Modern Topics in Microwave Propagation and Air-Sea Interaction

NATO ADVANCED STUDY INSTITUTES SERIES

Proceedings of the Advanced Study Institute Programme, which aims at the dissemination of advanced knowledge and the formation of contacts among scientists from different countries

The series is published by an international board of publishers in conjunction with NATO Scientific Affairs Division

A	Life Sciences	Plenum Publishing Corporation
B	Physics	London and New York
C	Mathematical and Physical Sciences	D. Reidel Publishing Company Dordrecht and Boston
D	Behavioral and Social Sciences	Sijthoff International Publishing Company Leiden
E	Applied Sciences	Noordhoff International Publishing Leiden

Series C – Mathematical and Physical Sciences

Volume 5 – Modern Topics in Microwave Propagation and Air-Sea Interaction

Modern Topics in Microwave Propagation and Air-Sea Interaction

Proceedings of the NATO Advanced Study Institute held at Sorrento, Italy, June 5–14, 1973

edited by

A. ZANCLA
Italian Meteorological Service, Rome, Italy

D. Reidel Publishing Company
Dordrecht-Holland / Boston-U.S.A.

Published in cooperation with NATO Scientific Affairs Division

First printing: December 1973

Library of Congress Catalog Card Number 73–91210

ISBN-13: 978-94-010-2683-3 e-ISBN-13: 978-94-010-2681-9
DOI: 10.1007/978-94-010-2681-9

Published by D. Reidel Publishing Company
P.O. Box 17, Dordrecht, Holland

Sold and distributed in the U.S.A., Canada, and Mexico
by D. Reidel Publishing Company, Inc.
306 Dartmouth Street, Boston, Mass. 02116, U.S.A.

CONTENTS

PREFACE

The increasing use of very high frequencies for radio communications, the demand for additional channels above 10 GHz and the use of new techniques and instrumentation for sounding the troposphere were the starting points for the choice of the topics to deal with at the NATO Advanced Institute held in Sorrento (ITALY), June 5 through 14, 1973.

Moreover the particular interest presently devoted to the propagation of microwaves along sea paths suggested the convenience to consider also the air-sea interface problems and the physical processes of evaporation from the sea.

The valuable contributions of high qualified lecturers as well as the enthusiastic participation of the students to lectures, workshops and informal meetings, assured the goal that was in the aim of the Institute.

I am pleased to express my gratitude to prof. D.R. HAY, President of the Radiometeorology NATO Panel, to dr. D. GJESSING, member of the Advisory Committee and to dr. T. Kester, executive officer of the Panel itself and head of the Advanced Institute Program, for the advice they offered me during the organization of the Institute.

My appreciation is addressed to lecturers and to directors of the conference sections and workshops.

I thank prof. A. SERRA and prof. G. CENA, respectively former and newly appointed chiefs of the Italian A.F. Meteorological Service that helped me with suggestions, and the Italian Air Force STAFF for the logistic support.

A. Zancla
Director of the Institute

Note: The numbers refer to the 'List of participants and contributors', pp. xi-xiii.

List of participants and contributors
(contributors indicated by an asterisk)

BELGIUM

(39) Prof. O. Godart	- Université de Louvain, Institut d'Astronomie et de Geophysique, Louvain-le-Neuve
(25) Mr. M.G. Schayes	- (as above)

CANADA

(7) Prof. R.W. Burling	- Institute of Oceanography, University of British Colombia, Vancouver
(2) Prof. D.R. Hay *	- University of Western Ontario, Faculty of Science, London, Ontario

DENMARK

(9) Prof. H. Bach	- Laboratory of Electromagnetic Thecry Technical University of Denmark, Lyngby
(3) Mr. J. Boisen	- Danish Defence Research Establishment, Copenhagen
(13) Dr. N. Klint-Hansen	- (as above)

FEDERAL REPUBLIC OF GERMANY

(8) Dr. H.W. Früchtenicht *	- Institut für Radiometeorologie und Maritime Meteorologie, Hamburg
(28) Dr. G. Kruspe *	- (as above)
(23) Dr. H. Jeske *	- (as above)
Dr. H.D. Seehars *	- (as above)
(20) Prof. L. Hasse *	- Meteorologisches Institut der Universitat, Hamburg

FRANCE

Mr. J. Battesti	- Centre National d'Etudes des Telec. Issy les Moulineaux
(38) Mr. M. Rooryck	- (as above)
(32) Dr. M. Coantic *	- Institut de Mécanique Statistique de la Turbulence, Marseille

GREECE

(31) Mr. D. Mavrakis * - Department of Physics of the University, Athens

HOLLAND

(40) Ing. M.J.M. Van Weert - Technische Hogeschool, Eindhoven

ITALY

Dr. A. Bernardini - Istituto di Comunicazioni Elettroniche dell'Università, Roma
(44) Prof. Ing. B. Peroni * - (as above)
Dr. M. Colacino - Consiglio Nazionale delle Ricerche I.F.A., Roma
(42) Prof. Ing. C. Capsoni - Istituto di Elettrotecnica ed Elettronica del Politecnico, Milano
(47) Prof. Ing. F. Carassa * - (as above)
(49) Dr. Ing. Drufuca * - (as above)
(48) Dr. Ing. A. Paraboni * - (as above)
(45) Dr. Ing. F. Fedi * - Fondazione Bordoni, Roma
(46) Ing. P. Mandarini - (as above)
(43) Dr. Ing. I. Ghidini - Radio Televisione Italiana, Roma
(1) Prof. G. Cena - Servizio Meteorologico A.M., Roma
(33) Dr. A. Nania * - (as above)
(10) Dr. G. Racca - (as above)
(5) Dr. A. Zancla - (as above)
(15) Mr. A. Pucciano - Centro Meteorologico Regionale Aeroporto Brindisi
(6) Prof. Ing. I. Ranzi * - Istituto Superiore Poste e Telecomunicazioni, Roma

N.A.T.O.

(11) Dr. T. Kester - Scientific Affairs Division, NATO-OTAN, Bruxelles

NORWAY

(36) Dr. T. Gjessing * - Defence Research Establishment, Kjeller
(30) Dr. K.S. McCormick - (as above)
(14) Mr. A. Tangerud - (as above)
(12) Mr. A. Moene - Meteorologisk Institut, Oslo

SWEDEN

(22) Mr. L.Ladell - Swedish Defence Research Establishment, Stockholm
(4) Dr. S. Wickerts * - (as above)

U.K.

Dr. D. Croom * - Radio and Space Research Station, Slough, Bucks

(41)	Dr. J.A. Lane	- Radio and Space Research Station, Slough, Bucks
(18)	Dr. D.T. Llewellyn-Jones *	- (as above)
(29)	Mr. P.A. Mattews	- Department of Electronics and Electrical Engineering, University of College, London
(34)	Mr. S. Rotheram *	- Marconi Company Ltd, Chelmford, Essex
(26)	Mr. Z. Warhaft	- (as Coantic)

U.S.A.

(27)	Prof. B.R. Bean *	- U.S. Department of Commerce, NOAA, Boulder
(37)	Dr. C.B. Emmanuel	- (as above)
(19)	Prof. R. Bolgiano *	- School of Electrical Engineering, Cornell University, Ithaca
(17)	Cpt. J.W. Goldey	- U.S. Air Force, Richards, Gebaur AFB
(24)	Dr. T. Laevastu *	- Environmental Prediction, Research Facility, Monteray
(35)	Mr. S.A. Stage	- Department of Atmospheric Science, University of Washington, Seattle
	Prof. A.W. Straiton *	- University of Texas, Austin
(21)	Dr. W. Vogel	- (as above)
(16)	Dr. M. Thompson *	- U.S. Department of Commerce, I.T.S. Boulder

OPENING SESSION

Introductory speech

A. Zancla

Director of the Advanced Study Institute

I am happy to meet you for the opening session of the NATO Advanced Study Institute 1973 on Radiometeorology.

My satisfaction is seeing you gathered here is double: first, because it crowns my two last years work, I carried out following up the engagements taken with the NATO Scientific Affairs Division, the Panel Members and the Chief of the Italian Meteorological Service; secondly, for it demonstrates that you have found valid and important the proposed matter that touches directly meteorologists and radio propagation engineers in the Mediterranean area. Here the meteorological parameters and the radio propagation phenomena assume a very important role because of the particular aspects of air-sea interaction and the rapid and frequent changes of the weather.

I am sure that the topics presented by very high qualified specialists, I thank to have accepted my invitation, will contribute to give us the explanation of many problems, and in any case, will excite our interest, new scientific researches and new collaborations.

I tried to take the highest care in both scientific and organizational aspects for a profitable success of the ASI.

In thanking the President and the members of the Panel, the lecturers and the scientists and the Deputy Chief of Meteorological Service, I am particularly honored to open the ASI 1973 on Radiometeorology.

Welcome address

G. Cena

Deputy Chief Italian A.F. Meteorological Service

Gentlemen, distinguished guests and colleagues,

this is the first time Italy has the honour of giving hospitality to a NATO Advanced Study Institute and I am especially pleased to be with you today and to participate in the opening session of it.

The variety and the importance of the lectures in the programme of the meeting are a guarantee that the maximum level of interest will be attained; many of them deal with fundamental problems of meteorology and physics-of-the atmosphere, as evaporation from sea, fields of temperature and humidity above the sea, determination of the wind stratification by propagation methods, visibility and atmospheric attenuation and so on. This means that the meeting will also riprisent a valuable opportunity for the Meteorological Services to improve their knowledge about some atmospheric processes. The presence of some riprisentatives of the Italian Meteorological Service in this meeting is due to the expectation, I may better say the certainty, of deriving benefits from the discussions such outstanding scientists will have during this session of the NATO ASI.

Gentlemen, hoping this meeting will be as profitable as your efforts deserve, I am very happy to express to you all, also on behalf of the Chief of the Italian Meteorological Service, my hearty welcome to Sorrento, my best wishes for your work and my cordial thanks for the help your studies will bring to my Service.

NTRODUCTORY LECTURE: "SOME ASPECTS OF MODERN RADIOMETEOROLOGY"
Summary

I. Ranzi

Istituto Superiore Poste e Telecomunicazioni,
Roma, Italy

Refraction, scattering from turbulence in large air volumes and from layers of enhanced turbulence, partial reflections from discontinuity layers, multipath effects, ducting propagation, focussing and defocussing effects, these are the main tropospheric effects which may influence microwave links; some of them are particularly important in the case of propagation over sea or along coastal boundaries.

While these effects are generally negligible in the case of earth-to-space links with high elevation angles, they may become very dangerous when the elevation angle is below 3° or 2°. In the X-band, fadings up to -30 dB and very harmful multipath effects are in fact observed.

Very powerful investigation methods, for the detection and the study of the reflecting and scattering layers in clear air conditions are now in use, by means of high power radars and of special FM-CW radars. The first kind of radar has the advantage to allow the observation of regions of enhanced turbulence over a large area; it is also very useful for the detection and tracking of the layered structures associated with the large scale billows due to Kelvin-Helmoltz instabilities and which produce dangerous turbulence to aircraft (CAT). But these radars have a limited resolution power. The FM-CW radars, generally used for vertical sounding, may detect the fine structure of the turbulent layers, the effects and the characteristics of acoustic-gravity waves, their breaking and their origin.

The combination of this kind of soundings with acoustic and laser sounding has already given very interesting results on the physics of the lower troposphere and of the boundary layer.

The remote determination of the temperature vertical profile by means of radiometric methods should also be mentioned.

As regards the microwave attenuation in a clear atmosphere, interesting results have been obtained in the study of the oxygen absorption band around 60 GHz. At heights greater than 10 km, the pressure and Doppler broadening effects rapidly decrease with the height and the band splits in about 40 separated lines. It is then feasible to establish communication links between high altitude aircraft or spacecraft on frequencies placed between two adjacent absorption lines, with a very high shielding degree, due to the lower atmosphere, with respect to noise and interference coming from the earth surface and also to interception at the ground. The oxygen microwave properties may be also used for radiometric remote sensing of some stratospheric data.

In the field of absorption and scattering of microwaves by rain hail and snow particles, studies are being carried out to predict these effects in the case of complex shapes and irregular distributions of the dielectric constant within the scatterer; the problems with wet hail and melting snow are also under study. The presently used methods for attenuation calculation should also be improved, especially for frequencies higher than 10 GHz, to take into account the presence of small drops and of multiple scattering.

As regards the experimental texts on the relationship between rainfall rate and attenuation, it has been put in evidence that errors may be introduced by the methods generally used for rainfall rate measurements, due to the high degree of inhomogeneity in a rain cell and to the influence of wind on the raingauges.

Also the problem of establishing cumulative curves of rainfall rate for very small percentages of the time on the ground of conventional meteorological data, was studies by various researchers but no satisfactory results have been obtained.

I GROUP

Air-Sea Interaction and Ducting:
Meteorological and Radiometeorological Aspects

UNE REVUE DES PROCESSUS PHYSIQUES DES INTERACTIONS OCEANS-ATMOSPHERE

Michel Coantic

Institut de Mécanique Statistique de la Turbulence
Marseille, France.

THE PHYSICAL PROCESSES OF AIR-SEA INTERACTIONS : A REVIEW.

Detailed Plan :

I - INTRODUCTION :

The importance of air-sea interaction as a geophysical phenomenon, and its consequences upon wave propagation.

II THE BASIC TRANSFER PROCESSES :

2.1. Radiation - Short-wave radiation, and absorption of solar energy in the sea. - Long-wave radiation, and radiative transfer processes in the atmosphere.
2.2. Water vapour transfer, evaporation, and latent heat transport.
2.3. Sensible heat transfer.
2.4. The global energy balance across the air-sea interface, and its consequences.
2.5. Kinetic energy transfer - Turbulent friction, and drift currents. - Wave generation by wind.

III THE RECIPROCAL INTERACTIONS BETWEEN THE VARIOUS MECHANISMS :

3.1. The fundamental effects of turbulence.
3.2. Stability effects.
3.3. Coupling between heat and mass transfer processes.

3.4. Interaction between turbulent and radiative heat transfers.

3.5. Effects related to the nature of the air-sea interface - Wave generation, modification of airflow structure, and of scalar transfer processes. - Two-phase effects at high velocities.

IV CONCLUSIONS :

Overall view of the present state of experimental and theoretical knowledge about air-sea interactions.

1. INTRODUCTION

Les échanges d'énergie qui se produisent sous différentes formes à travers l'interface air-mer constituent un phénomène géophysique d'une importance essentielle, aussi bien du point de vue de la météorologie que de celui de l'océanographie. A partir de la source initiale d'énergie que constitue le rayonnement solaire, ces échanges développent en effet des processus thermodynamiques et mécaniques dans l'atmosphère et dans les océans, sur une gamme d'échelles qui s'étend des circulations générales jusqu'aux structures les plus fines. Ces processus sont responsables de l'apparition des turbulences atmosphériques et océaniques, dont le rôle est déterminant pour le mécanisme et l'intensité des transferts dont les les deux milieux sont le siège.

Les interactions océans-atmosphère sont donc l'un des principaux facteurs qui conditionnent notre environnement, et ceci dans l'ensemble de l'atmosphère et de l'hydrosphère, et tout particulièrement dans les basses couches de l'atmosphère marine qui nous concernent ici. Elles contrôlent les variations de la pression barométrique, le régime des vents et des courants, l'évolution de la température dans l'air et dans l'eau, les distributions de l'humidité et de la salinité. Aux échelles les plus fines, les processus turbulents responsables des échanges de quantité de mouvement, de chaleur et de masse au voisinage de la surface marine s'accompagnent de fluctuations aléatoires de la vitesse, de la température, de l'humidité, de la pression, et enfin du niveau de l'interface.

Les interactions océans-atmosphère contrôlent donc directement les conditions de la propagation des ondes électromagnétiques dans la basse atmosphère marine. L'indice de réfraction de l'air dépendant de sa pression, de sa température, et de son humidité, les variations moyennes et les fluctuations turbulentes de ces trois variables affectent la portée et la qualité des communications. Par ailleurs, les déformations importantes que les vagues et la houle provoquent sur la surface marine, les embruns que le vent arrache de cette dernière, les gouttes d'eau de plus ou moins grandes dimensions qui résultent des condensations et des précipitations, les aérosols naturels ou artificiels diffusés par la turbulence, entraînent des réflexions ou des diffractions parasites dont l'importance est évidente.

La suite de cet exposé est consacrée à une présentation des principaux mécanismes physiques à petite échelle qui sont responsables des transferts au voisinage de l'interface, et qui affectent donc directement la propagation dans les basses couches de l'atmosphère. Les ouvrages cités dans la bibliographie permettent de compléter cette information, et de l'étendre aux caractéristiques à l'échelle moyenne et à l'échelle synoptique des interactions air-mer.

2. LES PROCESSUS DE TRANSFERT FONDAMENTAUX

Les échanges d'énergie entre l'atmosphère et les océans résultent principalement de quatre mécanismes physiques.

2.1. Le rayonnement

Il faut distinguer le rayonnement de courtes longueurs d'ondes ($< 3\mu$), d'origine solaire, et le rayonnement de grandes longueurs d'ondes (de 4 à 120μ) provenant de l'atmosphère et de la surface de la mer.

Le premier dépend de la latitude, de l'heure et de la saison, des diffusions et absorptions dans les couches supérieures, de la nébulosité. Sa valeur au niveau de la mer est, en moyenne, de l'ordre de 15 milliwatts/cm^2. Une fraction variable de ce flux est réfléchie à l'interface, mais la plus grande partie pénètre dans la masse marine où son énergie est absorbée dans les couches superficielles (voir Figure 1.). La mer augmente de température, et accumule donc de l'énergie sous la forme d'enthalpie sensible.

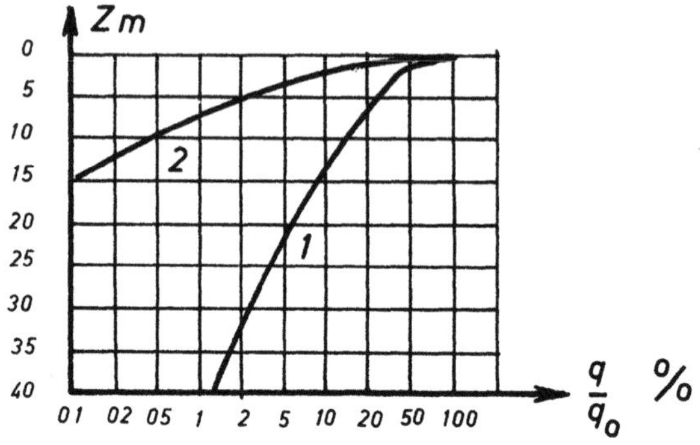

Fig. 1. Absorption du rayonnement solaire dans l'océan, d'après Sverdrup. Les courbes représentent le pourcentage d'énergie qui subsiste à une profondeur donnée. 1. Eau océanique moyenne. 2. Eau côtière moyenne.

Le mécanisme est très différent pour les grandes longueurs d'ondes, c'est-à-dire le rayonnement infra-rouge. Le coefficient d'absorption de l'eau est extrêmement élevé, et la surface marine se comporte pratiquement comme un corps noir. Dans l'atmosphère, la présence de vapeur d'eau et de gaz carbonique entraîne des processus de transfert radiatif très complexes, les différentes couches absorbant une certaine fraction des rayonnements qui les traversent et rayonnant simultanément une certaine quantité d'énergie, en fonction notamment de leur température, et de leur teneur en eau gazeuse ou liquide. Le flux net, qui résulte du bilan des rayonnements dirigés de la surface vers l'atmosphère et de celle-ci vers la surface, peut varier de façon appréciable dans les basses couches. Il est généralement ascendant, entraînant une déperdition d'énergie pour la mer qui est en moyenne de l'ordre de 7 milliwatts/cm^2.

2.2. Le transfert de vapeur d'eau, l'évaporation et l'enthalpie latente

L'on peut admettre, en première approximation, que la pression partielle de la vapeur d'eau en un point de l'atmosphère situé au contact immédiat de la surface marine est égale à la pression de vapeur saturante correspondant à la température de l'interface. La pression partielle sera, en général, différente dans les couches supérieures, ce qui entraîne un transfert (par convection et diffusion moléculaire et turbulente) de la vapeur d'eau depuis (ou vers) la surface et donc une évaporation (ou une condensation) à celle-ci. Il s'agit généralement d'une évaporation, dont la valeur moyenne est estimée à un mètre par an pour l'ensemble des océans.

Ce processus de transfert de masse s'accompagne d'effets thermodynamiques très importants. En effet, l'évaporation d'une certaine masse d'eau nécessite l'apport d'une importante quantité de chaleur, correspondant au produit de cette masse par la chaleur latente de vaporisation à la température de l'interface. Cet apport calorifique s'effectue nécessairement aux dépens des couches d'air, et surtout des couches d'eau, adjacentes à l'interface, dont la température s'abaisse. En termes thermodynamiques, le processus d'évaporation constitue donc un mécanisme de transfert d'énergie, celle-ci étant fournie sous la forme d'enthalpie sensible par la mer et transférée à l'atmosphère sous la forme d'enthalpie latente par la vapeur d'eau. La chaleur correspondante sera libérée sous la forme sensible dès que la vapeur se condensera. Ce mécanisme, particulièrement intense dans les zones tropicales, est d'une importance capitale pour la météorologie et entraîne une déperdition moyenne pour l'ensemble des océans de l'ordre de 7,5 milliwatts/cm^2. Il est schématisé sur la Figure 2.

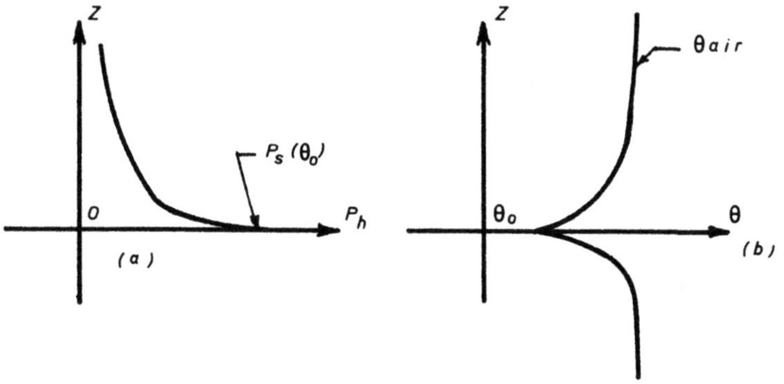

Fig. 2. Le transfert de vapeur d'eau, et ses conséquences thermodynamiques. a) Répartition de la pression partielle ; b) Effet sur la température (si θ air = θ eau).

2.3. Le transfert d'enthalpie sensible, et le bilan global d'énergie

Les températures de l'air et de l'eau de part et d'autre de l'interface sont généralement différentes, l'océan étant en moyenne plus chaud que l'atmosphère. Aux mécanismes précédents vient donc se juxtaposer un processus de transfert d'enthalpie sensible, par convection et diffusion moléculaire et turbulente au sein des deux phases. Sa valeur moyenne correspond aux 0,5 milliwatts/cm^2 nécessaires pour fermer le bilan.

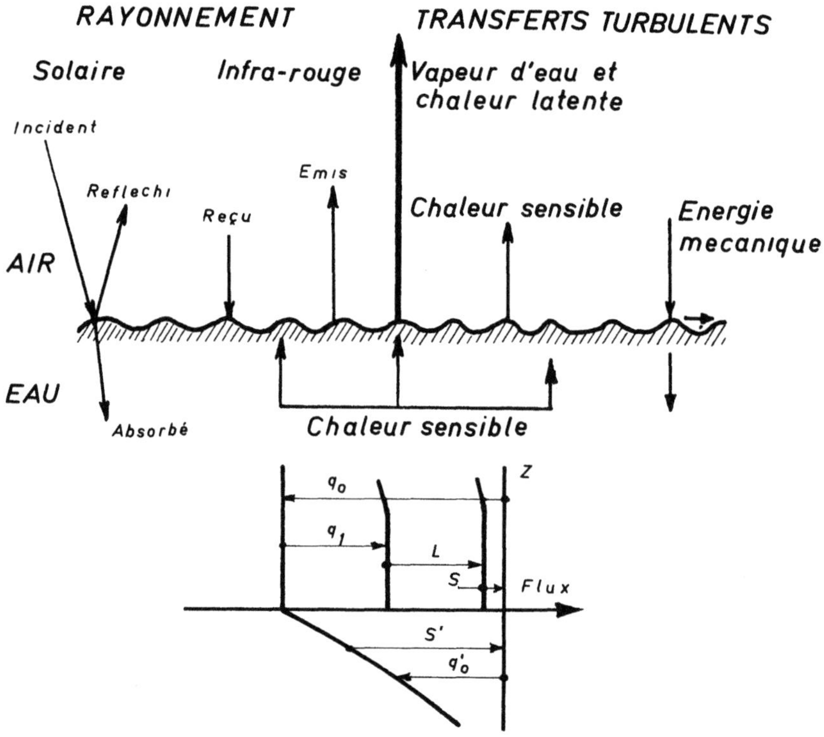

Fig. 3. Représentation schématique des mécanismes de transfert et de transformation d'énergie au voisinage de l'interface océan-atmosphère.

La Figure 3. schématise l'ensemble des flux énergétiques, et des apports de chaleur qui en sont la conséquence, dans une situation supposée stationnaire. L'énergie solaire est absorbée, en un premier temps, dans l'océan où elle apparaît sous la forme sensible, puis transférée, en un deuxième temps, vers la surface, par la convection turbulente. L'atmosphère est ensuite alimentée en énergie, depuis la surface, par les jeux combinés du rayonnement infrarouge, de l'évaporation, et de la convection thermique. Il est clair que la principale source d'énergie de la troposphère est la vapeur d'eau.

2.4. Le transfert d'énergie mécanique

Les inégalités à grande échelle, dans l'espace et dans le temps, des apports d'énergie et de masse que les mécanismes précédents opèrent dans l'atmosphère et dans l'hydrosphère entraînent des variations dans la répartition des densités de ces deux milieux fluides. Compte tenu de l'influence de la gravité, de la géométrie et de la rotation terrestres, ces inégalités engendrent à leur tour des mouvements relativement rapides de l'atmosphère (les vents), et des mouvements plus lents des océans (la circulation thermohaline). Le rendement thermodynamique de cette transformation d'énergie interne en énergie mécanique est évidemment très faible, mais ses conséquences sont pour nous d'une importance capitale, en raison de ses effets sur l'évolution de notre environnement.

A l'échelle qui nous intéresse ici, l'énergie cinétique est généralement transférée de l'atmosphère à l'océan. Le ralentissement imposé par la surface marine entraîne l'apparition d'une "couche limite planétaire", au sein de laquelle la répartition du vent en altitude suit, sous l'effet du frottement turbulent et de l'accélération de Coriolis, la "spirale d'Ekman" bien connue. Une fraction appréciable de l'énergie atteignant les basses couches est alors transférée à l'océan suivant deux mécanismes qui sont intimement liés:

- la génération des courants, qui existerait seule si la surface de la mer restait plane. Sous l'effet de la force d'entraînement variable exercée par le vent sur la surface, des forces internes exercées par l'accélération de Coriolis et par la turbulence, des conditions aux limites imposées par la topographie du fond et des côtes, divers types de courants sont engendrés.

- la génération des vagues, dont le mécanisme est extrêmement complexe, et encore mal connu. Il est généralement admis que les fluctuations turbulentes de la pression dans les basses couches excitent initialement, par résonance, certains modes d'oscillation de la surface. Dès que les déplacements de cette dernière atteignent une certaine amplitude, l'écoulement d'air se trouve modifié de telle sorte que les tensions normales et tangentielles exercées sur la surface travaillent de façon à transmettre à travers elle un flux important d'énergie mécanique. L'énergie s'accumule, sous les formes cinétique et potentielle, dans les couches superficielles et se trouve ensuite rayonnée par la houle, transférée aux courants, et dissipée sous forme de turbulence, suivant des mécanismes et dans des proportions encore très mal connus.

3. LES INTERACTIONS RECIPROQUES ENTRE LES DIVERS PROCESSUS DE TRANSFERT

L'une des principales difficultés du problème des échanges mer-atmosphère vient de ce que les mécanismes précédents sont en général étroitement couplés.

3.1. Le rôle fondamental de la turbulence

Dans les milieux fluides que nous considérons ici, la diffusion moléculaire constitue un mécanisme très peu efficace pour transférer les grandeurs telles que la quantité de mouvement, la chaleur, ou la masse. L'efficacité du mécanisme de diffusion turbulente est considérablement supérieure, dans un rapport qui est couramment de l'ordre de 10^4 et on peut atteindre des valeurs de l'ordre de 10^8. Les taux des transferts convectifs d'enthalpie sensible et d'enthalpie latente dans le système océans-atmosphère, et la répartition des variables telles que la température et l'humidité qui y fixe les conditions des transferts radiatifs d'énergie, dépendent donc principalement du niveau et de la structure des processus turbulents dans les deux milieux.

La turbulence dans les couches voisines de l'interface air-mer est avant tout d'origine dynamique. Bien que les quantités d'énergie qu'il met directement en jeu soient relativement faibles, si l'on se place du point de vue thermodynamique, le mécanisme de transfert d'énergie mécanique est donc d'une importance capitale pour l'ensemble des échanges océans-atmosphère, à l'égard desquels il joue en quelque sorte le rôle d'un catalyseur.

3.2. Les effets de la stratification

La température, l'humidité, et la salinité ne peuvent pas être considérées ici comme des "contaminants scalaires passifs", c'est-à-dire sans influence sur le processus dynamique. Les variations locales de ces grandeurs s'accompagnent en effet de variations de masse spécifique qui entraînent l'apparition de forces d'Archimède. Ces dernières travaillent de façon à accroître ou à diminuer le niveau des mouvements turbulents, suivant que le milieu est dans un état de stratification instable, ou au contraire stable. Plus précisément, l'on verra que des quantités d'énergie proportionnelles aux flux turbulents de chaleur et de masse sont apportées ou soustraites à la turbulence, dont l'intensité et la structure sont profondément modifiées lorsque le nombre de Richardson atteint des valeurs appréciables.

Les transferts d'énergie réagissent donc sur le mécanisme dont ils sont eux-mêmes la conséquence, et les échanges entre l'océan et l'atmosphère peuvent s'établir en définitive à des niveaux très différents de ceux qui correspondraient à un état de stratification "neutre". Le couplage entre les divers modes de transfert est donc bi-directionnel.

3.3. La liaison entre le transfert de masse, et les transferts de chaleur, par turbulence et par rayonnement

Nous avons signalé plus haut les importantes conséquences que le transfert de masse exerce, en raison de l'évaporation, sur les échanges de chaleur sensible. Réciproquement, ces derniers agissent sur la température de l'interface, et donc sur l'écart de pression partielle qui conditionne le transfert de vapeur d'eau. L'existence quasi-permanente d'un film froid superficiel à la surface des mers est notamment d'une importance non-négligeable pour le cycle de l'eau à la surface de notre globe.

Les interactions réciproques qui existent entre les transferts radiatifs de différentes longueurs d'ondes et les transferts turbulents de chaleur et de masse n'apparaissent pas seulement au niveau des conditions aux limites et des bilans globaux. La turbulence contribue en effet au contrôle des répartitions, non seulement de la température et de l'humidité, mais encore du gaz carbonique, des noyaux de condensation, et de divers aérosols, dans l'ensemble de la troposphère. C'est donc elle qui fixe les conditions de l'absorption et de la réémission des rayonnements solaire et infrarouge, déterminant ainsi les éléments du bilan radiatif des couches voisines de l'interface air-mer.

3.4. Les effets propres à l'interface air-mer

Une bonne partie de ce qui vient d'être dit au sujet des basses couches de l'atmosphère marine pourrait s'appliquer sans grandes modifications à la région correspondante de l'atmosphère terrestre près d'un sol suffisamment plat. Certaines dissemblances méritent toutefois d'être soulignées :

- La totalité de l'énergie solaire incidente est transformée en chaleur sensible, dans le dernier cas, à la surface même du sol où elle ne peut pénétrer que par conduction moléculaire. Elle n'intéresse donc qu'une mince pellicule, d'où elle sera restituée aisément, notamment par le rayonnement nocturne. L'océan est au contraire apte à accumuler l'énergie qu'il reçoit, en la répartissant sur une certaine profondeur. Les différences qui en résultent sont bien connues des météorologistes. A l'échelle qui nous intéresse, la diminution de l'amplitude des variations journalières de la température superficielle limite au-dessus des mers les effets d'instabilité diurne et d'inversion nocturne qui caractérisent la micrométéorologie terrestre.

- Même dans le cas de l'évapotranspiration d'une végétation fortement irriguée, divers mécanismes concourent à limiter les échanges d'eau entre le sol et l'atmosphère. Le rôle de l'évaporation, et son importance dans le bilan énergétique, est donc moindre dans le cas terrestre que dans le cas marin.

- La présence, à la surface de la mer, de vagues ou de houles qui ne dépendent pas seulement des conditions locales peut avoir sur le mécanisme des transferts dans l'air et dans l'eau des effets importants, mais qui restent encore mal connus. Ainsi, la géométrie de l'interface affecte les processus d'échange radiatif. Par ailleurs, l'importance relative des forces de pression et des forces de frottement tangentielles dans le transfert de l'énergie mécanique du vent aux vagues variera suivant le taux de développement de celles-ci. Le degré de similitude (analogie de Reynolds) entre les échanges turbulents de quantité de mouvement, de chaleur et de masse évolue donc au voisinage de la mer suivant des lois encore plus complexes que celles qui gouvernent ses variations à proximité d'un terrain de rugosité géométrique donnée.

- Lorsque la vitesse du vent augmente, les phénomènes de déferlement, d'arrachement d'embruns, de production de bulles et d'écume, prennent une importance croissante. Ces effets diphasiques vont modifier profondément les mécanismes de transfert, et très probablement accroître l'efficacité des échanges lors des tempêtes. D'après certains auteurs, leurs conséquences pourraient être primordiales pour les bilans synoptiques.

4. CONCLUSIONS

Cette brève revue des principaux mécanismes physiques par lesquels s'opèrent les transferts et transformations des différentes formes d'énergie au voisinage de l'interface océans-atmosphère avait pour but de situer l'importance et les difficultés du problème des interactions entre les deux milieux. L'on possède vraisemblablement à l'heure actuelle une conception qualitative globale qui est correcte de l'ensemble des mécanismes qui contribuent aux échanges. Par ailleurs, les conférences qui vont suivre montreront que l'on sait mettre en équations le problème d'une façon assez satisfaisante, que l'on dispose de méthodes expérimentales relativement précises pour des études à la mer et au laboratoire, et enfin qu'un ensemble important de recherches théoriques et expérimentales ont été menées à bien, particulièrement au cours des dernières années. Aujourd'hui, une prévision quantitative des ordres de grandeur des divers effets est possible dans le cadre de théories semi-empiriques, la description expérimentale des processus physiques a beaucoup progressé, et enfin certaines caractéristiques à petite échelle sont connues d'une façon relativement sure. Il reste cependant encore beaucoup à faire, et l'on ne peut espérer parvenir à une solution réelle du problème des interactions air-mer tant qu'une théorie générale des mécanismes turbulents n'aura pas été établie.

Certains aspects des conditions de la propagation des ondes électromagnétiques dans la basse atmosphère marine peuvent donc aujourd'hui être expliqués, et même approximativement prévus, mais l'on réalisera aisément que l'ensemble de la question est encore loin d'être connu d'une manière pleinement satisfaisante, puisque l'environnement naturel qui conditionne la propagation ne l'est pas lui-même.

BIBLIOGRAPHIE

Kitaigorodskii S. A. (1970). L'interaction à petite échelle entre l'atmosphère et l'océan (en russe). Nauka, Moscou.

Kraus E. B. (1972). Atmosphere-ocean interaction. Clarendon Press, Oxford.

Lacombe H. (1965). Cours d'océanographie physique, Gauthier-Villars, Paris.

Lumley J. L. et Panofsky H. A. (1964). The structure of atmospheric turbulence. Interscience, New-York.

Monin A.S. et Yaglom A. M. (1971). Statistical fluid mechanics (traduction). M. I. T. Press, Cambridge.

Munn R. E. (1966). Descriptive micrometeorology. Academic Press, New-York.

Phillips O. M. (1966). The dynamics of the upper ocean. Cambridge University Press, Cambridge.

Roll H. U. (1965). Physics of the marine atmosphere. Academic Press New-York.

DISCUSSION

QUESTION, by S. WICKERTS, Research Institute of National Defence, Sweden.

Will you please comment on the figures you mentioned concerning the relation between molecular and turbulent diffusion (10^4 - 10^8)

REPLY, by M. COANTIC.

The last figure applies to salt diffusion in the oceans: published values range from 1 to 10^3 cm2 sec^{-1} for vertical, and from 10^5 to 10^8 cm2 sec^{-1} for horizontal eddy diffusivities, while the molecular value is of order 10^{-5} cm2 sec^{-1}. The first figure is representative of the effect of turbulence in the surface layer of the atmosphere, where $K_m = \kappa U^* z$ is typically of order 10^{-1} -1 m2 sec^{-1}, while the kinematic viscosity coefficient is of order 10^{-5}.

COMMENTS, by L. HASSE, Meteorologisches Institut Universität Hamburg, F.R.G.

1/ With regard to the surface "cooling" by heat loss from the water surface to the atmosphere it may be noted that a warmer surface layer may exist when the input of solar radiation considerably exceeds the heat losses.

2/ A remark of J. MALKUS on the role of evaporation for the general circulation of the atmosphere was cited. In fact, the role of water vapor is a very important one. According to an argument of SANDSTRÖM, the atmosphere as a heat engine is most effective if heat is supplied at high pressure and withdrawn at low pressure. As evaporation uses up the energy available from solar radiation at high pressure and releases the latent heat at lower pressures, it reduces the overall efficiency of the atmosphere as a heat engine.

3/ One additional aspect of sea waves is that they take up part of the momentum given from the atmosphere to the ocean. This part or fractions of it may be handed over to currents or internal gravity waves without reappearing in the oceans surface layer as turbulence. The turbulence intensity or friction velocity available in the surface layer of the ocean for transport of heat, matter and gases is therefore reduced compared to that in the atmospheric surface layer.

REFERENCES :

L. HASSE, 1971 : The sea surface temperature deviation and the heat flow at the sea-air interface
Boundary-Layer Meteor. 1, 368-370.

J.W. SANDSTRÖM, 1916 :
Göteborgs kgl. Vetenskaps - och Vitterhets - Samhälles Handl., Fjärde Folgen 17, 2.

PRESENT STATE OF KNOWLEDGE ON EVAPORATION FROM THE SEA AND NUMERICAL ANALYSIS/FORECASTING OF WATER VAPOR PRESSURE IN THE SURFACE LAYERS OF THE AIR

T. Laevastu

Environmental Prediction Research Facility
Monterey, California

1. INTRODUCTION

This condensed review of evaporation from the ocean places emphasis on the practical aspects of evaporation which have a bearing on radar propagation in the lower layers of the atmosphere over the sea, creating the evaporation duct. Special emphasis is placed on the relation between evaporation and synoptic weather systems and the use of synoptic meteorological data for the computation of evaporation. Voluminous literature exists on the subject -- especially on the different aspects of computing evaporation. The length of this paper does not allow discussion and evaluation of all pertinent facts upon which the concluding evaluation has been based. References are selective and minimal. The figures included are meant to compliment the text in the form of examples and are not, as a rule, discussed further in the text.

The evaporation data, with respect to both analysis and prediction (direct evaporation measurements and recordings are impossible over the ocean) are used mainly for four types of activities. These are (1) analysis and prediction of the weather -- especially numerical analysis using primitive equation models where uptake, release and transport of moisture plays a determining role; (2) hydrological forecasts -- including forecasts of floods and droughts and the transport of moisture across continental boundaries. (This application is increasing in importance as better medium- and long-range forecasting models of the atmosphere come into use); (3) oceanographic forecasts -- where the main activity deals with forecasting the thermal structure in the surface layer of the ocean; and

(4) forecasting the so-called "evaporation duct" which affects radar propagation close to the surface.

2. SPACE AND TIME SCALES

Evaporation might be considered as a so called turbulent transfer process. Turbulence and turbulent transfer might be considered terms of convenience which help overcome some difficulties in following the details of movements and processes in Nature, where rapidly changing, apparently chaotic situations exit. The turbulent motions and processes must however, be considered on the basis of different space and time scales in order to bring some order into apparent chaos. This simple fact is, unfortunately, quite often overlooked. The conventional division of the space and time scales is, with respect to Nature, rather arbitrary and has been made on the basis of some conveniences and past approaches and applications.

With respect to space, four different scales may be used on some overlapping basis for both the atmosphere and the oceans. These scales are (1) hemispheric -- especially useful in dealing with processes and motions in upper atmosphere; (2) regional -- such as one ocean or part of the ocean (e.g., Norwegian Sea). (this scale permits consideration of surface atmospheric and oceanic features such as the various surface atmospheric systems, highs and lows, and related winds and currents; (3) local scale (or mesoscale) -- considers the surroundings of a given location to the extent of a few hundred kilometers in the horizontal; and (4) microscale -- concerns very local processes and distributions in the order of cm and meters to a few kilometers in the horizontal. It should be noted that different scales can apply to the vertical extent.

With respect to time, four scales can be recognized conveniently: (1) annual, seasonal and monthly scales; (2) weekly scale and the scale on the basis of so-called natural periods which have been lately recognized in the atmosphere as ranging from a few days to a maximum of a few weeks; (3) synoptic scale which in common usage varies from a few hours to a few days; and (4) sub-synoptic scale ranging from fractions of seconds to a few hours.

3. THE THEORY OF EVAPORATION AND FACTORS AFFECTING IT

3.1 The Theory of Evaporation

The existing knowledge of evaporation has not yet reached the stage where only one theory (or few slightly different modifications) is accepted by the majority of scientists concerned, and especially where the theory can be favorably compared with empirical measurements. Sverdrup (1951) reviewed the existing evaporation theories and concluded that different theoretical approaches (five considered) were useless until valid arguments could be advanced for accepting one of the many models. No progress seems to have been made in this subject in the last twenty years. Simularly, Marciano and Harbeck (1954) found that all theoretical evaporation formulae were unsatisfactory. A few years later, Deacon, Bristley and Swinbank (1958) described the practical difficulties encountered in the use of various theoretical approaches for calculating evaporation. Deacon and Webb (1962) summarized the situation of evaporation computation with the following statement which is still valid: "No vigorous theoretical approach of the evaporation is at present possible due to a lack of complete knowledge of the transfer characteristics in the lowest layer of the air near the water surface; and the perspective theories are based on different assumptions as to the nature of these characteristics."

Pesaresi (1969) has made the latest worthwhile contribution to an understanding of the processes involved in evaporation, in taking more into consideration the atmospheric stability conditions over the oceans than before. Earlier, Laevastu (1960) had made use of the stability in a simplified manner. Some additional contributions about the understanding and use of evaporation formulas were made by Schmitz (1963) who took into account the effects of variations of pressure and, by applying weighted means, the effects of fluctuative transport of mass, using the Austausch principle. Schmitz emphasized that with an unstable lapse rate there is an inequality of the eddy transfer coefficients of vapor and heat. The proof for the latter statement is not yet complete, and the principle that heat and moisture transfer are subject to the same laws and processes is still the common concept in estimating and computing evaporation.

Finally, Harris (1966) has upset many of the evaporation considerations which use a specific turbulent transfer approach. He showed that there exists a "wave driven wind" (in the case of progressive waves) and that this flow reached at least half a wave length, or about 14 wave amplitudes, above the mean water surface. This wave driven wind in the atmosphere is similar to mass transport by waves in the water as shown by Stokes.

3.2 Factors Affecting Evaporation

Some general information on the factors affecting evaporation from the oceans is given in Table 1.

TABLE 1. FACTORS AFFECTING THE EVAPORATION FROM THE OCEANS

MAIN FACTOR	ELEMENTS/CONDITIONS AFFECTING OR DETERMINING THE MAIN FACTOR	REMARKS*
Saturation vapor pressure of the sea surface	Sea surface temperature (SST)	SA, AL
	Thickness of "skin" and difference between "skin" temp. and SST	INA, SNA, greatly variable
	Surface salinity	Effect small, (SNA)
	Surface films (organic)	Effect small, (SNA)
Vapor pressure deficit of the air near the interface	Air temperature	SA, AL
	Dew point temperature	SA, AL
	Atmospheric pressure	SA
Turbulent transport of moisture from the surface layer of the air	Vapor pressure gradient with height	NA
	Thermal convection (Dimensions and intensity of "thermals")	NA
	Intensity of turbulence	NA
	Wind (speed, direction)	SA, AL
Nature of the sea surface	Roughness of the sea surface including wave parameters	SA, AL, related to wind effects
	Spray	NA
	Surface films	NA

* SA = Synoptic information available
AL = Accuracy low
INA = Information not available
SNA = Synoptic information not available
NA = Not available

The effect of spray on evaporation has not been fully resolved. While many workers emphasize its importance, others are convinced that its effect is minor indeed except at very high wind speeds.

Though some of the available theory and knowledge of factors affecting evaporation is quite reasonable, it might not always be applicable to the computation of evaporation in practical cases because of either the lack of proper measurements of desired parameters, or a lack of real verification of the quantitative relationship. The general approach in applying the available knowledge of the evaporation process as a turbulent transfer process for computations and estimations is via the following assumptions: (1) the vertical gradients are approximated by sea-air water vapor and temperature differences at defined heights and depths; (2) the turbulent transfer rate is, furthermore, made a function of wind speed at a defined level; and (3) stability conditions of the air affecting the rate of turbulent transfer is taken into consideration via temperature difference between the sea surface and air at a defined level.

Finally, it should be pointed out that microscale evaporation processes and associated turbulent transfer might not be the only consideration to be used. Latest observations and recordings of variability of various parameters related to evaporation indicate that "local" scale thermals might play a determining role in both the evaporation process *per se* and in the distribution of moisture (and temperature) in the lower layers of the atmosphere.

4. MEASUREMENT OF EVAPORATION AT SEA

Two basically different approaches have been tried for the direct measurement of evaporation at the sea: (1) evaporation pans aboard ships, and (2) floating evaporation pans on the sea surface. Both measurements are extremely uncertain due to a number of difficulties (a) the wind field and water vapor pressure gradient with height is considerably disturbed by the presence of the ship; (b) there is considerable splashing in the pan due to the movement of the ship or floating pan; (c) the rim of the pan (which must be considerably higher than that of an evaporation pan on land) shelters the free surface; (d) direct solar heating, nocturnal radiational cooling as well as cooling by wind advection and heating by the ship's presence cause temperature changes and vertical gradients which might not normally be present in the sea; and (e) there can be considerable splash and spray into the floating pan which can be used only in relatively calm weather.

As there is at present no reliable method for the direct measurement of evaporation from the sea, one must use for verification purposes the results of the pan measurements on land and small water body evaporation measurements (of which there is voluminous literature available) in order to relate quantitatively the factors affecting the evaporation from the water surface, and adapt these relations for synoptic work at sea. The best known evaporation measurements under a great variety of conditions were made by Rohwer (1931). Some additional later projects such as the Lake Mead Study (Marciano and Harbeck, 1954) have greatly confirmed the results of Rohwer's work. However, there still exists a need for simulation experiments in the measurement of evaporation under some natural conditions -- especially with respect to the effect of waves on the evaporation rate.

Two indirect methods have been tried for the measurements of evaporation from the sea surface using either (a) fixed construction in shallow water or (b) floating stabilized masts as instrument platforms. These are (a) the flux methods and

(b) the profile method.

The eddy flux method described by Deacon (1959) and others is based on:

$$F_s = (\overline{\rho w})s \sim \overline{\rho w's'}$$

Where F_s is the eddy flux, w is the vertical velocity component, ρ is the density, s the covariance of the flux of s, and (') notes departure from the mean and (-) notes average over a time period. With proper instrumentation, the eddy flux method gives reasonable results over land areas. Over the sea it runs into many difficulties, both with respect to instrumentation and to the basic theory of the computation of evaporation. It should be noted that this method does not explain the evaporation process nor does it relate these measurements to the synoptically observed weather elements affecting the evaporation. The main difficulties in obtaining correct turbulent flux measurements are (1) the separation between the instruments for recording w' and s' must be smaller than the smallest size of the eddies contributing to the water vapor flux; (2) the instruments must have the same response time to give representative covariances; (3) response time must be small so that the flux contributions of the smallest eddies will be measured; (4) the surface must be fixed or the instrument must be a constant distance from the surface (this requirement arises from Harris (1966) wave driven wind); and (5) the measuring period and/or area must be large enough to include both spacial and temporal variation, especially "thermals." Thus the eddy flux method is scarcely of any use for the determination of oceanic evaporation.

The profile method attempts to relate the turbulent vertical moisture transport to the vertical gradients of the properties and the transfer coefficient (K_e):

$$E = - \rho(D+K_e\ U_*Z)\ (\frac{\partial q}{\partial z})$$

Where E is the moisture flux, ρ is the density, D is the molecular diffusivity, U_* is the friction velocity, Z is the height and $\frac{\partial q}{\partial z}$ is the vertical moisture gradient. This method has been further developed by Roll (1963). With respect to measurements, it requires experimental determination of the values of U_*, K_e and especially the detailed moisture gradient with height. The impossibility of determining the friction velocity over a wavey surface, and specifically the lack of synoptic data on the detailed gradient of moisture with height, limits this method to the more empirical Dalton type approach to be discussed in later chapters.

5. COMPUTATION OF EVAPORATION OVER THE SEA

It is impossible to get humidity and wind speed profile measurements on a synoptic basis from voluntary observing vessels. Therefore, certain simplifications of the formulae must be made for computation of evaporation:

(a) The vapor pressure profile gradient is approximated by the difference between the saturation water vapor pressure at the sea surface and that measured at a higher level, preferably a "standard level" (10 m.).

(b) Wind speed is measured at the same standard level. Thus we derive the Dalton type formulae:

$$E = K(e_w - e_a)\ u$$

The coefficient K must be derived from empirical measurements. If e_w and e_a (water vapor pressure of the water surface and of air at a given height) are in mb and u (wind speed) in m sec^{-1}, and E (evaporation) in mm $(24h)^{-1}$, the available literature for a single factor K shows values from 0.095 to 0.157.

Several slightly modified forms of Dalton formulae have been used in the past. Some of them use air temperature as well as atmospheric pressure as additional parameters. The inclusion of the latter two parameters seems not be justified on the basis of empirical data.

The best type of modified Dalton type formula is:

$$E = (a + bu)\ (e_w - e_a)$$

where a and b are empirical factors and other symbols are the same as in previous formulas. The factor a will allow evaporation during calm weather in convective turnover. Furthermore, if omitted, it allows computation of condensation if $e_w - e_a$ is negative.

Laevastu (1960) has compared available evaporation formulas

and presented a revised formula of the Dalton type:

$$E = (0.26 + 0.077\ u)\ (0.98e_w - e_a)$$

The value 0.98 indicated that the water vapor pressure of the sea surface is about 2% less than that of fresh water. If $e_w - e_a$ is negative, the factor 0.26 is omitted.

For synoptic analytical computations of evaporation the following analysis/forecasts must be available (a) sea surface temperature; (b) water vapor pressure of the air; and (c) wind speed (and direction). From the above analysis the following additional parameters must be computed: (d) saturation vapor pressure of the sea surface; (e) vapor pressure difference of sea surface and surface air. An example of the density of synoptic observations is given in Figure 1 and examples of computed synoptic evaporation patterns are shown on Figures 2 and 3.

Some examples of the variability of the water vapor pressure of the air measured directly by Gonella (1964) is shown in Figure 4. Such local and micro-scale variability of parameters affecting evaporation will cause considerable fluctuations of evaporation on the same scales.

It can be expected that the evaporation in coastal regions is considerably higher than that further offshore -- especially when dry, cold (or also dry, warm) air moves over the water. In case of warmer air, considerable stability can be expected in the lower layers due to cooling, thus decreasing the transport of water vapor upwards. This problem of the humidity profiles under conditions of warmer air movement over colder sea are discussed in another paper (this seminar) by Hamilton and Laevastu. The sea and land breeze affects radio propagation (which is also one of the coastal phenomena) and is already well described by Bean and Dutton, (1966). Obviously the sea and land breeze also cause considerable variation of evaporation in coastal regions.

The surface properties of the air respond relatively rapidly to the properties of underlying sea surface as will be further explored in the next chapter.

It could be argued that the coastal influence can be traced in properties of both the surface atmosphere and sea surface. Indeed "the range of coastal influence" can be so demonstrated as shown in Figure 5. The range is usually from 10 n mi to 50 n mi offshore, the offshore being often demarcated by a thermal front on the sea surface. These so-called coastal fronts approximately

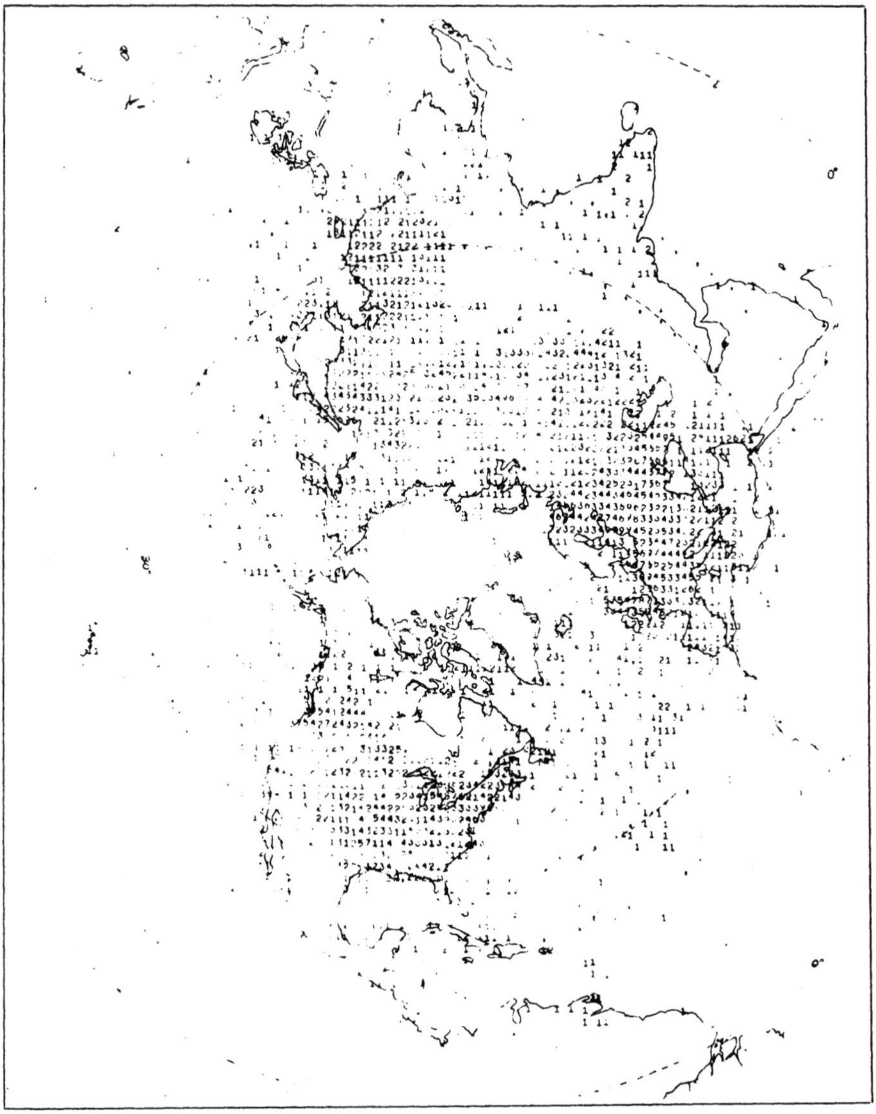

Fig. 1. Density of surface weather reports on 18Z 08 August 1966

coincide with either the continental slope or with other more pronounced depth changes in the case of wider continental shelves.

6. COMPUTATION OF WATER VAPOR PRESSURE OF THE AIR OVER THE SEA ON SYNOPTIC AND SUBSYNOPTIC SCALE

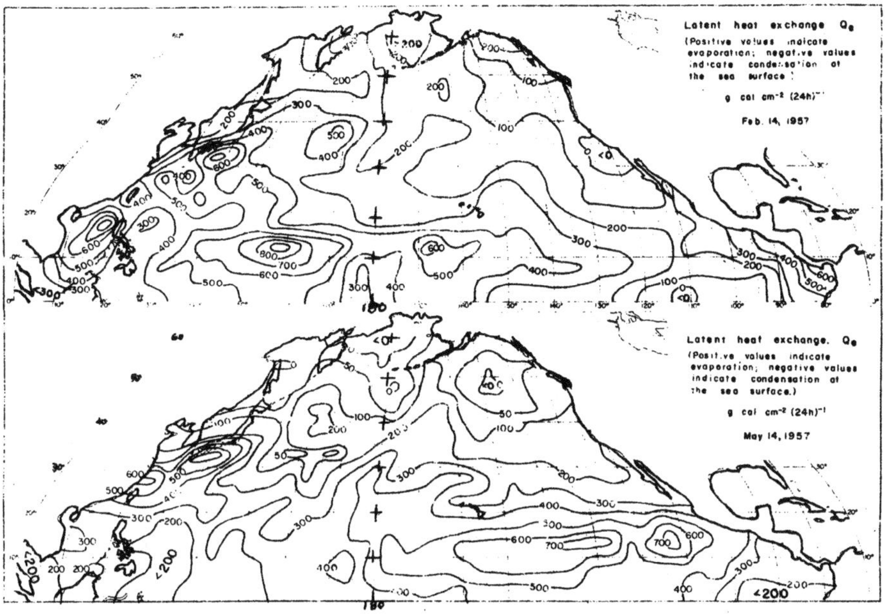

Fig. 2. Examples of synoptic evaporation charts for the North Pacific (given as latent heat exchange in g cal cm^{-2} $(24h)^{-1}$) (from Laevastu 1965)

A Norwegian theory originally reported by Åmot (1944) and later by Mosby (1957) and Bøyum (1962) is presented here with a few modifications. This theory is adaptable to numerical computations using parameters available in routine meteorological observations. In this theory the trajectory of the surface air is followed.

An air parcel moving over the sea surface has its water vapor pressure changed mainly by the convective turbulent transfer of moisture. If the saturation water vapor pressure of the sea surface (e_w) changes along the trajectory of the air, the corresponding individual change of sea-air water vapor pressure difference $\Delta e = e_w - e_a$ is:

$$\frac{d\Delta e}{dt} = \frac{de_w}{dt} - \frac{de_a}{dt} \qquad (1)$$

where e_a is the water vapor pressure of the surface air.

The change due to the effects of the saturation water vapor pressure of the sea-surface along the trajectory is essentially composed of two terms:

Fig. 3. Examples of synoptic hemispheric analyses of latent and sensible heat exchange, 00Z 21 June 1965

$$\frac{de_w}{dt} = \frac{\partial e_w}{\partial t} + \mathbb{V} \cdot \nabla e_w \tag{2}$$

The "local change" of sea surface saturation water vapor pressure $\frac{\partial e_w}{\partial t}$ is small (usually much smaller than 0.01 mb/hour) and can be neglected especially in comparison with the last term ($\mathbb{V} \cdot \nabla e_w$), which represents the sea surface saturation water vapor pressure change along the trajectory of the surface air. $\mathbb{V}$ is the velocity of the surface air parcel and can be assumed to be the surface

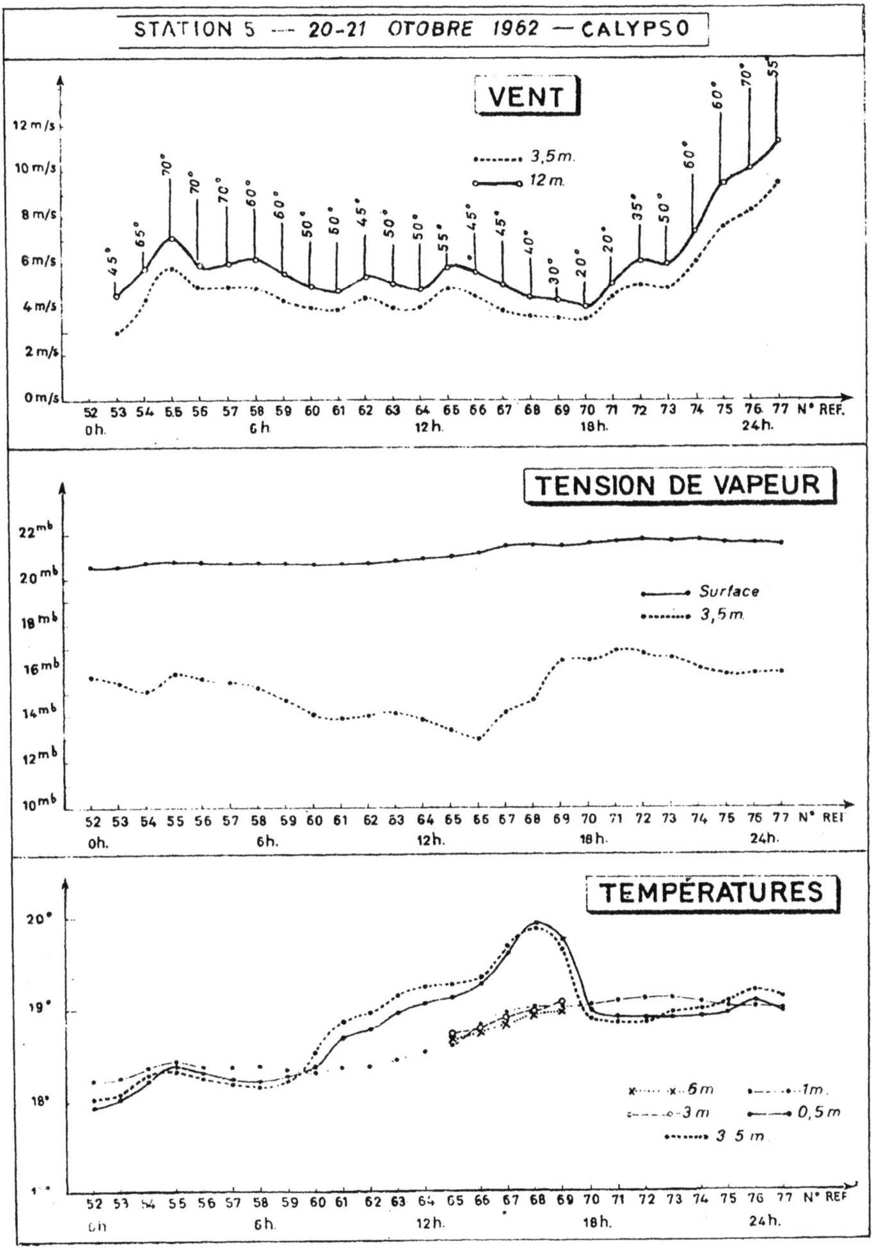

Fig. 4. Recording of wind, water vapor pressure and temperature at a station between Nice and Calvi, 20-21 October 1962 (from Gonella, 1964)

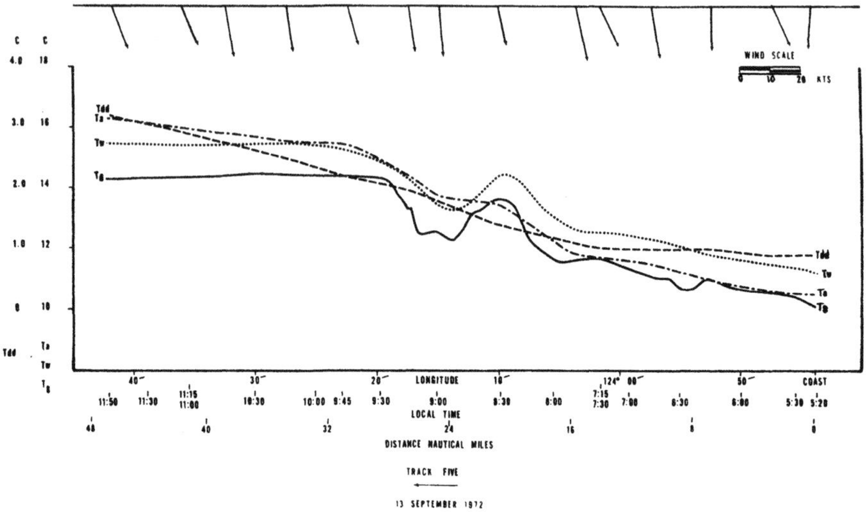

Fig. 5. Changes of sea surface temperature (Tw), temperature at 8m depth (T8), air temperature (Ta), dew point temperature (Td) and winds with the distance from the coast along a track off Point Arena, (north of San Francisco) on 13 September 1972

wind speed. Thus equation (1) reduces to:

$$\frac{d\ \nabla e}{dt} = \mathbb{V} \cdot \nabla e_w - \frac{de_a}{dt} \tag{3}$$

Synoptic sea surface temperature analysis (from which the saturation water vapor pressure is computed) as well as analysis of surface wind, are available routinely.

The convective transfer of moisture is a function of Δ_e (see Laevastu, 1960). Furthermore, Mosby (1933) and Bøyum (1962) found from observations that the changes of air temperature and in analogue the water vapor pressure of the surface air can be well represented as:

$$\frac{de_a}{dt} = -K\ (e_w - e_a)\ -C \tag{4}$$

where K and C are positive empirical constants. Substituting (4) into (3), re-arranging, and assuming $\mathbb{V} \cdot \Delta e_w$ to be constant, equation (3) can be transformed into the integrable form:

$$\frac{d\ \Delta\ e}{dt} + K \cdot \Delta e = C + \mathbb{V} \cdot \nabla e_w \tag{5}$$

which after integration gives:

$$\Delta e = \Delta e_o e^{-Kt} + \left(\frac{C}{K} + \frac{1}{K} V_r \frac{\partial e_w}{\partial r} \right) (1-e^{-Kt}) \qquad (6)$$

where r is distance. This is a useful formula for the computation of Δ e where Δe_o is at the time t_o. The empirical constants of C and K must be determined from observational data. The latest tuned values for the constants C and K are .35 and .10 respectively.

If the wind blows over the sea so that the rate of change of sea surface temperature is constant an equilibrium value of saturation water vapor pressure difference of sea surface and surface air will be reached in about 5 hours (see Figure 6). The equilibrium difference depends on the rate of change of sea surface temperature along a trajectory and implies a constant rate of removal of water vapor from the surface layers. However, there are considerable short time and distance variations of this removal as well as in other parameters used.

Thus the method cannot be used for local and micro-scale sub-synoptic computations. However, it is useful and the only method available for synoptic regional and hemispheric computations of water vapor pressure of the air in the surface layers.

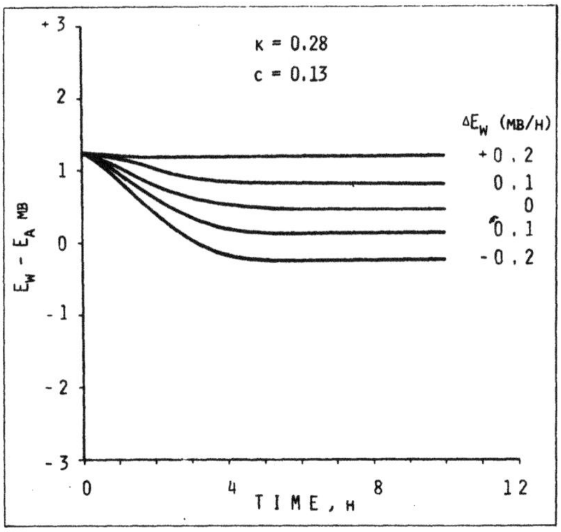

Fig. 6. Changes of Δe(=ew - ea) with time, using Formula 6 and different rate of sea surface temperature change under the trajectory (+0.2 to -0.2 mb per hours travel). Initial Δe=1.25 ^{o}C; C=0.13; K=0.28.

7. REFERENCES

1. AMOT, A., "On the Temperature Differences between the Air and Sea Surface and its Applicability in the Practical Weather Analysis", Met. Ann.,(Oslo) 1(10): 1-9 1955.

2. BEAN, B. R., and E. J. DUTTON, "Radio Meteorology" Nat. Bur. of Standard, Monogr.,92:435pp., 1966.

3. BØYUM, B., "A Study of Evaporation and Heat Exchange Between the Sea Surface and the Atmosphere" Geophys. Publ. (Geophys. Norv.), 22(7):15pp., 1962.

4. DEACON, E. L., C. H. B. PRIESTLY and W. C. SWINBANK, "Evaporation and the Water Balance" UNESCO Arid Zone Res., Climatology, 9-34., 1958.

5. DEACON, E. L., "The Measurement of Turbulent Transfer in the Lower Atmosphere" Advances in Geophys. 6:211-224., 1959.

6. DEACON, E. L., and E. K. WEBB, "Interchange of Properties Between Sea and Air, Small Scale Interactions." The Sea Vol. I (Interscience Publ. N. Y.) 43-87., 1966.

7. GONELLA, J., "Contribution a l'etude des echanges entre l'air et la mer en Mediterranee Occidentale" Cah. Oceanogr., 16(7):547-556., 1964.

8. HARRIS, D. L., "The Wave-Driven Wind" J. Atm. Sc., 23(¢):688-693., 1966.

9. LAEVASTU, T., "Daily Heat Exchange in the North Pacific, in Relation to Weather and its Oceanographic Consequences" Soc. Scient. Fennica. Comm. Physica-Mathem., 31(2):1-53., 1965.

10. MARCIANO, J. J. and G. E. HARBECK, JR., "Mass Transfer Studies, Water Loss Investigations, Lake Hefner Studies." U. S. Geol. Surv. Prof..Paper, 269:46-70.

11. MOSBY, H., "The Sea Surface and the Air" Scient. Res. Norv. Antarct. Exp., 1927-28, No. 10 140pp., 1933.

12. PESARESI, R., "Air-Sea Interaction: The Fluxes in the Atmosphere Boundary Layer" SACLANT ASW Res. Centre. Techn. Rpt. 153 (mimeo), 1969.

13. ROLL, H. U., "On the Application of Turbulence Theory to the Determination of Marine Evaporation" Geofisica, Pura e. Applicata, 56(3):150-158., 1963.

14. ROHWER, C., "Evaporation From Free Water Surfaces" Agriculture Techn. Bull., 217:96pp., 1931.

15. SCHMITZ, H. P., "Erweiterte Grundgleichungen zur Bestimmung der Verdunstung auf dem Meer aus dem klein-und grossräumige Feuchteaustausch" Dtsch. Hydrogr. Zeitschr., 16(3):105-136., 1963.

PRELIMINARY OBSERVATION OF THE EVAPORATION DUCT IN THE I.M.S.T. AIR-SEA WIND TUNNEL

R. Bolgiano, Jr. and Z. Warhaft

School of Electrical Engineering, Cornell University
Radio and Space Research Station, S.R.C.

I. INTRODUCTION

Several of the preceding lectures have noted the difficulties attendant upon attempts to observe the detailed structure of refractivity, temperature, and humidity in the actual atmosphere and to determine from such measurements the manner in which this structure is influenced by the background meteorological conditions. In the first place, it is difficult to place your rapid-response, fluctuation-sensitive instruments at points at which the structure is interesting. Secondly, it is seldom if ever possible to measure adequately the environmental wind, temperature, and humidity so as to describe precisely the conditions under which the observed fine-scale structure occurs. Finally, it is in no sense possible to control these conditions: i.e., it is not possible to "repeat" an experiment, changing the parameters in a known and controlled manner. Consequently, our ability to examine the detailed structure or to discover its relation to the mesoscale or synoptic meteorology is severely restricted.

With these limitations in mind, the I.M.S.T. Air-Sea Interaction Facility appeared to offer unique opportunities for exploration of the underlying mechanisms responsible for the development of boundary-layer structure under various stability and shear conditions, especially in the presence of a free water surface. Therefore a program of investigation has been undertaken, cooperatively between the Radio and Space Research Station, l'Institut de Mécanique Statistique de la Turbulence, and the senior author, to study humidity variability, and its transport, together with the consequent refractivity structure, under temperature and velocity conditions that simulate a range of atmospheric circumstances. In this facility it is possible to achieve and maintain given background conditions, and reproduce the same conditions at later times, then to examine the profiles and fluctuations of the relevant variables with high precision over a wide frequency spectrum.

II. THE FACILITY

The I.M.S.T. Air-Sea Interaction Facility (Coantic and Favre, 1973) consists of a large, low-speed(U_{max}=14 m sec^{-1}) wind tunnel in which the floor of the working section consists of a water channel. The complete air circuit is about 100m in length, the working section is 40m long by 3m wide by 1.4m high and the canal is about 0.8m deep. The fan is in the return part of the air path, above the working section; just ahead of the settling chamber, where the mean air speed is considerably reduced, are the condensing, heating, and moisturizing elements of what is essentially an industrial air-conditioning system. The temperature of the water, which is circulated slowly to insure homogeneity, is also

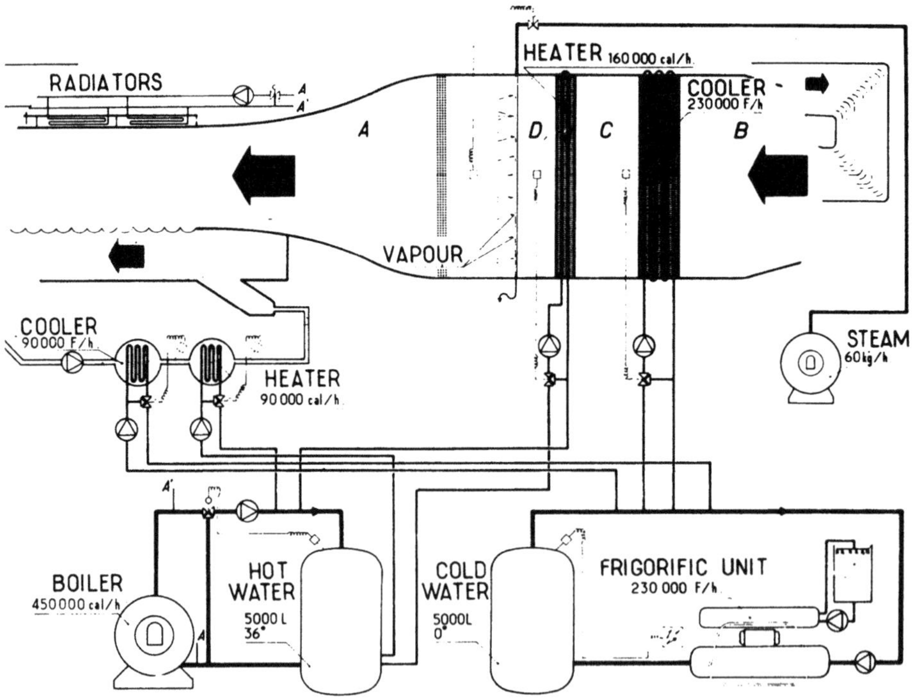

Figure 1. I.M.S.T. Air-Sea Interaction Facility control systems.

controlled; and a wave-making pump is installed at the input end of the canal so that various degrees of sea-state, in addition to that produced by the air motion alone, may be simulated.

All of the controls are brought to a central control panel where they may be connected in closed-circuit feedback systems, with recording indicators. Thus, wind speed, air temperature, specific humidity of the incoming air, and water temperature all may be set independently so as to approximate an extensive range of ocean-atmosphere circumstances. Figure 1 shows, diagrammatically, the wind tunnel complete with its control systems.

The measurement system is depicted in Figure 2. Mean and fluctuation components of air speed, temperature, and humidity are each measured by the various means indicated. One of the most difficult quantities to sense adequately is the rapid variation of humidity. The microwave refractometer (Birnbaum type), fitted with 10μ Wallastan wire temperature-sensing elements at both ends of the open cavity, served as a rapid response (up to 100 Hz) humidity fluctuation sensor. By connecting refractive index and temperature outputs into a simple electronic analogue computing circuit humidity variations were made directly available. Sample

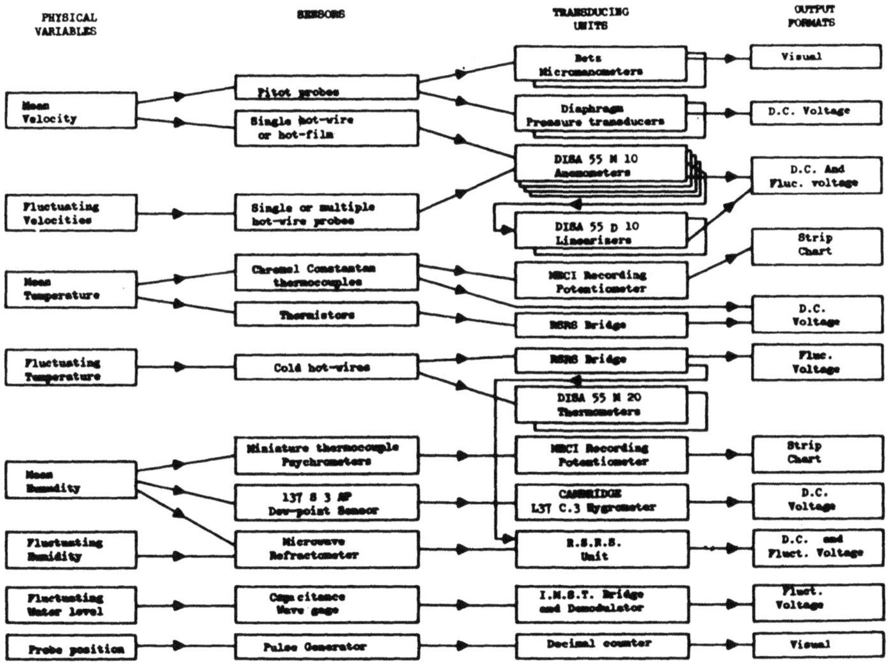

Figure 2. I.M.S.T. Air-Sea Interaction Facility measurement system.

strip-charts were taken of the various fluctuating quantities on a U-V oscillograph and time series were recorded on magnetic tape for later analysis.

III. THE MEASUREMENT PROGRAM

The initial plan called for detailed measurement of mean profile information by I.M.S.T., not necessarily at the same time as the examination of the fluctuating terms. Temperature fluctuations θ', refractivity fluctuations N', humidity fluctuations q', vertical velocity w', and variations of horizontal velocity u', were all to be measured cooperatively by R.S.R.S. and I.M.S.T. under a variety of background conditions involving differing degrees of stability and shear.

Inasmuch as the differences in temperature and specific humidity, between the input air stream and the air in contact with the water surface, determine the difference in density $\Delta\rho$ and, since water current is negligibly small, the velocity differential is essentially the mean speed U of the input air, the ratio of Archimedean forces to inertial forces in the equation of conservation of momentum is given by

$$(-g\Delta\rho/\rho_0)(L/U^2),$$

in which g is accelaration due to gravity, ρ_0 is reference density, and L is a length scale characteristic of the vartical dimension of the mean flow pattern, say, the boundary layer thickness. This ratio, in analogy with Reynolds Number, should bear a unique relation to transition from laminar to turbulent flow, or vice versa. It may be considered a bulk Richardson Number that is given, in terms of temperature difference ΔT and humidity difference Δq, by

$$R_i = 3.4 \times 10^{-3} (\Delta T + 0.18\Delta q)\ gL/(\rho_0 U^2),$$

the units of q being g/kg. The fluctuational character of the flow was explored for values of this parameter ranging from -0.7 (strongly convective) through zero (neutral) to +0.8 (strongly stable). Particular emphasis was placed on the case $R_i \sim +0.3$, since theory indicates this is approximately the critical value that should separate turbulent and non-turbulent regimes. Most of the variation of R_i was realized by changes of wind speed, although clearly the magnitude of ΔT had to be adjusted to realize neutrality and its sign had to be reversed to change from the stable to the convective case.

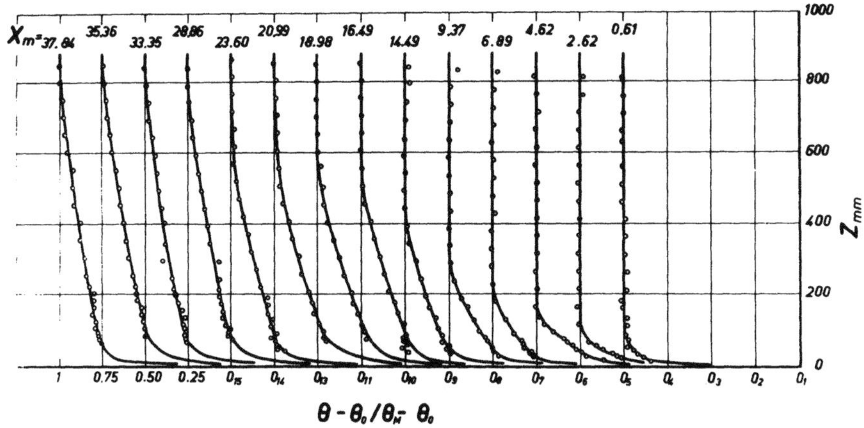

Figure 3. Normalized temperature profiles.
T_{air}=30°C, T_{wat}=20°C, T_{sat}=10°C, U=10m sec^{-1}.

IV. RESULTS

Figures 3 and 4 show mean profiles of normalized temperature and of normalized humidity for a statically stable situation with wind speed 10 m sec^{-1} ($R_i \sim 0.001$). The numbers at the tops of the curves tell the distance down-stream from the tunnel throat at which the profiles were measured. Of particular interest is the

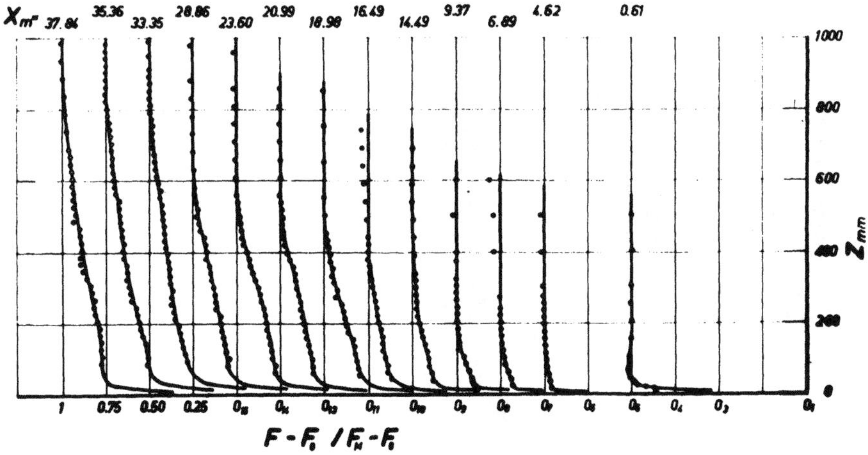

Figure 4. Normalized humidity profiles.
T_{air}=30°C, T_{wat}=20°C, T_{sat}=10°C, U=10m sec^{-1}.

development of the boundary layer with increasing fetch. Note that near the entrance, where wave height was still sufficiently small to permit measurements close to the surface, the steep gradients of the logarithmic profile are in evidence. (These portions of the curves further down stream have been added by hand and do not represent valid data.) The linear portion of the profile (from a few centimeters upward) maintains its slope as we move further from the input, simply extending further and further into the undisturbed air stream. Similar data have recently been collected for other conditions, but many more remain to be examined.

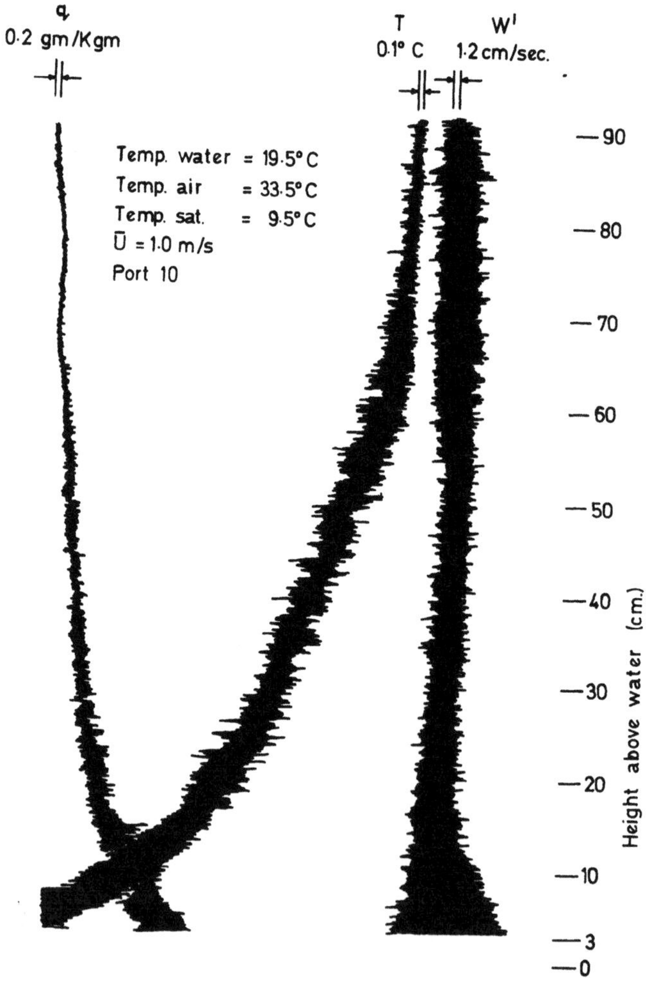

Figure 5. Fluctuation profiles, near neutral case ($R_i \sim +0.1$).

Figures 5, 6, and 7 are coarse profiles made from the fluctuation measurements by recording the variables on strip chart as the sensing elements were transported verically at a uniform rate of about half a meter per second. Figure 5 is for $R_i \sim +0.1$, Figure 6 for $R_i \sim +0.8$, and Figure 7 for $R_i \sim +0.3$. All were taken at a point approximately 25m from the entrance.

A significant increase in turbulent velocity fluctuations in the 10cm nearest the water surface is evident in Figure 5. The temperature and moisture traces indicate the variation of the mean as well as the increase in variance as the surface is

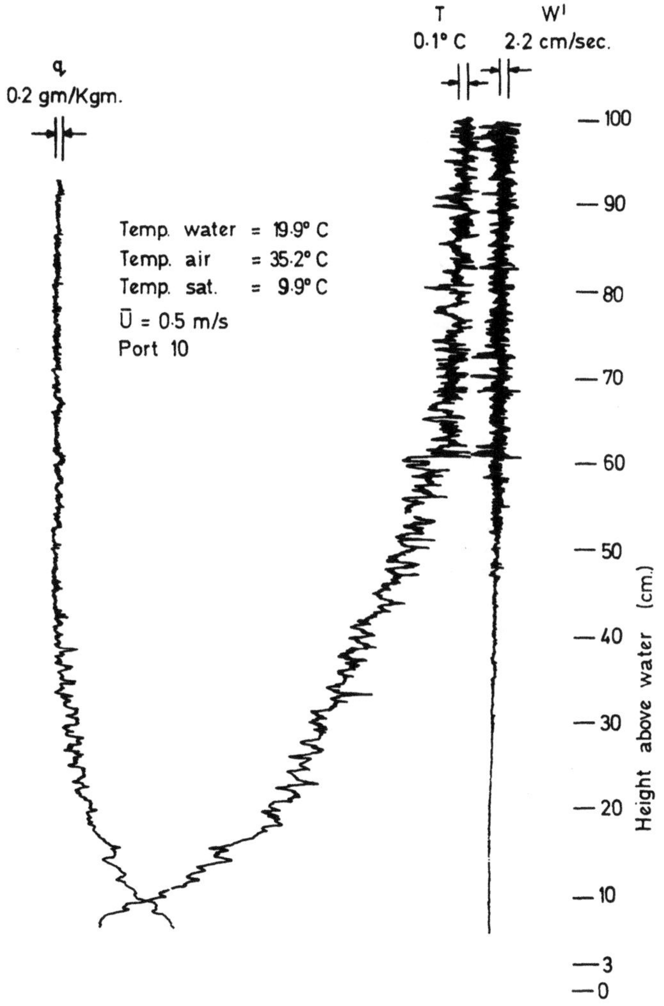

Figure 6. Fluctuation profiles, highly stable case ($R_i \sim +0.8$).

approached. Unfortunately the size of the refractometer cavity precluded taking data closer than 3cm above the water. Nonetheless some evidence of the logarithmic portion of the profile can be seen below 10cm in all three figures.

In the highly stable case, Figure 6, nearly total suppression of the turbulence can be seen below about 40cm. (Turbulence of the boundary layer at the entrance was insured by the presence of large roughness elements on the lower lip of the throat.) On the other hand, the temperature and humidity fluctuations, while clearly much slower in this non-turbulent zone, exhibit as large

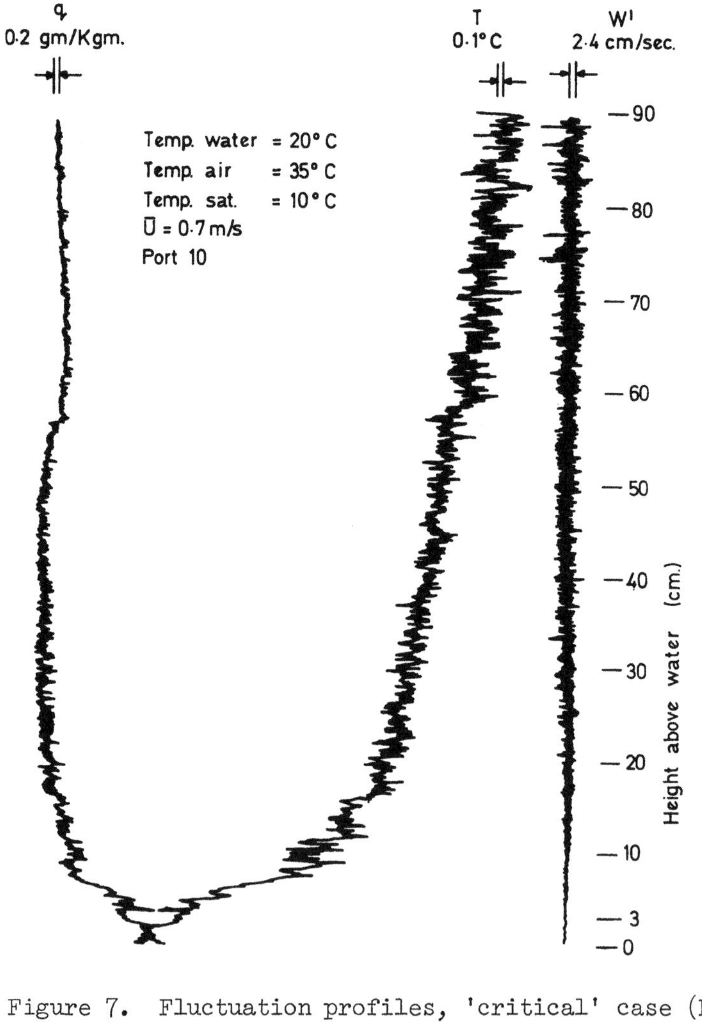

Figure 7. Fluctuation profiles, 'critical' case ($R_i \sim +0.3$).

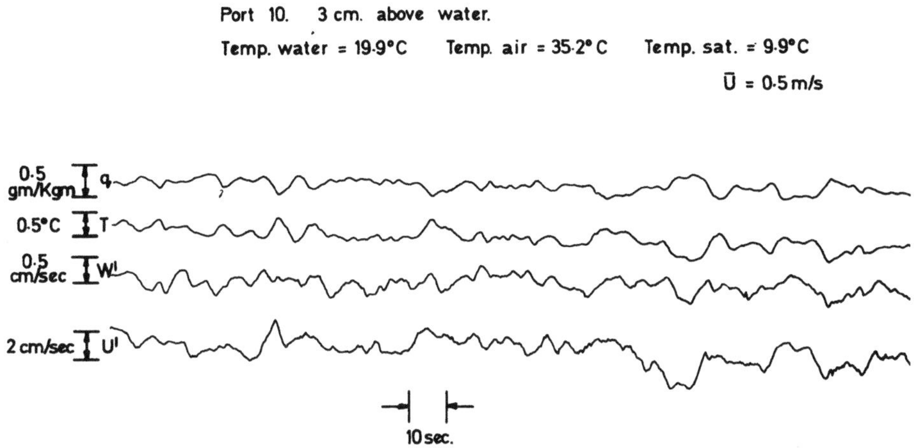

Figure 8. Time record of u', w', T', and q' at 3cm height, highly stable case ($R_i \sim 0.8$).

or larger amplitudes than in the near neutral case. The 'critical' case shows a similar character. Figure 8 is a time record of the fluctuating variables at 3cm altitude for the highly stable case, showing the very slow yet apparently random nature of the variations. Figure 9 is a corresponding record for the critical case. It includes an example of a turbulent burst, a phenomenon observed

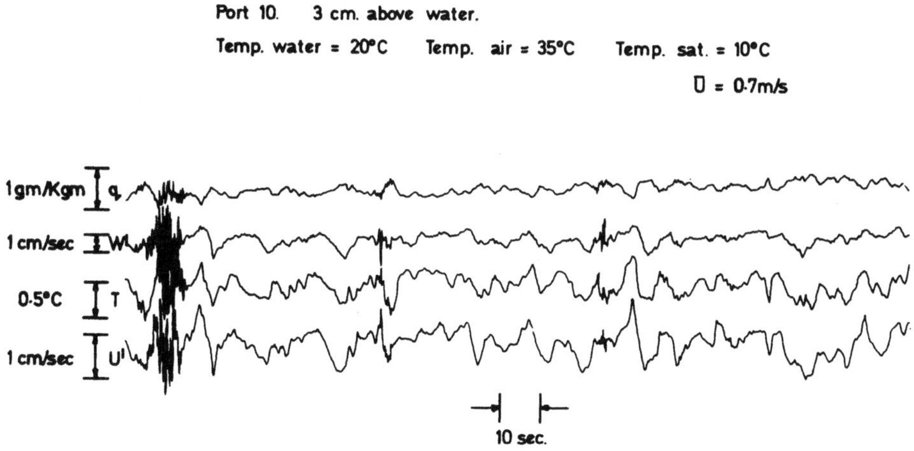

Figure 9. Time record of u', w', T', and q' at 3cm height, 'critical' case ($R_i \sim +0.3$).

only for conditions close to 'critical'. This seeming redevelopment of turbulence, after it has been largely if not totally suppressed, would appear to resemble the process often associated with the formation of a low-level jet just above a radiation inversion, in which turbulence reappears spontaneously following its suppression by the stable density gradient. The amplitude of low frequency fluctuations of q, T, and w is still higher than in the stable case.

Normalized power density estimates of the auto-spectra for the fluctuating quantities have been calculated. This has been achieved by computing the auto-correlation function from a lagged-product sample time series, then cosine-transforming the result and smoothing by means of a Hanning window. In the neutral case the spectra show the usual -5/3 form extending to the neighborhood of 10Hz for Θ' and q' (which are mechanically filtered by the large size of the sampling volume) and at least to the neighborhood of 100 Hz for the velocities. In the highly stable case, as shown in Figure 10, the spectra drop much more steeply (slope about -3.5) and exhibit very little energy above a few tenths of a Hertz. Note the error bar giving the 80% confidence limit for the q'-spectrum.

For the 'critical' case spectral analyses have been made of the data in the slowly varying regions between bursts and also within a burst, both at 3cm height. Figure 11 is an example of the former, showing a very steep drop off of all spectra at about 0.1 to 0.2 Hz, followed by a somewhat less steep form at higher frequencies. It is postulated this latter effect may be a reflection of the occasional insipient bursts that appear between fully developed bursts. The spectra within a burst (those of u' and w' are shown in Figure 12) are somewhat different, both in shape and frequency range. They extend out to tens of Hertz (those of Θ' and q' drop sharply above about 5 Hz in consequence of the reduced mean velocity as well as the smoothing caused by the cavity size); but those of u', Θ', and q' fall off rather more rapidly than in the neutral case. On the other hand, the spectrum of w' exhibits a marked leveling off in the range of 1 to 5 Hz, then also falls with the same slope as u'. Measurements made at 10cm height, although showing no intermittency, do exhibit similar spectral character to that in Figure 12.

From the analyses carried out thus far neither the character of these bursts nor their dynamical significance is fully delineated. Their frequency and, to some extent, their nature is exceedingly sensitive to the value of R_i. A change of as little as 10% in the value of mean wind speed U can alter their pattern completely. Also, although this bears much more careful checking, it may be that they are affected significantly by evaporation. Very recently some comparative tests have been made between

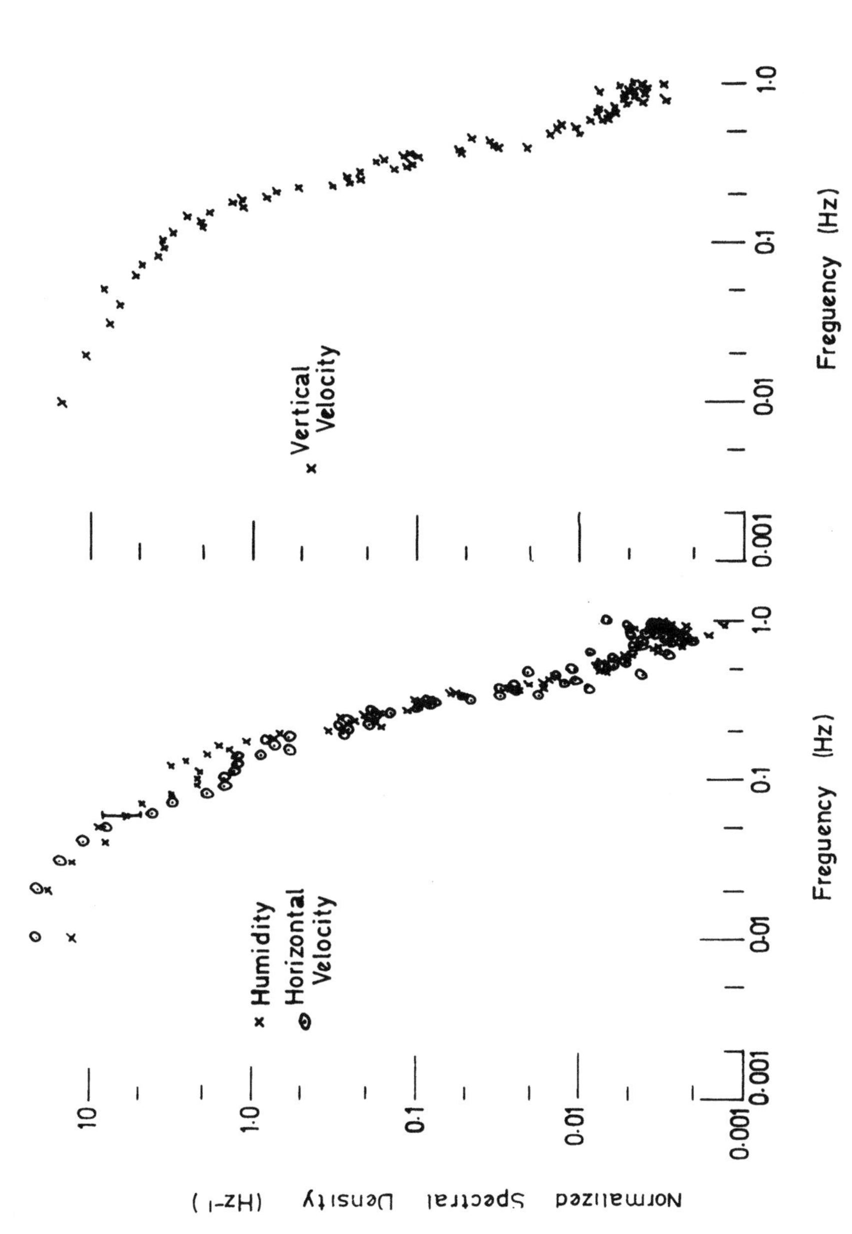

Figure 10. Normalized auto-spectra of u', w', and q' at 3cm height, highly stable case ($R_i \sim +0.8$).

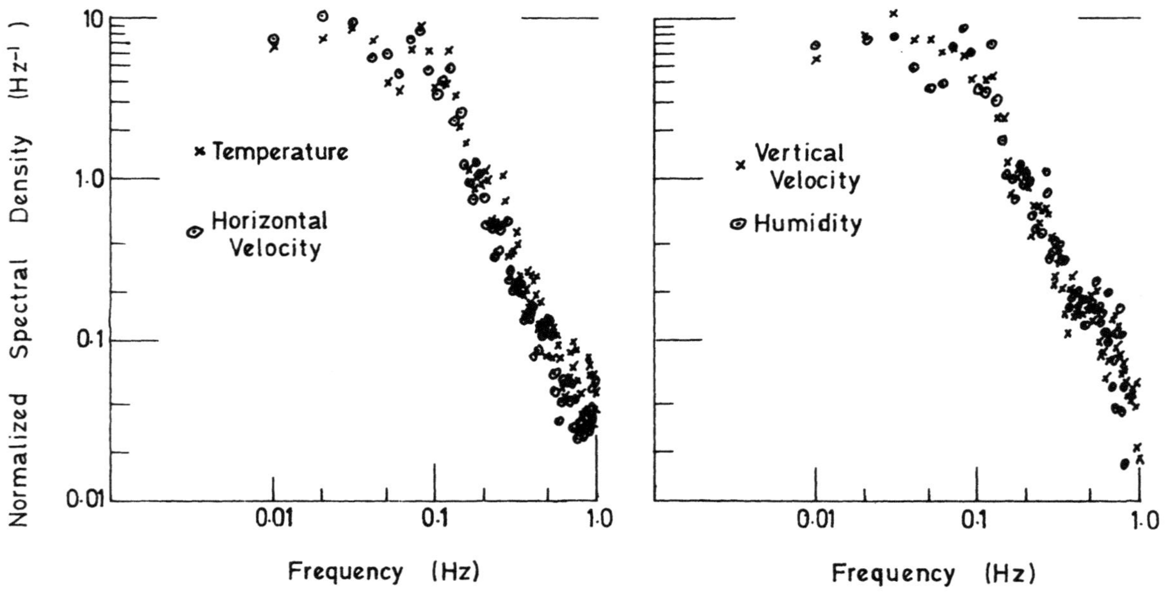

Figure 11. Normalized auto-spectra of u', w', T', and q' at 3cm height between bursts, 'critical' case ($R_i \sim +0.3$).

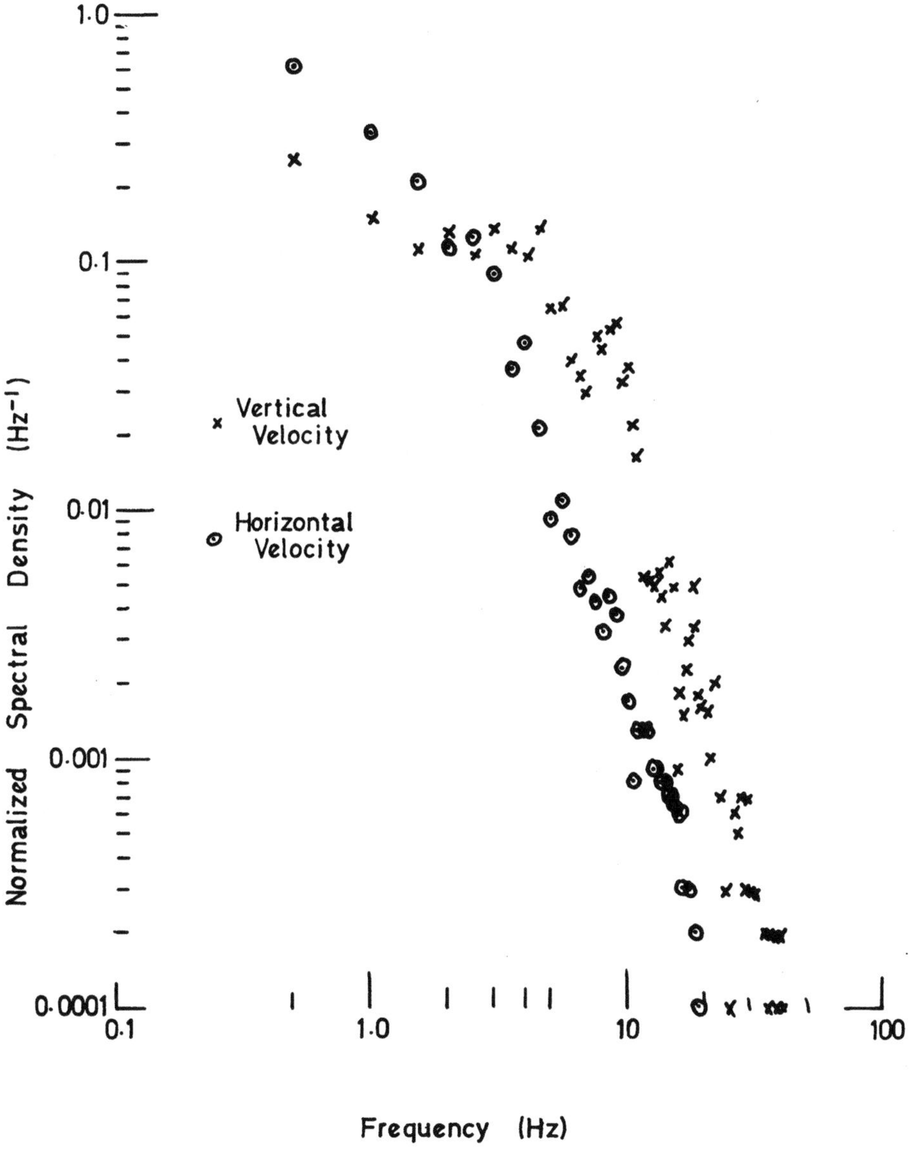

Figure 12. Normalized auto-spectra of u' and w' at 3cm height within a burst, 'critical' case ($R_i \sim +0.3$).

situations in which R_i was maintained essentially constant while evaporation was turned ON and OFF by adjusting both ΔT and Δq. The results have yet to be analyzed in detail; nonetheless there do appear to be some readily discernable differences.

Considerably more study is needed if an understanding of this phenomenon is to be gained. In addition the slow variations in the highly stable situation, and between bursts in the 'critical' case, require careful investigation to ascertain if they are waves or turbulence.

Finally, in specific reference to the radio evaporation duct, profiles of N similar to those of T and q depicted in Figures 5 - 7 show gradients of refractive index that are, in the linear portion of the profile, about 100 times stronger than needed to effect ducting. This extends up to the order of a meter above the water for a fetch of 25m. In the logarithmic region, up to 10cm height, the refraction may well reach 1000 times earth curvature. Such values of refractive gradient were observed for R_i values from +0.1 to +0.8 and mean velocities from 2 to 0.5 m sec^{-1}. A very highly refractive layer of this nature, even though it may be only centimeters thick, may exert a significant effect on microwave propagation so long as the surface of the water remains relatively smooth. Further studies are needed to determine the full extent of this effect. Also additional experimental exploration of the structure of the evaporation duct under a wider range of environmental conditions should be made in the I.M.S.T. Air-Sea Interaction Facility.

V. REFERENCE

Coantic, M. and A. Favre, Activities in and preliminary results of air-sea interactions research at I.M.S.T., paper presented at Second IUTAM-IUGG Symposium on Turbulent Diffusion in Environmental Pollution, Charlottesville, Va. 8-14 April 1973 (in press).

VI. ACKNOWLEDGMENT

This program of investigation has been made possible by the cooperative effort of personnel at l'Institut de Mécanique Statistique de la Turbulence and the Radio and Space Research Station, and through extensive use of the facilities of both laboratories. Financial support has been provided by: N.A.T.O., A.G.A.R.D., Centre National pour l'Exploitation des Océans, Science Research Council, Délégation Générale à la Recherche Scientifique et Technique, and Cornell University.

THE DYNAMICS OF WATER VAPOR FLUX IN THE MARINE BOUNDARY LAYER

B. R. Bean and C. B. Emmanuel

U. S. Department of Commerce, NOAA/ERL, Office of Weather Modification, Boulder, Colorado, U.S.A.

INTRODUCTION

Evaporation from a water surface is defined as the amount of water vapor carried aloft from unit area per unit time; hence,

$$E = \rho_W w \tag{1}$$

where ρ_W is the water vapor density [g/m³], and w is the vertical wind [m/s]. Normally, the determination of the evaporation implies a time average of both ρ_W and w. Consequently, we can express the instantaneous values of both of these quantities as composed of a mean value (denoted by a bracket, $< >$) and a fluctuating quantity (denoted by a prime), i.e.,

$$\rho_W = < \rho_W > + \rho_W'; \; w = < w > + w', \tag{2}$$

then

$$E = < \rho_W'w' > + < \rho_W > < w > , \tag{3}$$

where the first term on the right is the eddy flux. For sufficiently long averaging time, $< w > \equiv 0$, then

$$E = < \rho_W'w' > , \; [g/m^2 \, s]. \tag{4}$$

The short-term fluctuations in the radio refractive index are caused, to a large extent, by variations in the water vapor content of the atmosphere. Thus, a rapid response microwave refractometer may be used to measure the spectral characteristics of water vapor. The addition of a vertical wind sensor allows one to acquire the product $< \rho_W'w' >$, which can be obtained either by electronic multiplication directly in the field or by the use of a computer at a later time. We now discuss results obtained on water vapor flux during three separate field experiments.

BARBADOS OCEANOGRAPHIC AND METEOROLOGICAL EXPERIMENT (BOMEX)

During the BOMEX experimental period measurements were generally made at heights of 18-, 45-, and 152-m above the ocean surface. The aircraft used was a DC-6 instrumented with a gust probe system referenced to an inertial platform which contained high performance accelerometers. The water vapor density fluctuations were determined by a microwave refractometer and a thermistor, both mounted on the gust probe boom. The vertical flux of water vapor was calculated from Eq. (4) for either a 5- or 10-min average of the data which were sampled at 20 times per second. At the typical airspeed of 93 ms^{-1}, this represents a flight path length of 28 or 56 km.

Figs. 1, 2 and 3 present records of w', ρ_W' and $\rho_W'w'$ for upwind, downwind, and crosswind runs made on 11 May 1969. A comparison of the alongwind and crosswind ρ_W' traces shows a striking difference in that the alongwind traces have a ramp-like character while the crosswind traces have a "square wave" signature. Note

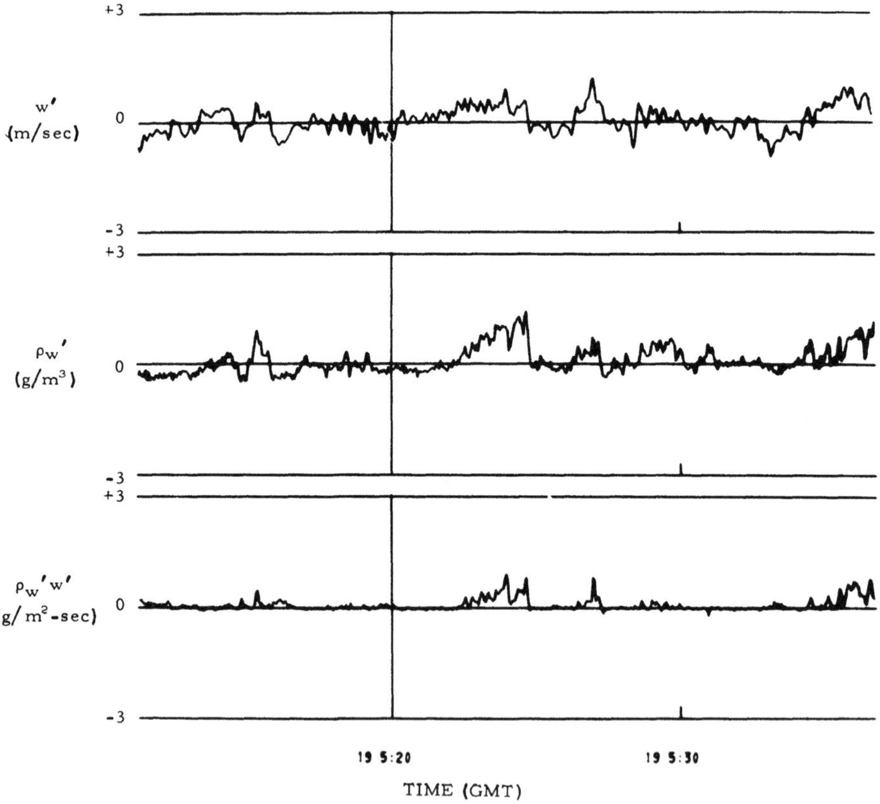

Fig. 1. Sample of time series of w', ρ_W' and $\rho_W'w'$ taken on an 18 m run upwind on 11 May 1969 during BOMEX. Data were taken at 20 samples sec^{-1}.

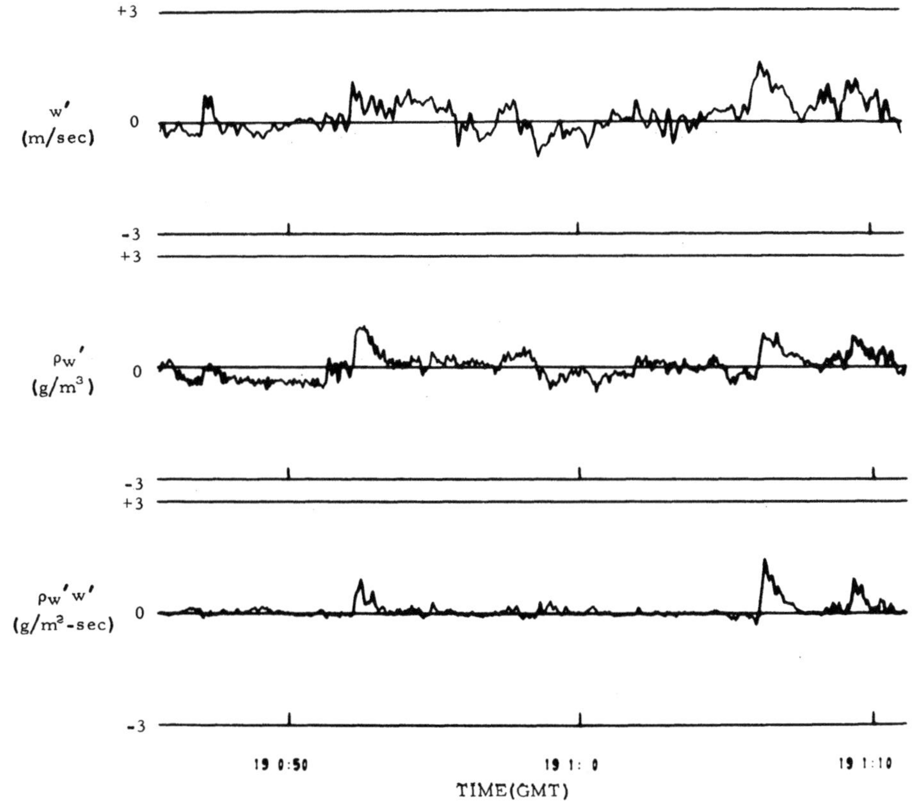

Fig. 2. Same as Fig. 1 except for downwind at 46 m.

also that the ramp signature reverses in going from an upwind to a downwind flight. Thus, a convective element displays a sharp decrease of ρ_w' to windward of the maximum and a slow decrease of ρ_w' to leeward of the maximum (see Fig. 2). Figure 4 demonstrates the peak variance of water vapor which tends to move to lower frequencies with increasing height, i.e., large eddies are more effective in transporting humidity with increasing height while near the sea surface the small eddies contribute more to the flux of water vapor.

A composite of all the vertical wind, water vapor, and water vapor flux spectra are shown in Fig. 5 taken from 10-min samples. The water vapor spectra reveal a peak at 0.024 Hz, irrespective of time, position, heading or altitude at the frequency of the phugoid motion of the aircraft. Note, however, that the phugoid motion does not appear in the final $< \rho_w'w' >$ spectra since it was removed from the w' spectra by the gust probe - INS system.

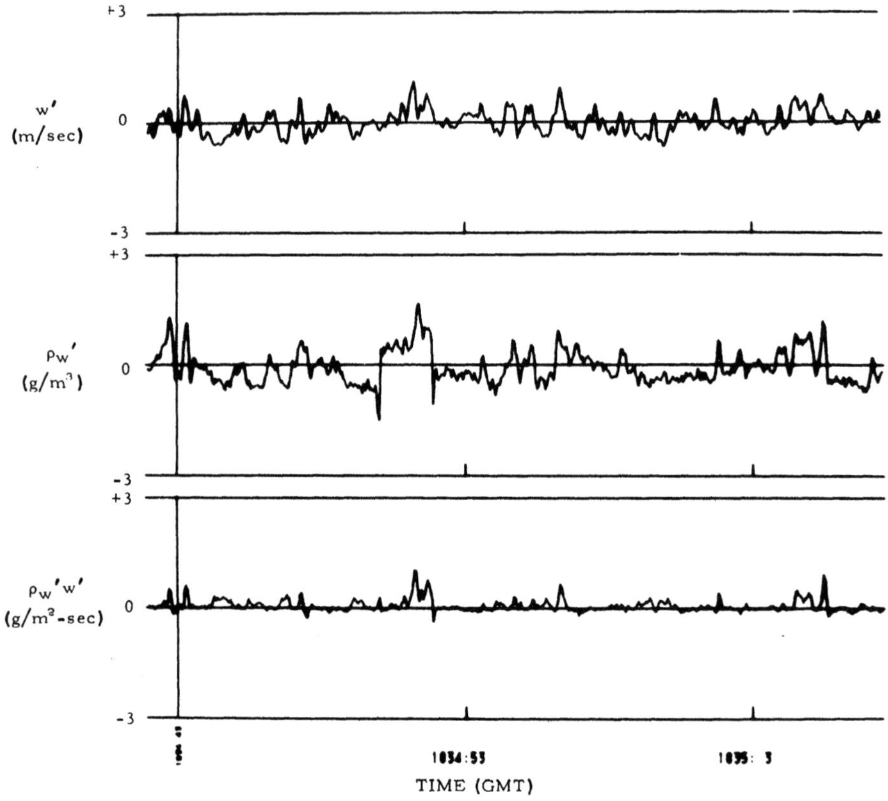

Fig. 3. Same as Fig. 1 except for crosswind at 18 m.

No significant difference could be found in the water vapor flux below 150 m, implying a constant water vapor flux layer. Fig. 6 presents the mean and extremes of E (both 5- and 10-min samples) as measured throughout the BOMEX period, and shows rather clearly that E is quite variable both within a day and on a day-to-day basis.

THE HAWAIIAN ISLANDS TRADE WIND INVERSION STUDY

The need to monitor and measure the pertinent parameters of the ever changing atmospheric structure has led to the development of remote sensing techniques. One such remote sensing tool has been the FM-CW radar [Richter, 1969]. The FM-CW radar was first used to study atmospheric structure in the San Diego area [Gossard and Richter, 1970] and later on the high plains of eastern Colorado with equal success [Bean et al. 1971]. Recently, it was used to study the Pacific trade wind inversion in the area of the Hawaiian Islands

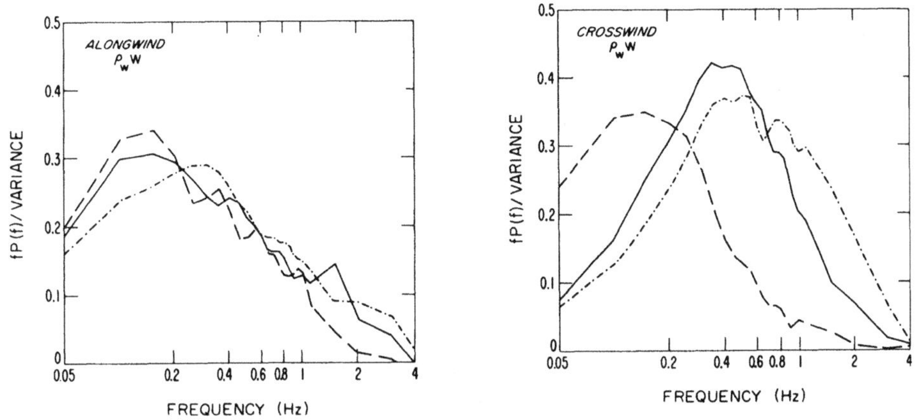

Fig. 4. Composite of all along- and cross-wind water vapor flux spectra taken near FLIP during BOMEX at three altitudes (same as Fig. 1). The frequency-weighted spectral density is normalized by the average variance of all the samples, the 90% confidence limits are very close to the curves presented.

The Hawaiian Islands, Fig. 7, are situated about 21° N latitude and are under the influence of the trade wind inversion. Warm, dry air subsides from above over the moist cooler air from the ocean surface. The interface between these two air masses is characterized by a positive gradient of the temperature with height and by a sudden drop in humidity. The region of the inversion can often be seen in the sheared-off tops of the clouds. The prevailing wind is from the northeast resulting in cloud structures on the same side of the mountains along the island chain due to the lifting of moist air.

The FM-CW radar was located near the southwestern coast on the island of Lanai at an elevation of approximately 400 m. Two 3-m parabolic antennas were used in the backscatter mode. The isolation between the transmitting and receiving antennas was in excess of 85 dB.

In-situ measurements were made by an aircraft instrumented with a microwave refractometer as well as a thermistor for measurements of the temperature field. Fig. 8 illustrates the refractivity structure of the trade wind inversion as a function of position along the wind and across the island of Lanai. Over the open ocean the inversion level was found to be quite stable between 1.3 and 1.5 km above the sea surface but lifted in the vicinity of the island. As the air passed through the Pailolo Channel between Maui and Molokai, the layer strengthened and rose to more than 2 km, returning to its initial height only after it cleared the channel.

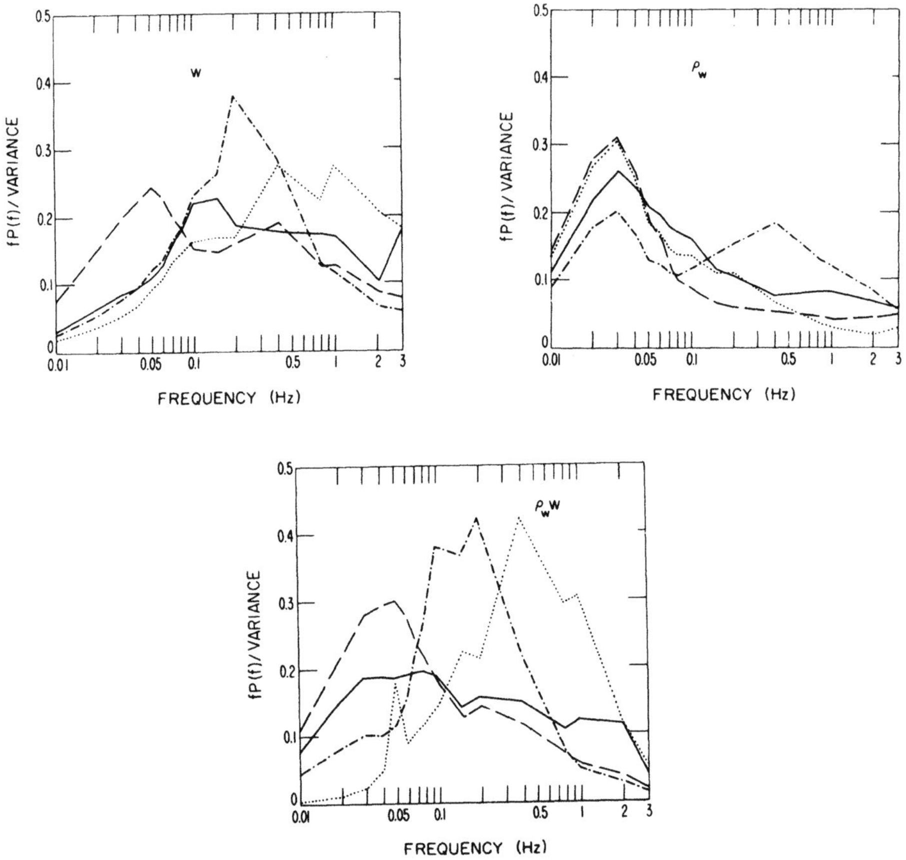

Fig. 5. Composite of all vertical wind, water vapor, and water vapor flux spectra taken from 10-min sample lengths. The frequency-weighted spectral density is normalized to the average of all the particular variances used in the composite. There are approximately four runs at each altitude and heading. Along-wind spectra: 152 m (long dashes) and 33 m (solid line); cross-wind spectra: 152 m (dash-dotted line) and 33 m (dots).

Fig. 9 depicts the FM-CW radar observations on 18 and 20 June 1971. Note that the inversion level is highly variable both in position and intensity. At times the layer would move from above 2 km to less than 1 km within a time period of 5 minutes, while its thickness would change rapidly from a few tens of meters to more than 100 meters. The wave-like structure is presumably induced by the mountain on the east side of the island. The layered structure is

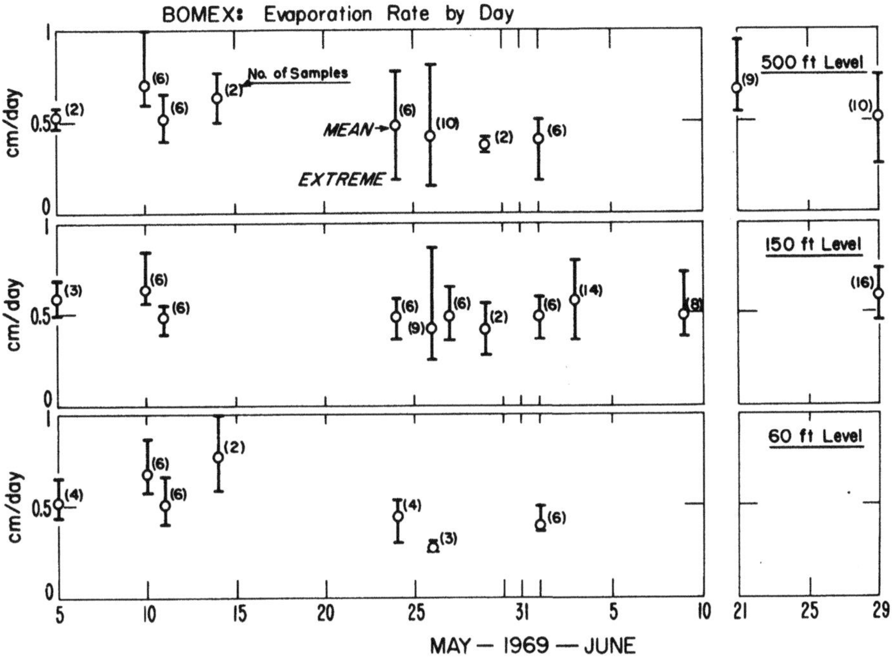

Fig. 6. Daily averages and extremes of water vapor flux taken during BOMEX. Note that day-to-day variations seem to follow at all altitudes.

observed in regions where there exist sharp breaks in the refractivity profile and where the temperature gradient turns sharply positive, i.e., inverts. Most layers exist just below the point where the profile turns back to "normal." A rather interesting observation was noted on 19 June 1971 when the aircraft observations indicated the existence of two layers, one at 2.0 km and the other at 1.8 km. However, during this time, the FM-CW radar indicated only a single layer descending rapidly from 2.0 km to 1.8 km. The aircraft apparently had traversed through the same layer twice.

In a previous study on the high plains of southeastern Colorado [Bean et al., 1971], the FM-CW radar returns were compared to refractivity measurements taken on a nearby tower. Defining the structure function of refractive index as

$$\overline{(\Delta n)^2} = C_n^2 \, r^{2/3} \, , \qquad (6)$$

where $\overline{(\Delta n)^2}$ is the mean square of the difference in refractivity at two sensor points separated by a distance r, and C_n is the structure function constant. Expressing the radar reflectivity, η, as

$$\eta = 0.38 \, \lambda^{-1/3} \, C_n^2 , \qquad (7)$$

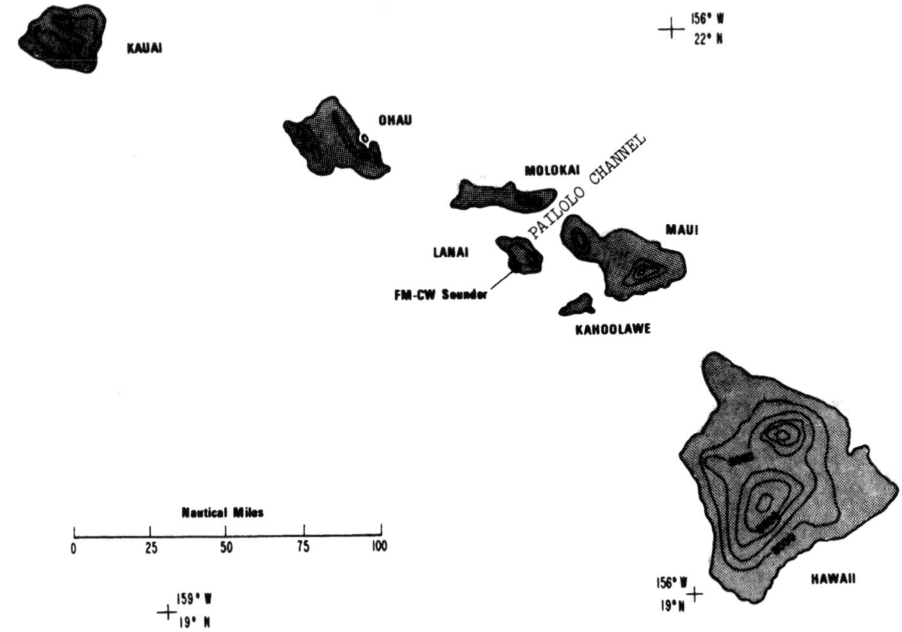

Fig. 7. The Hawaiian Islands.

where λ is the wavelength, we have

$$\eta = 0.38\ \lambda^{-1/3}\ r^{-2/3}\ \overline{(\Delta n)^2}\ , \qquad (8)$$

i.e., a relationship linking the radar reflectivity and the structure function of the refractive index (n). In the Colorado experiment, although the reflectivities varied over two orders of magnitude (10^{13}- 10^{15}), the radar measurements and the tower measurements agreed to within a factor of two, a discrepancy that may well be accounted for by the separation between tower and radar (240 m). The same experiment also found that the largest values of $\overline{(\Delta n)^2}$ occurred in regions where the refractivity (and temperature) profiles exhibited sudden changes in the gradient. Hence, although the radar does not detect the gradient, it does detect the turbulent fluctuations associated with sharp changes in the gradient.

It is possible to compare the relative contributions to the refractivity profile from the temperature and humidity. At one kilometer (Ottersten, 1969), the difference of refractivity, N, is

$$\Delta N = \Delta(n - 1)\ 10^6 = -\ 1.09\ \Delta T + 4.72\ \Delta e + 0.28\ \Delta P, \qquad (9)$$

where ΔT in °C, Δe and ΔP in mb are the differences in temperature, vapor pressure and total pressure, respectively. The results of this study are presented in Table I. The dominant gradient results from humidity which produces the very strong gradients of refractivity and, presumably, the fluctuations of N which give rise to the

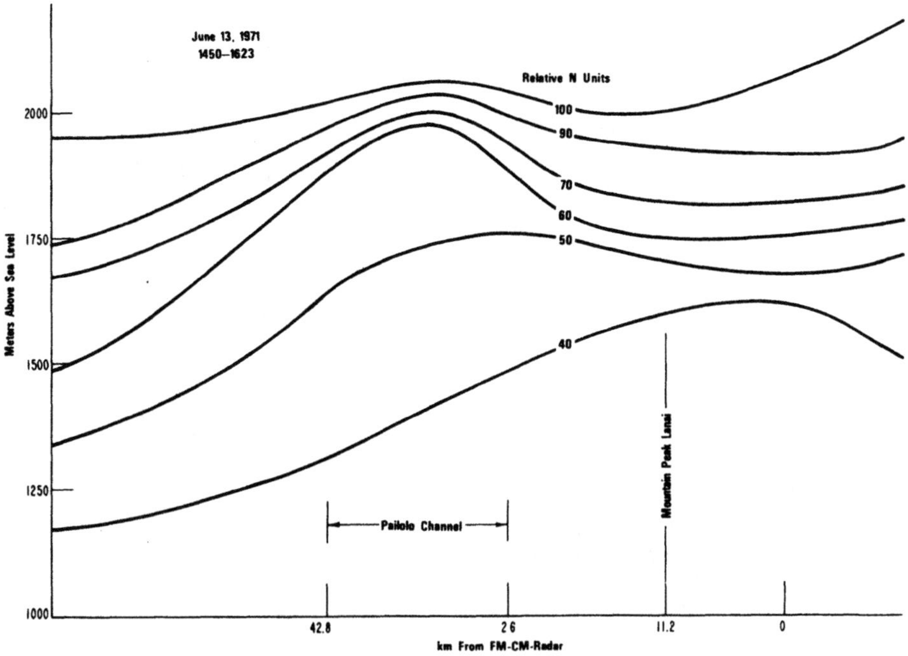

Fig. 8. Refractivity structure along the wind across island of Lanai.

radar return. It also helps to explain the flat sheared-off tops of clouds as they evaporate in the dry warm region of the inversion.

THE INTERNATIONAL FIELD YEAR FOR THE GREAT LAKES (IFYGL)--MEASUREMENTS OF EVAPORATION FROM LAKE ONTARIO

Airborne measurements of evaporation from Lake Ontario, Fig. 10, were made from a DC-6 aircraft equipped with a gust probe system and a microwave refractometer as during BOMEX. These measurements were taken during five intensive periods (each of about 10 days in length) spanning the "evaporation year" from May to November 1972.

IFYGL was a cooperative effort by scientific teams from Canada and the United States. It was an intensive study aimed at providing knowledge and understanding of lake processes to improve management of the quantity and quality of lake water. Consequently, the biology, chemistry, meteorology and water movement and balance of the lake water were investigated. Here we discuss the airborne measurements of evaporation and heat flux. Table II gives the monthly averages of the surface temperature, the temperature at 100 ft above the surface, the water vapor, the vertical flux of water vapor and the vertical temperature flux. Note that in May and June the water surface temperature (T_o) is much less than the air temperature at

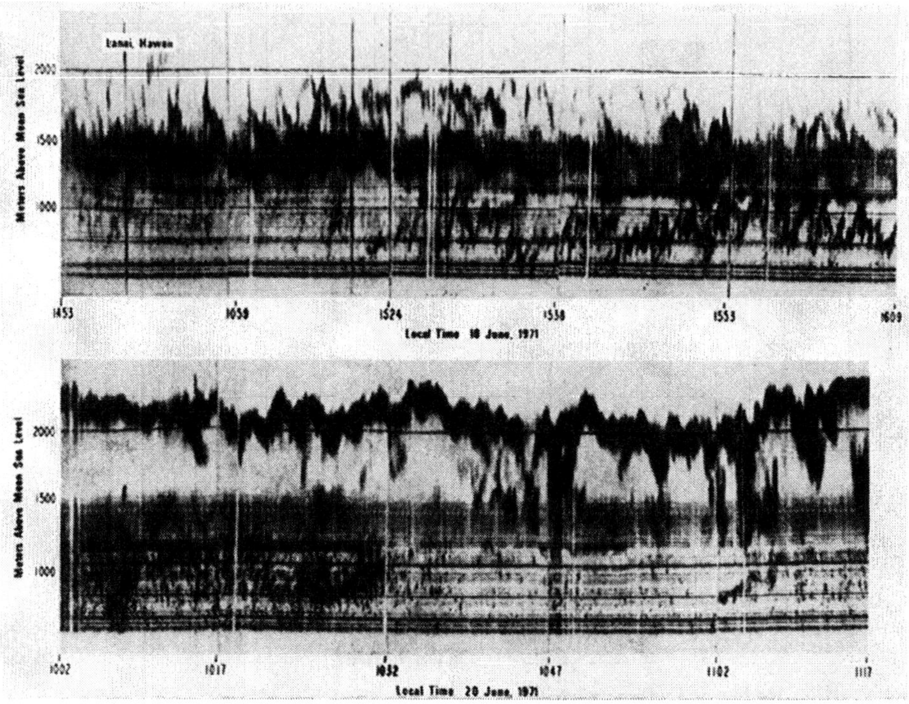

Fig. 9. FM-CW radar returns from clear atmosphere.

100 ft. During this time we found strong temperature inversions to be rather common, with the fluxes of water vapor and temperature to be negative or zero indicating that water vapor was condensing on the lake and that heat flow was from air to water. In August, the surface temperature of the lake and the temperature at 100 ft were almost in equilibrium resulting in a negligible evaporation rate.

The peak in the evaporation cycle appears in October when the surface temperature was consistently higher than the overlying air temperature. Both the water vapor and temperature flux were positive. The same results were also found during the month of November, although there is a slight shift toward a lower evaporation.

Table III presents the results for a two-day period in October during which time there was a Polar Continental outbreak. This phenomenon produces remarkable bursts in evaporation. Cold dry air from Canada moved across the lake at 25 knots. The lake surface temperature remained unchanged but the air temperature at 100 ft dropped from 12° to 6°, while the evaporation increased to almost 1 cm/day (October 9). The following day, October 10, was marked by an unusual event. The wind, quite brisk, was from the north blowing the warm surface water to the south shore of the lake with upwelling of cool water on the north (this, to some extent, was also observed

TABLE I

The Gradients of Meteorological Parameters Across the Pacific Trade Wind Inversion Over the Island of Lanai

June 18, 1971 1450 - 1600

Altitude ≈ 1 km

$$N = -1.09\ \Delta T + 4.72\ \Delta e + 0.28\ \Delta P^*$$

Layer Height Interval	Measured Differences			Contributions to N			Derived Humidity Difference
ΔH	ΔP	ΔT	ΔN	ΔN_t	ΔN_p	ΔN_e	Δe
300	- 31 mb	2.4°	- 46	- 2.6	- 8.7	- 34.7	7.4 mb
300	- 31	1.5	- 42	- 1.6	- 8.7	- 31.7	6.7
100	- 10	1.5	- 2.2	- 1.6	- 2.8	- 21.6	4.6
300	- 31	1.8	- 42	- 2.0	- 8.7	- 31.3	6.6
60	- 6	2,6	- 28	- 2.8	- 1.7	- 23.5	5.0

*Bean and Dutton (1968)

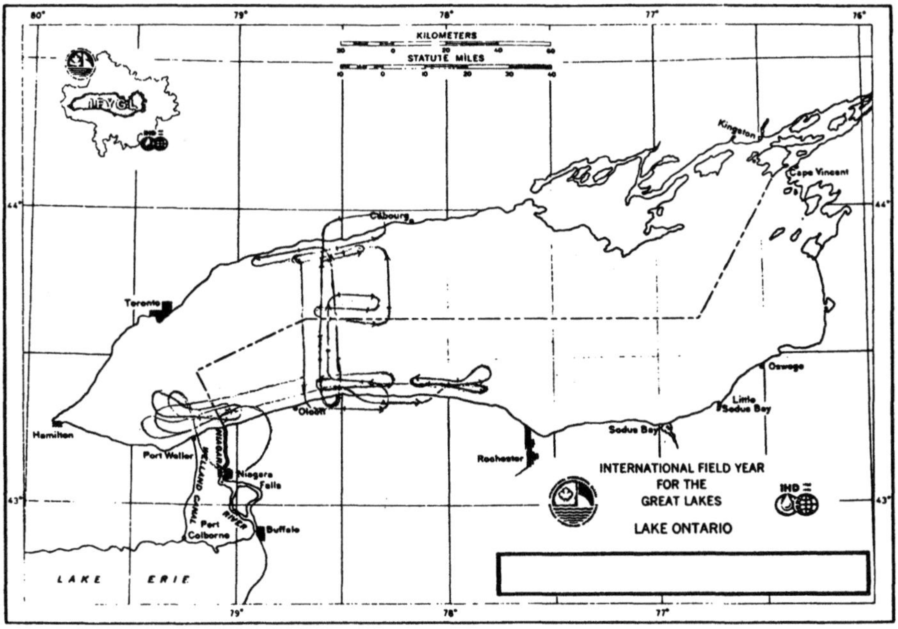

Fig. 10. Lake Ontario with typical flight pattern.

TABLE II

IFYGL Average Values 32.8 m (100 ft.). All Data.

	T_o(°C)	T(°C)	ρ_v(gm/m³)	$f_{\rho v}$* (cm/day mw/cm²)	f_T† (°m/sec mw/cm²)
May	2°	10°	5	0	0
June	9°	17°	11	- 0.013 - 0.376	- 0.008 - 0.965
August	19°	18°	12	0.040 1.16	0
October	14°	12°	8	0.315 9.104	0.025 3.015
November	6°	0°	4	0.20 5.78	0.037 4.46

*Vertical water vapor flux
†Vertical temperature flux

TABLE III

IFYGL Average Values 32.8 m (100 ft.)

	T_o(°C)	T(°C)	ρ_v(gm/m³)	$f_{\rho v}$ (cm/day mw/cm²)	f_T (°m/sec mw/cm²)
October 9	14°*	6°	4	0.95 27.4	0.092 11.1
October 10	14°*	6°	5	0.38 11.0	0.040 4.82

*Average surface temperature. North shore surface temperature was 6°C.

on the 9th of October). Fig. 11 shows the lake surface, the white streak separates the cooler and warmer waters. North of the line the surface temperature is 6 °C, south of the line 12 °C. The white streak is interpreted as the mixing region between the two water regions of markedly different temperatures. Fig. 12 represents the change in the surface temperature (∿ 2 °C) of the lake as the aircraft traversed across the streak line, which is only a few meters in width.

Fig. 11. The interface between warm and cold water masses.

CONCLUSIONS

It is found that airborne measurements of water vapor and vertical velocity yield a method of determining evaporation for both the "steady-state" conditions of the trade winds and the rapid variations associated with air-lake interactions. It is further shown that orographic effects are large and may lead to substantial error in interpretation of such measurements. The FM-CW radar serves as an invaluable guide to track elevated structure and interpret the aircraft measurements.

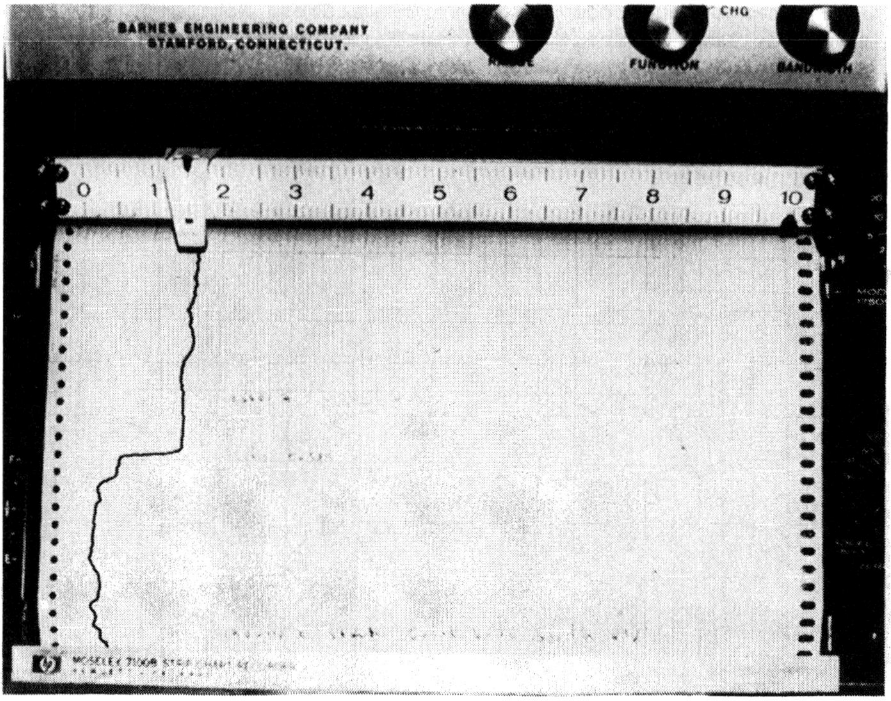

Fig. 12. Temperature across water interface.

REFERENCES

Bean, B. R., R. Gilmer, R. L. Grossman, R. E. McGavin, and C. Travis (1972): An analysis of airborne measurements of the vertical water vapor flux during BOMEX, J. Atmos. Sci., 29, No. 5, pp. 860-869.

Bean, B. R., R. E. McGavin, R. B. Chadwick, and B. D. Warner (1971): Preliminary results of utilizing the high resolution FM radar as a boundary layer probe, Boundary Layer Meteor., 1, pp. 466-473.

Gossard, E. E. and J. H. Richter (1970): The shape of internal waves of finite amplitude from high resolution radar sounding of the lower atmosphere, J. Atmos. Sci., 27, pp. 971-973.

Ottersten, H. (1969): Radio back scattering from the turbulent clear atmosphere, Radio Sci., 4, No. 12, pp. 1251-1255.

Richter, J. H. (1969): High resolution tropospheric radar sounding, Radio Sci., 4, No. 12, pp. 1261-1268.

THE STRUCTURE OF THE ATMOSPHERIC BOUNDARY LAYER AT SEA

Lutz Hasse

Universitat Hamburg,
Meteorologisches Institut

SUMMARY

Recent experimental results of the parametrization of the turbulent fluxes of momentum, heat and moisture in the atmospheric surface layer at sea are compared. Parametrization by the bulk aerodynamic approach in terms of surface layer variables was in general agreement for both observations from the tropics and midlatitudes. For the small sensible heat flux in the tropics parametrization was poor. This may be due to different mechanism of vertical fluxes of heat and moisture which are demostrated by the dissimilarity of spectra and the different behaviour of time-cross-sections of temperature and humidity. For the determination of surface layer parameters in terms of Planetary Boundary Layer external parameters the resistance law and experimental knowledge of the grostrophic drag coefficient are discussed.

1. SURFACE LAYER

The theory of the surface layer of the atmosphere has been reviewed in this meeting already by Dr. Coantic and Prof. Bolgiano. Concerning the near water surface layer this paper therefore is restricted to a short discussion of recent experimental results.

During the past years the main effort of meteorological experimentation in the surface layer at sea has been aimed on the determination of the turbulent fluxes of momentum, heat and water vapor. With a knowledge of these fluxes we obtain at the same time through the so called flux-gradient relationship the knowledge of the gradients of water vapor and temperature, which are important for e.m. wave propagation.

The turbulent vertical fluxes are customarily related to simple surface layer variables, say mean wind speed, mean air sea temperature difference, mean air sea water vapor difference. This description, which is now called parametrization, is an important tool for several reasons. As the fluxes are highly variable, in order to compare results obtained at different times and areas it is necessary to relate them to other variables. Parametrization with aid of surface layer variables is the simplest way to do so, since the mean parameters can be obtained with good accuracy. On the other hand, parametrization is more or less the only way to determine the turbulent vertical transports at the air sea interface on a large scale, for example for numerical or experimental studies of air sea interaction or wave propagation forecasts.

In the following, some results of parametrization are discussed. It is assumed that it will not be necessary to detail the methods which were applied to determine turbulent transports. These methods are known as

eddy correlation or direct method
(from measurements of the fluctuating quantities)
aerodynamic of profile method
(from vertical mean profiles)
dissipation or spectra method
(from the -5/3 region of the spectrum where the spectral energy is related to the dissipation).

Most earlier results of parametrization were obtained in mid latitudes. The ATEX and BOMEX Expeditions of 1969 gave a welcome possibility to compare these with results from low latitudes. The variance of the bulk transfer coefficients within one experiment and from one author to the other still amounts to 15% or 20%. Therefore it is especially welcome to check what factors, which are not included in the parametrization, are important in the sense that parts of the variance can be traced back to their influence. This finally may lead to a better parametrization. Compared to middle latitudes, in low latitudes we have a Coriolis parameter smaller by, say, a factor of 4 and a higher water vapor content with its influence on stability. As far as the influence of waves is concerned there was a really long fetch. There may be differences in the surface films too.

Momentum flux: The rusults of direct measurements are given in fig. 1, both from ATEX and BOMEX (Pond et al. 1971). It is evident that there is no systematic difference. The stress from our measurements can be described through an average drag coefficient c_D^2 of 1.4×10^{-3}. During both expeditions the observations were taken at slightly unstable conditions and the neutral drag coefficient would be a little smaller. Since a typical value of the neutral drag coefficient at higher latitudes is 1.3×10^{-3}, there is general agreement, that is to say, the influence of variation of the Coriolis parameter, if any, is small compared to the scatter.

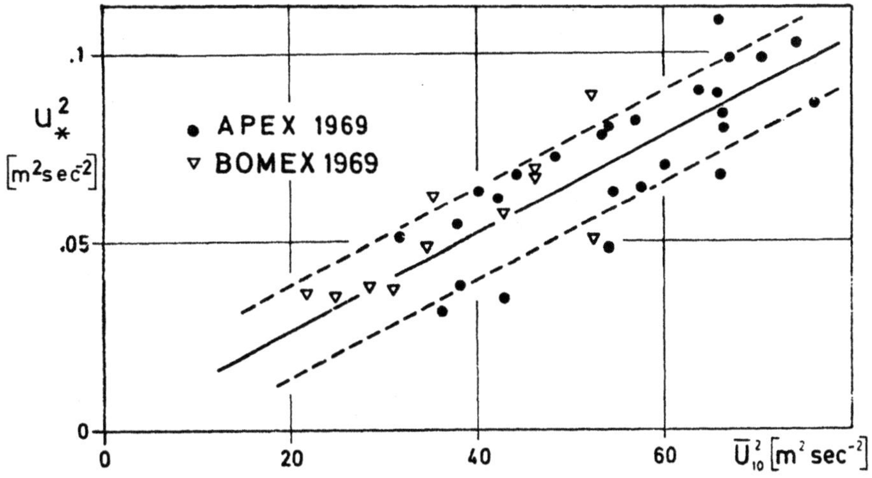

Fig. 1: Friction velocity u_* versus mean wind speed from direct measurements during BOMEX (Pond et al. 1971) and ATEX (Dunckel et al. 1973).

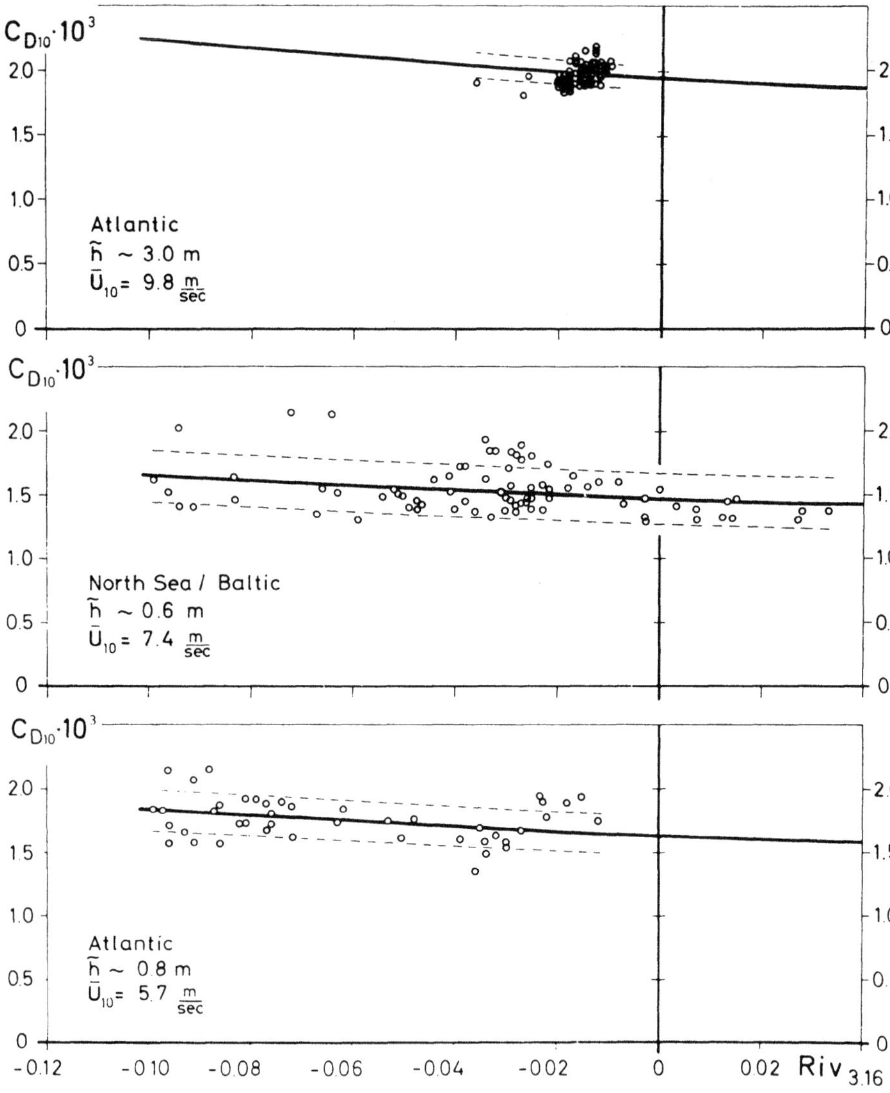

Fig. 2: Apparant drag coefficient c_D2 (referred to 10 m height) from profiles under different conditions of sea state. $\tilde{h}$ is estimated wave height (Dunckel et al. 1973).

An interesting phenomenon is that the determination of the friction velocity from profiles between 1.2 m and 8 m yielded a higher drag coefficient, around 1.9×10^{-3}. Krügermeyer, (Dunckel et al., 1973) could show that this is an influence of the greater wave heights encountered at the Atlantic (fig. 2). As we worked with a surface following buoy, we were able to measure the wind speed at a height (1.2 m) which was sometimes lower than the wave heights (double amplitude up to 3 m estimated). If the levels below 2 m are disregarded the drag coefficient from profiles approaches the value from direct measurements (of cause, the scatter is increased as there are fewer levels left for the determination of the profile). It has been hypothesized that the total stress above the wave τ_0 is splitted into two parts, one purely turbulent τ_t, one wave induced τ_w. The wind profile would be determined by τ_t. In case of swell, waves would feed momentum upward and increase τ_t as τ_0 is approximately constant. It seems that the figure gives a welcome experimental evidence of wave induced stress at the open ocean.

I have been asked to comment on the roughness parameter zo at sea. It may be here a convenient point to do so. Personally, I do not like z_o. It is much in use now because it has the dimension of a length and is welcome as a length scale in the surface layer for the use in dimensional arguments. But, if we look in the physical interpretation of z_o we have to go back to Prandtl who 40 years introduced it as an artefact. z_o is defined only for neutral conditions through the logarithmic profile

$$u(z) = \frac{u_*}{k} \ln \frac{z}{z_o}$$

The logarithmic profile can be derived in several ways, but almost all assume that u_* is at least approximately constant with height. This is not true at sea. The stress is partially taken up by the waves. Obviously, as mentioned, the profile is deformated in the lower part of the "constant stress layer". The physical significance of z_o therefore is rather doubtful.

Since $$c_D = \frac{u_*}{U} = \frac{k}{\ln z/z_o}$$

under neutral conditions z_o is not independent of c_D, rather ln z_o varies like c_D. It is preferable to use c_D, since this does not imply extrapolation down to zero wind speed, but rather refers to the momentum flux $\tau = \varrho c_D^2 U^2$ at some height above the waves, say 10 m. c_D is approximately constant, it varies only very little with wind speed as shown by Brocks and Krügermeyer (1972) through careful analysis of a selection of strictly neutral wind profiles. The Charnock formula $z_o = u_*^2/g$ which indicates a variation of z_o with wind speed was derived by dimensional reasoning as a simple dimensionally correct relation, but obviously does not describe the physical process.

Water vapor flux: Determination of water vapor flux from profiles during ATEX and from eddy correlation technique during BOMEX (Pond and Phelps 1971) agree very well. The mean water vapor transport coefficient $c_E = E/L \rho U \Delta q$ is 1.3×10^{-3}. This is in very good agreement with the early profile measurements of Brocks (e.g. 1963) at higher latitudes.

It can therefore be concluded that parametrization of momentum and water vapor flux in terms of surface layer variables does not show differences between low and middle latitudes which would exceed the known experimental scatter.

Vertical flux of sensible heat: Earlier determination showed that the sensible flux may well be parametrized by the product $U \Delta \theta$ (e.g. Kruspe 1972). During BOMEX and especially ATEX sensible heat fluxes were small and energetically uninteresting. They were peculiar in the sense that they could not be well parametrized. Holland found from the BOMEX data that the vertical sensible heat flux correlates well with $U \cdot \Delta q$. The implication is that the high moisture content of the air in the tropics noticeably influences the buoyancy. Wet air will tend to ascend. If wet correlates with warm, this gives raise to an enhanced sensible heat flux. In case of the ATEX data, the bulk parameter product $U \Delta q$ did not vary much, therefore the correlation between H and $U \Delta q$ was poor. This neither confirms nor contradicts Holland's findings. The influence of moisture in our case cannot explain the poor parametrization. Wucknitz (Dunckel et al. 1973) showed that the correlation of the covariance $w'T'$ with the bulk parameter product $U \Delta \theta$ is different in different frequency bands. The correlation was high (.87) for the frequency band 0.005 to 0.25 Hz and poor outside of this band.

We are not yet able to explain this in the sense that it may be traced back to other characteristics of the boundary layer to allow a parametrization. It is felt that there is at least a connection to the findings of Phelps and Pond (1971) during BOMEX who demonstrated a dissimilarity in the shape of the spectra of temperature and humidity.

A corresponding dissimilarity of spectra of temperature and humidity has been demonstrated for middle latitudes too (fig. 3, Kruspe 1973). Compared to the vertical velocity, the temperature shows more spectral energy at higher frequencies and the humidity at lower frequencies. This difference in the spectra is carried over into the co-spectra (fig. 4).

From the fact, that the dissimilarity of spectra has been observed in mid-latitudes too it can be concluded that the reason cannot lie in the influence of the humidity on buoyancy, since in our latitudes the water vapor content is by a factor of, say, 3 smaller than in the tropics.

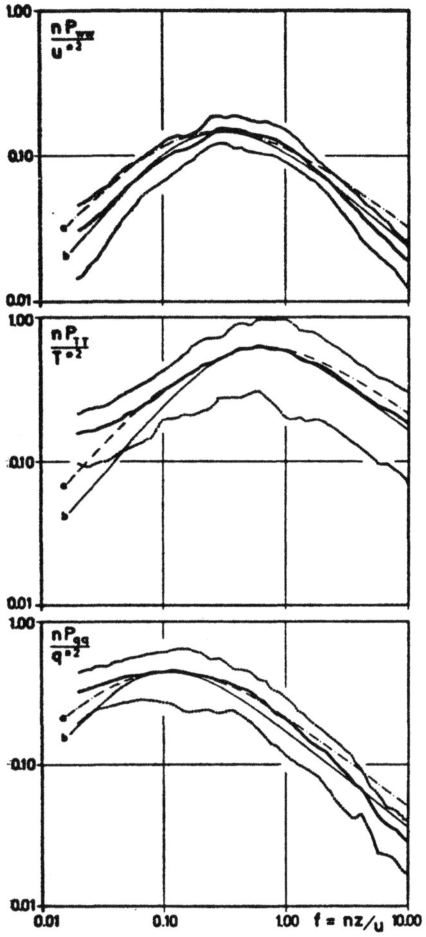

fig.3 Average and range of spectral energy for vertical velocity (above), temperature(center) and specific humidity (below) from the North Sea (Kruspe 1973).

A different behaviour of temperature and humidity at somewhat longer periods has also been detected in time cross-sections from temperature and water vapor profiles, where the water vapor shows rather more variation than the temperature, if both are compared to their average gradient in the same layer. It may well be that this is a result of different vertical variation up to somewhat greater heights. The air in higher layers is dryer than in the surface layer, while the potenzial temperature does not differ considerably. If through convection air from the surface layer is replaced with air from above, more variations of humidity than of temperature can be expected.

In case of more intense convection its footprints can be observed in the surface layer. In the time cross-sections sudden drops of temperature were discovered, which sometimes were accompanied by an also sudden change of water vapor. The sudden drop

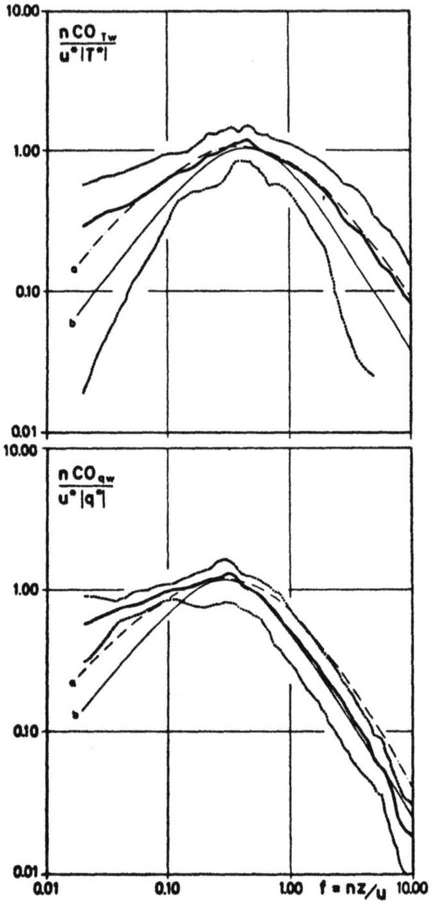

fig.4 Average and range of cospectral energy for vertical fluxes of sensible heat and moisture from the North Sea (Kruspe 1973).

was followed by a slow recovery to the prior temperature level. The relation between depth and length of these "breaks" is given in fig. 5 by Prümm (1973). Analysis of equivalent potential temperature reveals that in this cases the surface layer air had been replaced by air from considerably greater heights (several hundred meters). It is quite interesting to see with aid of the time cross-sections from the profile measurements that these exchanges of air go right down to the sea surface.

It is obvious that for a determination of the interaction of the different scales involved meteorologists are in need of suitable instrumentation. It would be necessary to determine unbiased temperature, humidity and possibly wind components up to some 1000 m height in an area of some 10 km diameter. Radiometeorology could provide such meso scale instrumentation.

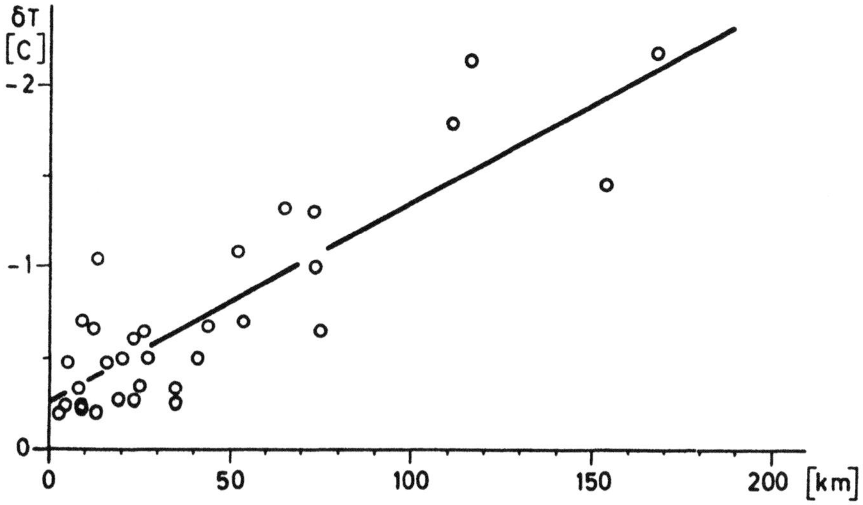

Fig. 5: Depth of temperature drop δ T versus spatial scale of the event during convective activity (Prümm 1973)

2. PLANETARY BOUNDARY LAYER

For description of the planetary boundary layer there are three main roads of approach available:

i numerical modelling
ii exchange or mixing length hypothesis with various modifications
iii dimensional analysis

(i) It is not yet possible with todays computers to fully simulat the turbulent processes in the atmosphere with their frequency range of 3 or 4 decades. Numerical modelling therefore usually has to revert to some sort of parametrization e.g. by one of the other two devices.

Method (ii) will not be treated in any detail in this context for the following reason. It has been known from laboratory experiments for about 40 years that the exchange or mixing length and related hypotheses work well, as long as there are strong gradients, e.g. in the center of pipe flow. Such approaches are useful in the lower layer of the planetary boundary layer, but will fail as soon as the transport by mixing is dominated by other transport mechanism, e.g. water vapor transport by convection and momentum transport by internal waves.

(iii) Dimensional analysis has been applied with considerable success in hydrodynamics, especially in those cases, where a closed solution cannot be given. If the characteristic variables of a problem are known, there must be a relation between these variables. This relation, as every meaningful physical relation, must be dimensionally homogeneous. If the problem is seen as one between dimensionless products of variables, this constraint usually leeds to a smaller number of independent products.

Lettau in 1959 defined a geostrophic drag coefficient c_g by

$$\tau = c_g^2 \varrho G^2$$

where the geostrophic wind speed G represents the pressure gradient as the driving force of the flow. He then hypothesized that c_g should depend on the dimensionless surface Rossby number Ro = = $G/f\cdot z_o$ and could demonstrate an empirical resistance law. A theoretical resistance law was derived by Kasanski and Monin (1961) for a neutral barotropic atmosphere by dimensional arguments and the postulate, that in the surface layer the wind profile should approach the logarithmic profile:

$$\ln Ro = M - \ln c_g + \frac{k}{c_g} \cos \alpha_o$$

$$\sin \alpha_o = \frac{c_g}{k} \cdot N$$

where the coefficients M and N are independent of the surface Rossby number Ro.

Blackadar (1967) as well as Monin and Zilitinkevich (1967) showed that in a diabatic, barotropic atmosphere the coefficients of the resistance law depend on a dimensionless stability parameter (Planetary Richardson Number)

$$\mu = \frac{k^2}{f} \cdot \frac{g}{T} \cdot \frac{H/c_p \varrho}{u_*^2}$$

Similar approaches are possible for the fluxes of heat and water vapor.

The disadvantage of dimensional arguments is, that they do not give the numerical values of dimensionless constants. Our empirical knowledge of the coefficients is still rather insufficient (see e.g. Monin and Zilitinkevich 1967, Wippermann 1972 a,b). Besides, empirical determinations are taken mainly over land and their relevance at sea is doubtful as the mechanism of momentum transfer over sea is different from that over land due to the presence of waves.

Since empirical evidence at sea so far is few, I suggest to devide the problem into two steps:
let us write

$$c_g = c_D \cdot \frac{U}{G} = \frac{u_*}{U} \cdot \frac{U}{G}$$

In this way we can make use of the existing knowledge of the variation of c_D with other parameters. Additional, there is an experimental advantage which stems from the different nature of $c_D = u_*/U$ and of U/G. The friction velocity is a local variable, and its determination needs some experimental effort. G is a mesoscale parameter, usually determined from smoothed gradients over a horizontal scale of order 50 to 200 km. It is therefore convenient to determine the geostrophic drag coefficient and its variation in several steps:

(1) The pressure field is determined from synoptic stations. The observations are fitted by a second order surface by method of least squares.

(2) The relationship U/G from wind observations at light-vessels in the middle of the area is obtained using air sea temperature difference as a stability parameter.

(3) z_o and c_D are assumed to be known as a function of wind speed under neutral conditions (Brocks and Krügermeyer 1972). Together with U/G this gives c_g as function of either wind speed or surface Rossby number $Ro = G/fz_o$. The observational evidence shows that the surface Rossby number at sea does not vary much with wind speed, as the increase of z_o (although the increase of c_D is small) cancels that of G. At any rate, due to the high Rossby number, the variation of c_g with Ro should be small. In fig. 6 the variation of c_g, composed from observations of c_D and U/G, with wind speed is plotted. If the pressure gradient is strong enough to dominate local circulations, c_g is independent of wind speed, as predicted.

(4) For a moderate wind speed, where both c_D and U/G are fairly well experimentally determined, the dependence of c_g on stability can be evaluated. A variation of c_D with stability is known from observations (Hasse 1968), but as there is considerable scatter, the integrated flux gradient relationship has been used (Panofsky 1963, Webb 1960, 1970). The variation of c_g with stability is shown in fig. 7. At the same time this gives the observed average variation of the coefficient M of the resistence law as a function of stability. Comparison of observations over land with those over sea shows a discrepancy both in the level of c_g for neutral conditions and for the stability dependence. Since the interferences

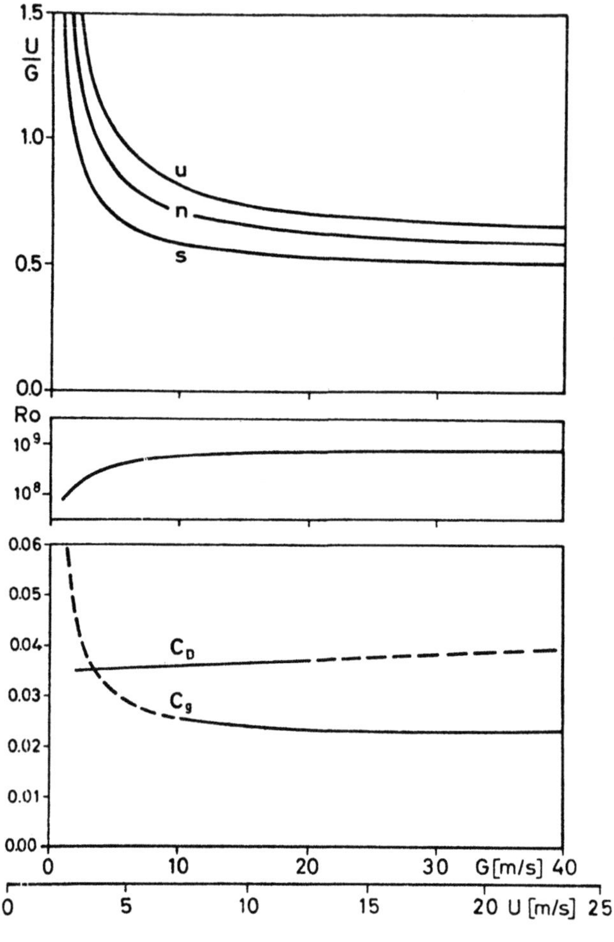

Fig. 6: Variation of surface to geostrophic wind speed ratio, surface Rossby number and geostrophic drag coefficient with stability (Hasse 1973).

for the Planetary Boundary Layer at sea are based on a considerable amount of reasonably good data, it is hoped that they can be used as a check on the applicability of the resistance law and its physical implications.

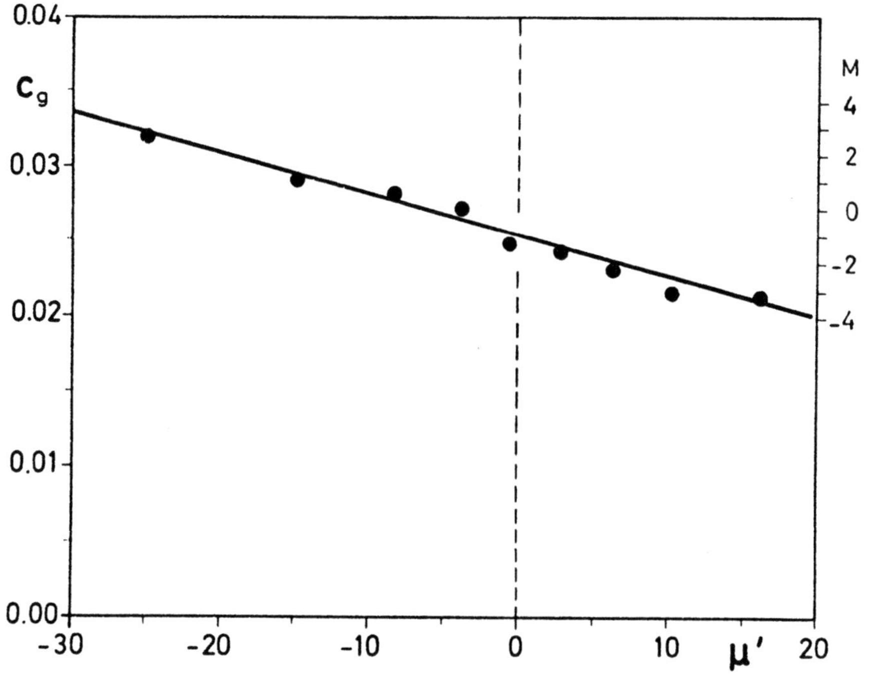

Fig. 7: Variation of geostrophic drag coefficient and the coefficient M of the resistance law with stability (Hasse 1973).

ACKNOWLEDGEMENT

In preparation of this paper I have drawn heavily on forthcoming papers from the Meteorologisches Institut der Universität Hamburg and the Institut für Radiometeorologie und Maritime Meteorologie an der Universitat Hamburg, Institut der Fraunhofer - Gesellschaft. This research was supported partly by the Deutsche Forschungsgemeinschaft, Schwerpunktprogramm Meeresforschung, lateron Sonderforschungsbereich Meeresforschung Hamburg, and by the Fraunhofer-Gesellschaft.

I acknowledge the contributions of my collegues at both institutes and their permission to use their results.

REFERENCES

Blackadar, A.K., 1967: External parameters of the wind flow in the barotropic boundary layer of the atmosphere, Rept. of the Study Conf. on GARP, Appendix IV, 11 pp.

Brocks, K. and Krügermeyer, L., 1972: The hydrodynamic roughness of the sea surface, in: Gordon (ed.) Studies in Physical Oceanography Vol. 1, 75-92, Gordon and Breach New York.

Dunckel, M., Hasse, L., Krügermeyer, L., Schriever, D., and Wucknitz, J., 1973: Turbulent fluxes of momentum, heat and water vapor in the surface layer at sea during ATEX, to be published in: Boundary-Layer Meteorology, Brocks Memorial Volume.

Hasse, L., 1973: On the surface to geostrophic wind relationship at sea and the stability dependence of the resistance law, to be published in: Boundary-Layer Meteorology, Brocks Memorial Volume.

Kasanskii, A.B. and Monin, A.S., 1961: On the dynamic interaction between the atmosphere and the earth surface, Izv.Geophys. Ser. 5, engl. ed., 514-515.

Kruspe, G., 1972: Autocovarianzspektren von Brechungs-index, vertikaler Windgeschwindigkeit, Lufttemperatur und -feuchte, Cospektren des vertikalen Wärme- und Feuchteflusses über See, Berichte des Instituts für Radiometeorologie und Maritime Met. an der Univ. Hamburg, Nr. 20.

Kruspe, G., 1973: Measurements of fluctuations of vertical wind velocity, temperature and radiorefractive index above the sea at a height of 30 meter, to be published in: Boundary - Layer Meteorology, Brocks Memorial Volume.

Lettau, H.H., 1959: Windprofile, surface stress and geostrophic drag coefficients in the atmospheric surface layer, Academic Press, New York, Adv. in Geophys. 6, 241-257.

Monin, A.S. and Zilitinkevich, S.S., 1967: Planetary boundary layer and large-scale atmospheric dynamics, Report on the study conference on GARP, Appendix V, 37 pp.

Panofsky, H.A., 1963: Determination of stress from wind and temperature measurements, Qu.J.R.Met.Soc. 89, 85-94.

Phelps, G.T. and Pond, S., 1971: Spectra of Temperature and Humidity Fluctuations and of the Fluxes of Moisture and Sensible Heat in the Marine Boundary Layer, J.Atm.Sci. 28, 918-928.

Pond, S., Phelps, G.T., Paquin, J.E., McBean, G. and Stewart, R.W., 1971: Measurements of the Turbulent Fluxes of Momentum, Moisture and Sensible Heat Over the Ocean, J.Atm.Sci. 28,901-91

Prümm, D., 1973: Zeitliche Variationen meteorologischer Größen in der wassernahen Luftschicht des atlantischen Nordost-Passats. Dissertation Meteorolog. Inst. d. Univ. Hamburg.

Wippermann, F., 1972 a: Baroclinic effects on the resistance law for the planetary boundary layer of the atmosphere, Beitr.Phys. Atmos. 45, 244-259.

ABOUT THE NEED OF A STANDARD TECHNIQUE TO COMPUTE ATMOSPHERIC REFRACTIVE INDEX FROM RADIOSONDE DATA

A. Nania and A. Pucciano

Italian Meteorological Service

INTRODUCTION

In radiometeorology there is still a problem, among others, for which a reasonable solution has not yet been found. That problem concerns with the computation of atmospheric refractive index and its vertical gradient in the studies on microwaves propagation. In order to utilize the computations for scientific and operational purposes, they ought to present hardly combined requirements: accuracy and economy. Accuracy is requested since small errors in the computation of the vertical gradient of atmospheric refractive index may infer large modifications on the theoric definition of radar coverage, but at the same time economy must be considered when we request a large number of measurements over extended geographic area and for long periods of time. The above said hard conceivability of accuracy and economy has determined two different categories of methods to obtain the requested data. In the first category there are the more precise and sophisticated instruments, such as absolute or differential refractometers, but they are very expensive, need to be handled by very specialized personell and frequently imply also very expensive means of transportation. In the other category there are the conventional radiosonde ascents performed in many stations twice a day as a routine. In this work we will analyze the procedures by which refractive index may be computed from radiosonde data and how to minimize errors due to measures and lectures of instrumental registrations.

REFRACTION INDEX

Pressure, temperature and humidity data are computed by the

relationship, valid in the field of radar wavelenghts

$$N = 77.6/T \cdot (p48I0 \cdot e/T) \qquad (I)$$

where N is the refractive index, T is the air temperature in °K,p is the total atmospheric pressure in mb,e is water vapor pressure in mb. The use of numeric values of constants, recomanded by SMITH and WEINTRAUB, causes an error of about ±0.5% N certainly less than that caused by the imprecision in measuring of T,p,e.

The calculation of N index can be made by numeric or graphic way. For numeric calculation some tables are used where, modifying the (I) in

$$N = Ap + Be \qquad (2)$$

with $A = 77.6/T$ and $B = 77.6 \cdot 48I0/T^2$. The values of A and B are tabulated for every 0.I degree of temperature from +40.0°C to -80.0°C. The accuracy in use of tables is good because the error is less than ±0.I N unit. The best accuracy, together with a very high computation speed, is possible by computers. The graphic computation is made by nomograms; a nomogram is used, in Italy, solving separately the Ap and Be terms of (2), and the addition by a slider is operated. The use of nomogram is faster than by tables, and accuracy is about ±0.2 N unit.

THE RADIOSONDE

As above said the N computation errors caused by numeric constants of the fundamental relationship or by computation rules, are of few decimals per N unit, good enough for operational purposes. The same accuracy is not achievable when the pressure, temperature and humidity data are obtained by radiosondes.

The radiosonde transmits alternately temperature and humidity data in a sequence scanned by a baroswitch. Every fifth humidity signal is replaced by a reference signal. A mean pressure variation of about I4 mb (pressure field I050 ÷ 700 mb) corresponds to a baroswitch contact,i.e. every a complete temperature-humidity sequence. The sensor features are: PRESSURE (aneroide capsule) accuracy $\simeq \pm 4$ millibars; TEMPERATURE (ceramic thermistor) accuracy $\simeq \pm 1$°C, time lag ≤ 4 sec.; HUMIDITY (lithium chloride or carbon) accuracy $\simeq \pm 5$% R.H., time lag $\simeq 7$ sec., greater for Low temperature.

At end of the sounding, in order to broadcast the results according with W.M.O. rules, a report which describes as faithfully as possible temperature and humidity profiles, as a function of pressure, is prepared. These profiles are broken lines limited by points, relative to mandatory levels (fixed pressure values) and significant levels. In order to select these last levels, following rules are valid:

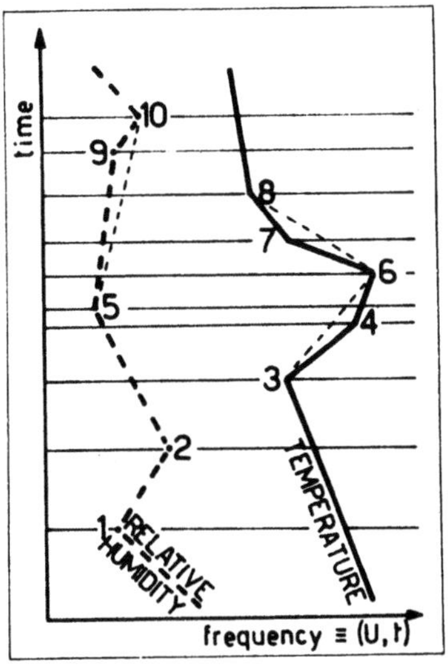

Fig. 1

the points of temperature or humidity profile showing a strong change are selected (e.g. in Fig. I, the levels relatives at points 3,6,8 for the temperature and I,2,5,IO for the humidity);
other significant levels are choiced, eventually, bearing in mind what follows:
Temperature - level selection is such that temperature value, read on union-line between two consecutive significant levels differ from recorded temperature more than I°C (under 300 mb or at tropopause level), (e.g. in Fig.I, between levels relative at points 3 and 6, and point 6 and 8);
Humidity - level selection is such that humidity value, read on union-line between two consecutive significant levels differs, at least, IO% from recorded humidity (e.g. in Fig. I, between levels relative at points 5 and IO);
a humidity significant level is neglected if close to a temperature significant level already selected. In such case temperature significant level is coupled to the humidity significant level, at the maximum change (e.g. in Fig. I, the value relative at point 5 is coupled to level relative at point 4).

ACCURACY AND ERRORS IN MEASUREMENT WITH RADIOSONDE

Launching and transmitting system

The radiosonde, due to its running feature, is not quite suitable for very precise measurements of refractive vertical gradients. First, the balloon carries up the radiosonde along the resultant between lifting power and wind, so forth sounding is not strictly vertical. Second, temperature and humidity measurements are not transmitted continuously. The pressure difference every a baroswitch contact is of about, I4 mb, as said before. According to ICAO atmosphere, within 3000 meters, this corresponds to an altitude mean change of I36 meters. During half of this space a temperature datum is recorded and in other half a humidity datum.

So, every 68 meters, there is a temperature or humidity datum loss alternately. Furthermore, as a consequence of transmitting system, every fifth contact the humidity signal is replaced by a reference (or calibration) signal, as a consequence every 680 meters there are about 204 meters without humidity data (Fig. 2). Time of a five contacts sequence consistes of 50 per cent temperature data, 40 per cent humidity data and IO per cent reference data. These discontinuities of the informative flux are acceptable for general synoptic purposes, but not for measurement of vertical refractive gradients when it is necessary to analyse atmospheric thicknesses as thin as possible. This disadvantage is difficult to be eliminated but it is possible to reduce partially its consequences by means of captured sounding and by a detailed analysis of instrumental registrations. This detailed data reduction technique will be discussed in further.

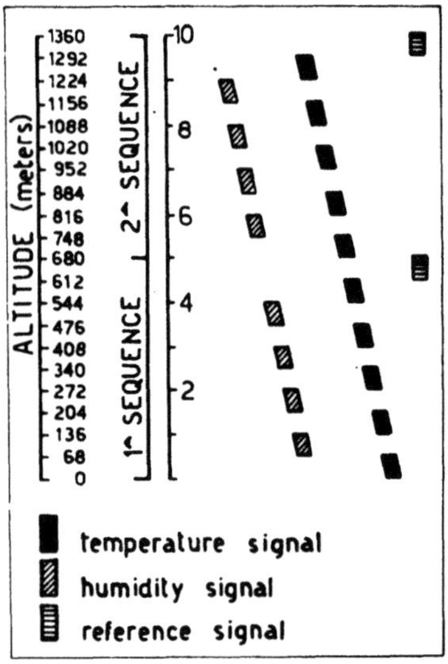

Fig. 2

Errors caused by sensors features

Checking the sensors features of a radiosonde it is easily seen that it is possible to have errors whether by accuracy or by quickness. Obviously it is not possible to eliminate errors caused by instrumental imprecision of present radiosondes. One can evaluate only their weight in the whole of errors relative to computation of N index. The entity of N change as consequence of errors in temperature, humidity and pressure measurements as a function of p,T,e changes is obtained from (I)

$$dN = (\delta N/\delta T)dT + (\delta N/\delta e)de + (\delta N/\delta p)dp \qquad (1)$$

Resolving partial derivatives in standard conditions (I5°C, IOI3 mb, 60% relative humidity) one has

$$\Delta N = - 1.27\Delta T + 4.50\Delta e + 0.27\Delta p \qquad (2)$$

The table I contains N changes relative to more general environmental conditions. From table I it follows that the smallest changes are caused by imprecision in pressure measurements, followed by imprecision in temperature measurements (when relative humidity is very high) and imprecision in relative humidity measurements (when temperature is high).

Errors \ Temperatures (°C)		+ 40	+ 20	0	- 20
± 1 mb of pressure		± 0.25	± 0.26	± 0.28	± 0.31
± 1°C of temperature	1000 mb	± 0.8	± 0.9	± 1.0	± 1.2
	850 mb	± 0.7	± 0.8	± 0.9	± 1.0
± 1°C of temp	relat. hum. 20%	± 2.6	± 1.4	± 0.5	± 0.1
	» » 60%	± 8.1	± 4.2	± 1.5	± 0.3
	» » 100%	± 13.5	± 7.2	± 2.4	± 0.6
± 10% relative humidity		± 28.1	± 10.1	± 3.1	± 0.7

Table I

Errors caused by time lag may have a partial compensation. This is very important considering that their weight is heavier than that relative to imprecision errors. In order to estimate the deviation between environmental value of a given quantity Θe (temperature or humidity) and the indicated value Θi caused by time lag, there is the relationship

$$\theta_{e,2} = \theta_{i,2} + R\lambda \cdot (\theta_{i,2} - \theta_{i,1}) / (h_2 - h_1) \qquad (3)$$

where R is the balloon ascension rate (meter/second); λ is the time lag (seconds); h_1 and h_2 two given levels (meters); indices 1 and 2 refer to two given levels. For a mean height difference between the end of a temperature (or humidity) signal and the beginning of the next, equal to 68 meters (Fig. 2) and for a mean ascension rate of 5.5 meters/second, the (3) becomes,

$$\theta_{e,2} = \theta_{i,2} + 0.081 \cdot \lambda \cdot \Delta\theta_i \qquad (4)$$

As far as the temperature is concerned the highest value of time lag of sensor (thermistor) is assumed (λ=4 seconds). From (4) deviations corresponding to various ΔT, relative to a 68 meters thickness, are as follows

ΔT (°C)/68 m	Deviation
±0.5	±0.16
±1.0	±0.32
±1.5	±0.49
±2.0	±0.65

From these values maximum deviation (±0.65) is relative to a very steep temperature lapse rate (±2°C per 68 meters). However this maximum is assumed, considering that in radiometeorology high temperature gradients are very important.

Humidity sensor time lag is higher than thermistor's and more difficult to be determined. According to other authors a value of IO seconds is assumed, more pessimistic than 7 seconds as said above, but closer, certainly, to mean working conditions of sensor in various environmental conditions. With λ = IO, by (4), the deviations corresponding to various U relative to a 68 meters thickness, are as follows

ΔU (%)/68 m	Deviation
± 5	± 4
±IO	± 8
±I5	±I2
±20	±I6

Following similar considerations about temperature also for humidity, the deviation value (±I6) is assumed corresponding to a vertical gradient of humidity (20 per cent of relative humidity variation per 68 meters).

Introducing these arbitrary two values into table I at temperature of 40°C and I00% relative humidity, the largest N unit changes are ±8.78 and ±45 respectively. Of course these values are arbitrary, since they correspond to unusual environmental conditions, and their comparison is very rough, nevertheless it is evident that N changes caused by time lag of thermistor are negligible compared to those caused by time lag of humidity sensor. As a consequence it is suggested, together with other authors, to consider only the latter.

Errors caused by significant level selection

The selection of significant levels of temperature and humidity with approximation of I°C and I0% respectively causes a maximum N unit error of ±I3.5 and ±28.I (table I). Also in this case the humidity is prevalent. Humidity influence is even stronger considering that some humidity significant level might be neglected if close to a temperature significant level. It is very difficult to fix the entity of this error; here, for practical purposes, a mean value of about I5% of relative humidity is assumed, keeping in mind that it is possible, in special conditions, to have a larger error.

Errors caused by reading approximation

Usually the sounding is performed by very trained operators but it may happen, anyway, to have errors caused chiefly by the use, for measurements, of work implements not extremely exact. The error values are included in "READING" part of table II. The approximations of pressure, temperature and humidity measurements are summed up in table II, where the totals for each element are reported since the errors feature permits to. From table the strong weight of errors relative to humidity comes out compared to the moderate importance of errors due to pressure and temperature; even considering that some arbitrary values are adopted. Furthmore most humidity errors are caused by the time lag of sensor and by reduction data techniques.

Dealing with the same subject, some authors like to evalue the errors weight in terms of number of ducts neglected by approximations, compared to ducts really existent. Obviously this comparison is possible only when proper instrumentation (refractometer) is available, and this is not yet the case for the authors of the present work.

USE OF RADIOSONDE DATA TO MEASURE N INDEX

All problems, discussed before, suggested to set up a suitable technique in order to eliminate, partially, some errors of measurement, at least the more significant, and to obtain more affidability for researches in radiometeorology. Another aspect of this problem aiming to increase the accuracy of N index measurement is the common use to represent the atmospheric refractive conditions with a reference to N unit changes per kilometer of altitude. As the data measurement is done relative to small atmospheric thicknesses, the reference to kilometer causes an increase of the errors (e.g. some measurements relative to a 50 meters thicknesses determine errors 20 times bigger).

In this paragraph two methods of using radiosonde data to measure N index will be discussed: the GRAPHIC INTERPOLATION METHOD and the NUMERIC INTERPOLATION METHOD, before going to treat the DETAILED REDUCTION METHOD. Technical literature about this aspect of the problem seems to be very poor.

Graphic interpolation method

The thermodynamic diagram (e.g.a STÜVE diagram) elaborated by radiosounding operator is used. The pressure, temperature and humidity data every 50 meters are read referring to the altitudes

		P (mb)	T (°C)	RH (%)
SENSORS	precision	4	1	5
	time lag		0.65	16
SIGNIFICANT LEVELS	selection		1	10
	selection for humidity			15
READING	baroswitch calibration diagram	1		
	record chart		1	2
	calculation ruler		1	3
TOTALS		5	4.65	51

Table II

curve, in the right side of diagram.

This method presents a very high subjective component and a rough accuracy, keeping in mind the poor definition of the diagram. Refractive profiles relative to the same soundings, obtained by this way, are similar but not equal. The graphic interpolation method, which seems widely diffused, is to reject because bears too many imprecision.

Numeric interpolation method

It is based on the use the data contained in an usual TEMP report. The way is as follows: the temperature and dew point values relative to mandatory levels are selected; by interpolation the altitude relative to significant levels are computed; pressure, temperature and dew point data every 50 meters are then obtained by interpolation.

The method is founded on the assumption that the vertical pressure change is to be considered as linear. That is quite true when interpolation is relative to small steps. The advantages of method are the objectivity of data treatment and the possibility of using the TEMP reports directly. The disadvantages are the same as for the graphic interpolation and, furthmore, the rounding of values due to cipher necessity (the temperature values to a fifth of degree and some dew point values to full degree).

Detailed reduction method

It is based on radiosounding record chart; the calibration diagram of baroswitch and the calculation ruler are ancillary.

The pressure values from start to finish of every a temperature signal, are obtained from the calibration diagram of the baroswitch. The frequency values relative to start and finish of every temperature and humidity signal are obtained from records chart, after making due corrections for chart-drift and reference-drift. Every a fifth contact, when a reference signal is recorded,

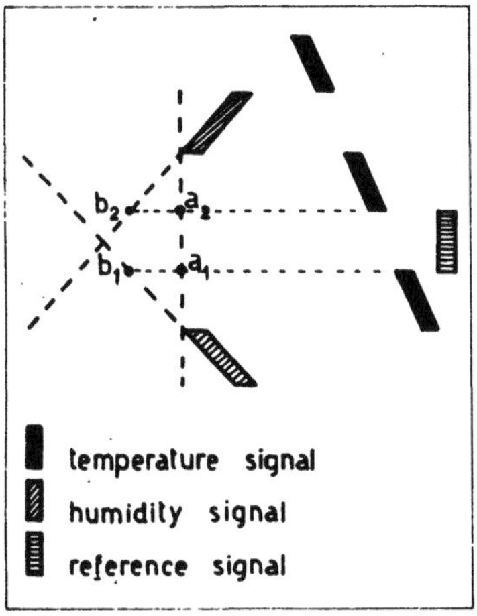

Fig. 3

the humidity value is lost, then it must be obtained by interpolation. As a rule, the interpolation is made along a line joining the upper left edge of lower signal to the lower left of upper signal (in Fig. 3, the point a_1 and a_2). On the contrary, here, it is suggested, when it is possible, to interpolate following the tendency detected by signals (in Fig. 3, the points b_1 and b_2). By the calculating ruler then pressure, temperature and humidity data for every an half baroswitch contact, are available. These values are used to compute N index and altitude for each level.

Table III gives a sample of complete calculation for the first 1500 meters of altitude.

No contact	t(°C)	U%	p(mb)	N	ΔN	h(m)	Δh(m)	ΔN(km^{-1})
1.3	15.2	80.0	1026.0	338.1	4.7	3.0	24.7	190.7
1.5	13.3	81.0	1023.0	333.4	2.5	27.7	64.5	38.2
2.0	12.9	82.0	1015.2	330.9	3.4	92.2	63.8	53.1
2.5	12.0	83.0	1007.5	327.5	3.3	156.0	62.7	52.6
3.0	11.4	83.0	1000.0	324.2	3.9	218.7	68.8	58.1
3.5	10.6	83.0	991.8	320.3	3.1	287.5	76.8	39.9
4.0	10.3	83.0	982.7	317.2	2.1	364.3	72.3	29.0
4.5	10.4	83.0	974.2	315.1	2.2	436.6	69.7	31.8
5.0	10.4	83.0	966.1	312.9	12.9	506.3	77.0	166.7
5.5	9.6	67.0	957.2	300.0	7.2	583.3	79.3	91.6
6.0	9.3	59.0	948.1	292.8	4.2	662.6	63.9	64.7
6.5	8.8	56.0	940.8	288.6	2.4	726.5	67.2	35.4
7.0	8.4	56.0	933.2	286.2	2.1	793.7	70.9	30.3
7.5	7.9	57.0	925.2	284.1	1.3	864.6	61.9	21.3
8.0	7.4	59.0	918.3	282.8	2.3	926.5	65.5	34.2
8.5	6.7	60.0	911.0	280.5	0.8	992.0	66.1	12.6
9.0	6.4	63.0	903.7	279.7	2.5	1058.1	71.8	35.2
9.5	5.5	64.0	895.8	277.2	3.2	1129.9	70.6	45.5
10.0	5.3	62.0	888.1	274.0	4.0	1200.5	73.9	53.3
10.5	5.3	58.0	880.1	270.0	3.1	1274.4	82.9	37.8
11.0	4.9	57.0	871.2	266.9	7.2	1357.3	68.7	104.7
11.5	4.5	45.0	863.9	259.7	0.4	1426.0	64.3	6.8
12.0	4.1	49.0	857.1	259.3		1490.3		

Table III

Table IV gives the same calculation of the foregoing sample, when the relative humidity is corrected by time lag according to (3) . The corrected values are emphasized by underlined italic types. It is easily seen that the small changes in humidity caused some important changes in the N index and gradients measurements. Within 3 ÷ 27.7 meters the gradient of 190.7 N (km^{-1}) is reduced to 128.9 N (km^{-1}). The upper duct (at about 500 meters) is identified with a higher value [from 166.7 to 249.3 N (km^{-1})] . Another duct (not detected by the different computation methods) with 161.6 N (km^{-1}) is revealed at about 1400 meters. The time lag corrections, so important on N index computation, are practically negligible when computing altitudes. This can be seen noting that the h and Δh values of table III are equal to correspondent values of table IV even when the humidity value is changed.

The advantages of method are the detailed use of all data sent by radiosonde during the ascent and the more accuracy in computation. The method can furnish a N value about every 70 meters compared to 50 meters of other methods. But the smaller definition is just apparent since a better agreement to actual thermodynamic conditions is obtained. A disadvantage is the the complexity of the method. But it must be said that using proper tables, reduces considerably the operations. Data reduction from record chart is obviously harder.

N	ΔN	h(m)	Δh(m)	ΔN(km^{-1})
338.1		3.0		
	3.2		24.7	128.9
334.9		27.7		
	3.4		64.5	53.0
331.5		92.2		
	3.4		63.8	53.6
328.1		156.0		
	3.9		62.7	61.8
324.2		218.7		
	3.9		68.8	58.1
320.3		287.5		
	3.1		76.8	39.9
317.2		364.3		
	2.1		72.3	29.0
315.1		436.6		
	2.2		69.7	31.8
312.9		506.3		
	19.2		77.0	249.3
293.7		583.3		
	-2.0		79.3	-2.5
295.7		662.6		
	8.4		63.9	131.5
287.3		726.5		
	1.1		67.2	16.4
286.2		793.7		
	1.7		70.9	24.0
284.5		864.6		
	0.8		61.9	12.9
283.7		926.5		
	2.8		65.5	42.7
280.9		992.0		
	0		66.1	0
280.9		1058.1		
	3.4		71.8	47.3
277.5		1129.9		
	2.7		70.6	38.3
274.8		1200.5		
	6.5		73.9	87.9
268.3		1274.4		
	1.7		82.9	20.5
266.6		1357.3		
	11.1		68.7	161.6
255.5		1426.0		
	5.2		64.3	80.9
260.7		1490.3		

Table IV

CONCLUSION

This paper was intended to be an initial tentative to treat the problem relative to standardize methods of utilization of radiosonde data for radiometeorology. It was intended to furnish a small contribution to the solution of this problem. Standardization becomes a necessity when we move from the field of local and isolated researches to that, certainly more interesting, concerning more extensive researches and studies, in time and space, and the comparison of results obtained in different countries. Authors hope that Sorrento meeting would furnish an occasion to promote further interest in this field considering also the present increasing development of the scientific and operational studies in radiometeorology.

REFERENCES

(1) BEAN, B.R. and DUTTON, E.J. - Concerning Radiosondes, Lag Constants and Radio Refractive Index Profiles. - J. Geoph. Res. v. 66, n. 11, Washington, D.C., 1961.

(2) SANTOMAURO, L. - Tables for quick calculation of atmospheric refractive index for radio microwaves propagation. - CENFAM Roma, 1962.

(3) PUCCIANO, A. - Calcolo grafico dell'indice di rifrazione atmosferica per le microonde. - Riv.Met.Aer., v. XXVII, n. 4, Roma 1967.

(4) ITAV, SERVIZIO METEOROLOGICO - Istruzioni tecniche per l'uso della radiosonda IA/AMT-1680B - Autovox, Roma, 1968.

(5) CENA, G. e ZACCOLO, A. - Osservazioni e misurazioni meteorologiche strumentali e a vista. - Scuola Applicazione A.M., Firenze, 1967.

(6) WAGNER, N.K. - An Analysis of radiosonde Effects on the Measured Frequency of Occurrence of Ducting Layers. - J. Geoph. Res. v. 65, n. 7, Washington, D.C., 1960.

(7) NANIA, A. - Esperimenti di sondaggi atmosferici con palloni frenati in prossimità del suolo. - Riv.Met.Aer., v. XXVIII, n. 3, Roma 1968.

(8) KERR, D.E. - Propagation of short radio Waves. - Dover Publications, New York, 1965.

(9) YERG, D.G. - The importance of water vapor in Microwave Propagation at Temperatures Below Freezing. - Bull.Am.Met.Soc., v. 31, n. 5, Boston 1950.

(10) BUNKER, A.F. - On the Determination of Moisture Gradients from Radiosonde Records. - Bull.Am.Met.Soc. v. 34, n. 9, Boston 1953.

(11) MORRIS, V.B. and SOBEL, F. - Some Experiments on the Speed of Response of the Electrolytic Hygrometer. - Bull.Am.Met.Soc. v. 35, n. 5, Boston 1954.

(12) SION, F. - Time Constants of Radiosonde Thermistors. - Bull. Am.Met.Soc. v. 36, n. 1, Boston 1955.

(13) WAGNER, N.K. - The Effect of the Constant of Radiosonde Sensors on the Measurement of Temperature and Humidity Discontinuities in the Atmosphere. - Bull.Am.Met.Soc. v. 42, n. 5, Boston 1961.

(14) BEAN, B.R. and DUTTON, E.J. - Radio Meteorology. - Dover Publications, New York, 1968.

(15) NANIA, A. e PUCCIANO, A. - Misura indiretta dell'indice di rifrazione atmosferica con l'impiego dei dati rilevati da radiosonde - Riv.Met.Aer., v. XXXII, n. 2, Roma 1972.

DETERMINATION OF THE WIND STRATIFICATION BY SCATTER PROPAGATION METHODS ON 7 AND 16 GHZ

H.D. Seehars

Institut für Radiometeorologie und
Maritime Meteorologie, Hamburg

A. GENERAL ASPECTS OF SCATTER PROPAGATION

In the last two decades a large number of investigations has been published, which deals with scattering of electromagnetic waves inside the inhomogeneous troposphere. By the conception "inhomogeneous troposphere" we mean the fact that the refractive index which is a physical parameter relevant to propagation, is a random function in time and space. All theories coincide with the following problems and objectives:

1. to give a sufficiently exact mathematical description of the turbulent medium,
2. to combine this description with the equation of propagation,
3. to try to solve the in each case inhomogeneous differential equation of propagation (i.e. the wave equation).

A general solution of the partial differential equation of propagation has up to now not been tried in order to avoid the enormous mathematical difficulties. A rather simple way to solve the equation of propagation is the separation of the temporal turbulent processes on the one hand and the spatial turbulent processes on the other hand. As a consequence of this stationary electromagnetic receiving fields and stochastically varying receiving fields are separately considered, i.e., a frozen turbulent structure is defined and introduced into the wave equation. As a solution you are obtaining an expression for the mean scattered field at the receiver. This field is not subject to temporal variations according to the presupposition, but only a function of the given spatial turbulent structure. The physical meaning of this model is covered by the title "scattering of wave energy in direction space".

On the other hand a turbulent refractive index field is supposed which contains as well a drift (i.e., a mean wind speed) as a temporal turbulent component superimposed to the drift. The motion of the turbulent inhomogeneities of refractive index and their temporal variability produce small frequency shifts of the incoming signal which are perceived in a view to the receiving field as very fast fluctuations of field-strength. The physical meaning of this model is covered by the title "scattering of wave energy in frequency space".

B. DEFINITION OF THE PROBLEM

Furthermore we shall deal with the second method, i.e. the assumption of drifts and temporal turbulent refractive index components. We shall describe here an indirect experimental method which allows to obtain results about the two-dimensional wind field by the aid of an analysis of the fine-structure of receiving signal, i.e. the stochastical fieldstrength variations. As a hypothesis we use the assumption that the received signal which has propagated along a well defined propagation path, i.e. a transhorizon link, is exclusively caused by scattering of the transmitted energy inside the "common volume".

As we have mentioned above we are intersted in results about the wind field, therefore we have to use an experimental set-up which allows to measure the spatial variations of the fluctuating receiving signal.

Such a system used by us is the beamswinging bi-frequency antenna system. Among other indirect methods this system has mainly three advantages in comparison with traditional wind measurements:

1. in a simple way it allows a survey of the spatial wind field.
2. Moveable measuring instruments and equipments are not necessary.
3. The measuring data can be recorded continuously.

C. PHYSICAL BASIS

In the following section we briefly want to discuss the physical basis inasmuch as it is of some interest for our problem.

All investigations are based on the Doppler effect which is caused by the motion or refractive index inhomogeneities inside the "common volume".

The common volume of a propagation path is defined as that volume which can be seen as well by the transmitting as the receiving antenna. In each case a monochromatical high frequency signal scattered by moving refractive index structures suffers a Doppler shift. Suppose the simplest case that two waves with small differences in frequency interfere at the receiver. Then from this a beat frequency results the envelope of which will likewise be a stochastical quantity according to the stochastical nature of the turbulent structure of refractive index. In the terminology of radio wave propagation the envelope is called "scintillation fading". According to the theory of the Doppler effect in case of single scattering the frequency shift is given by

$$\begin{aligned} \Delta v = v - v_0 &= \frac{1}{2\pi}(\vec{k}_1 - \vec{k}_2) \cdot \vec{V} \\ &= \frac{1}{2\pi}(\vec{K} \bullet \vec{V}) \\ &= \frac{1}{2\pi}|\vec{K}| \bullet |\vec{V}| \bullet \cos(\vec{K},\vec{V}) \end{aligned} \tag{1}$$

v_0 transmitting frequency, $\vec{k}_1$, $\vec{k}_2$ wavenumber vectors of the incoming and scattered wave, resp., $\vec{V}$ wind vector.

The difference vector $\vec{K}$ is defined according to the formula(1):

$$\vec{K} = \vec{k}_1 - \vec{k}_2 \tag{2}$$

The geometrical meaning of formula (2) can be seen from figure (1). Both vectors $\vec{k}_1$ and $\vec{k}_2$ form the scattering angle Θ . The magnitute of $\vec{K}$ can instantly be derived from fig. (1) and can obviously be interpreted by Bragg's condition. In analogy to the reflection of X-rays inside a crystal the scattering of electromagnetic waves is interpreted as reflection at netplanes of the same state of turbulence. According to Bragg's condition an amplification of the resulting amplitute of two waves scattered at two netplanes occurs in that case, if the phase difference of the corresponding rays is an integer multiple of the wavelength. Simply it follows that magnitute of $\vec{K}$ is given by

$$|\vec{K}| = K = 2k \bullet \sin\Theta/2 = \frac{4\pi}{\lambda} \sin\frac{\theta}{2} \tag{3}$$

According to equation (1) the Doppler shift is proportional to the scalar product between the difference vector $\vec{K}$ and the wind vector $\vec{V}$. From the definition of the scalar product it follows that a Doppler shift occurs only in that case if $\sphericalangle (\vec{K},\vec{V}) \neq 0$ and simultaneously $|\vec{K}|$ and $|\vec{V}|$ are $\neq 0$. Inasmuch as we are only dealing with transhorizon links, the case $|\vec{K}| = 0$ is of purely hypothetical nature. $|\vec{K}| = 0$ means that the scattering angle becomes approximately 0, to say the transhorizon path turns into a line-of-sight path.

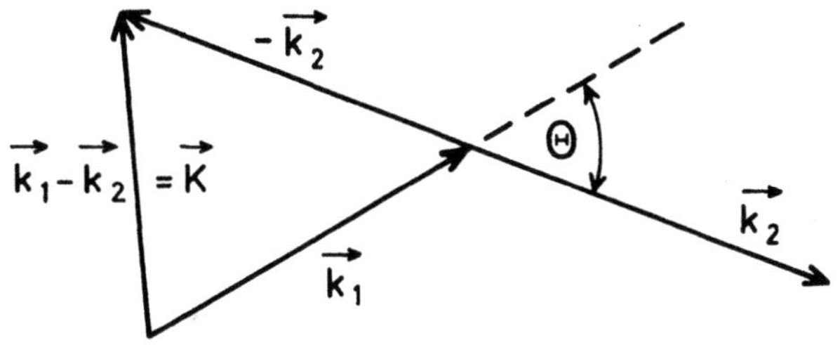

Fig. 1: Vector diagram of the electromagnetic and spectral wave-number component

This case is not of interest. Measurements of the Doppler frequency allow conclusions about the wind direction ($v \geq v_0 \rightarrow \Delta v \geq 0$ for $0 \leq \sphericalangle(\vec{K},\vec{V}) \leq \pi/2$ and $v \leq v_0 \rightarrow \Delta v \leq 0$ for $\pi/2 \leq \sphericalangle(\vec{K},\vec{V}) \leq \pi$, resp.) and wind velocity, measurements of the fading only conclusions about the magnitude of the wind vector. Generally the fading speed described by the fading frequency or the 1/ewidth of the autocorrelation function (AKF) can be defined by the magnitude of the Doppler shift. According to formula (1) the different constellations of $\vec{K}$ and $\vec{V}$ are able to produce strong Doppler shifts and an intensive fading, resp. The further considerations contain the following basic assumptions:

1. the influence of the vertical wind component is considered as negligible.
2. The refractive index inhomogeneities drift with the mean wind across the scattering volume.
3. The additional turbulent wind component has only a negligibly small influence on the fading.

Combining formula (1) and (3) the fading frequency and 1/e-width of the AKF are given by

$$f \sim \frac{1}{2\pi}\left|\frac{4\pi}{\lambda}\sin\Theta/2\right| \cdot |\vec{V}| \cdot \cos(\vec{K},\vec{V}) \qquad (4a)$$

$$\tau \sim \frac{\lambda}{2 \cdot \sin \theta/2} \cdot \frac{1}{\vec{V} \cdot \cos(\vec{K},\vec{V})} \qquad (4b)$$

In what follows that a fixed wavelength λ a variation of the fading frequency or the 1/e-width is connected with a variation of the scattering angle θ and the wind speed $\vec{V}$.

I point to the fact, that normally (i.e. symmetrical antenna constellations) the vector $\vec{K}$ is directly vertically, i.e. perpendicular to $\vec{V}$. Hence theoretically we cannot expect any fading in this case. Nevertheless Doppler shifts are observed which result from the finite scattering volume the lateral regions of which produce inclined $\vec{K}$-vectors. Generally the fading becomes the more intensive, the wider the scattering volume, i.e. the larger the horizontal antenna apertures. Evidently the fading is especially influenced by that wind component which is directed normal to the propagation path.

Hence equation (4) shows that if the scattering angle and the direction of the vector $\vec{K}$ remain constant, the fading intensity is only dependent on the wind. If the scattering volume is shifted into the horizontal or vertical direction, the variations of the fading intensity describe exactly the horizontal or vertical wind shear.

D. DESCRIPTION OF THE EXPERIMENTS

We have to distinguish between two experiemnts, which are denoted by the capital letters A and B. During experiment A the slightly varying scattering volume is shifted into the vertical (fig. 2). Both antennas are turned into the same direction by the same angle. The both horizontal axes of rotation penetrate the plane of the figure perpendicular to SE. The beamswinging mechanism is in principle the same for experiment B as for experiment A with the exception of the fact that both antennas are turned by the same angle into the opposite direction (fig. 3). Hence it is only a question of the shifts of the $\vec{K}$-vector in the vertical plane above the connection line between transmitter and receiver.

The direction of the vector $\vec{K}$ remains constant during experiment A, whereas the variation of the scattering angle causes slight variations of the magnitude of $\vec{K}$. During experiment B the vector $\vec{K}$ points in that direction which is given by the perpendicular line on the curve shown in fig. 3. That means the vector $\vec{K}$ varies slightly in direction, but the magnitude of $\vec{K}$ remains constant.

In fig. 4 a schematic three-dimensional presentation shows once more the different shifts of $\vec{K}$ during both experiments.

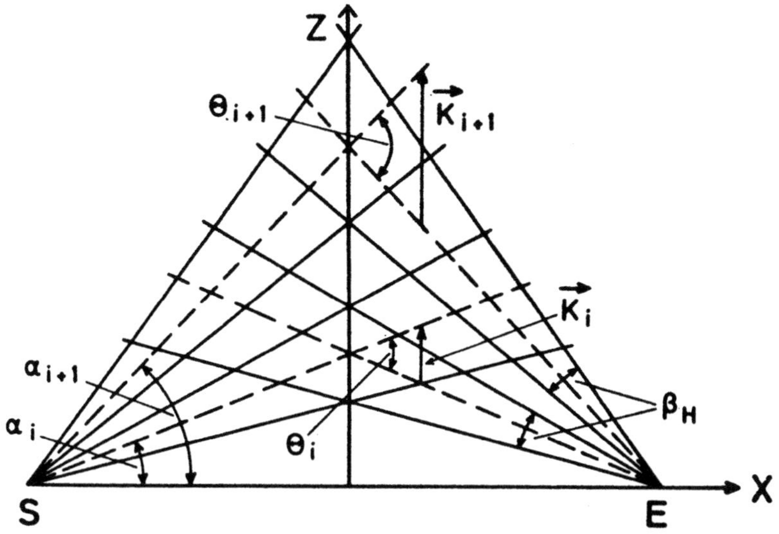

Fig. 2: Antenna elevations during experiment A

α_i, α_{i+1} vertical elevations of the antennas

θ_i, θ_{i+1} scattering angle

K_i, K_{i+1} wavenumber vectors for the ith and $i+1^{th}$ adjustment (i=0, 1,2,.....m)

β_H half beam width of the antennas

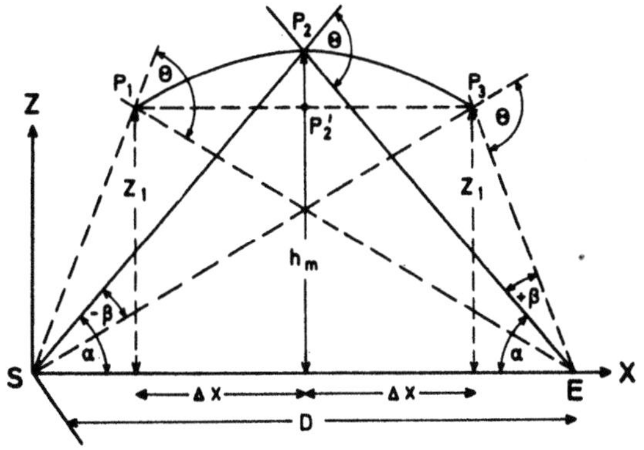

Fig. 3: Antenna elevations during experiment B. The following loci and antenna elevations are corresponding with oneanother:

P_2 $(D/2, h_m) \rightarrow \alpha_t = \alpha_r = \alpha$

P_1 $(D/2-\Delta x, z_1) \rightarrow \alpha_t = \alpha + \beta$; $\alpha_r = \alpha - \beta$

P_3 $(D/2+\Delta x, z_1) \rightarrow \alpha_t = \alpha - \beta$; $\alpha_r = \alpha + \beta$

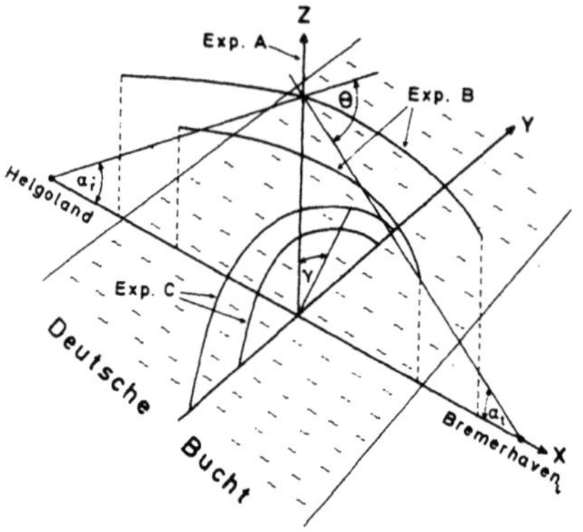

Fig. 4: Three-dimensional presentation of the variation in position of the scattering volume.

Now the height corresponding to the fading intensity which is measured during the beam-swinging experiments, is unknown. What we measure at the ground is a physical process which is caused within the scattering volume itself. To describe the relationship between wind speed and fading intensity we need the heights which are connected with the fading intensities. In order to take into account the refraction we determine the coordinates of the centre of the scattering volume during the experiments A and B by the aid of the general ray equation

$$\frac{d}{dx}\left(\frac{dz}{dx}\right) + f(z) \bullet \left(\frac{dz}{dx}\right)^2 - g(z) = 0 \qquad (5)$$

x horizontal coordinate, which is directed along the connection line between transmitter and receiver.
z vertical coordinate

$$f(z) = \frac{dn}{dz} + \frac{2}{a+z}; \qquad g(z) = \left[\frac{1}{n}\frac{dn}{dz}\left(1+\frac{z}{a}\right) + \frac{1}{a}\right] \bullet \left(1 + \frac{z}{a}\right) \quad (6)$$

a radius of the earth; n refractive index.

Equation (6) can be simplified by the following assumption:

$$\frac{2}{a+z} \simeq \frac{2}{a} \quad \text{and} \quad \frac{z}{a} + 1 \simeq 1 \tag{6a}$$

$$n \simeq 1$$

In what follows that

$$f(z) = \frac{dn}{dz} + \frac{2}{a} = C_1 \quad ; \qquad g(z) = \frac{dn}{dz} + \frac{1}{a} = C_2 \quad ;$$

Assuming a linear vertical refractive index profile equation (5) can be solved by quadrature. In this case the parameters C_1 and C_2 are constants. The general solution is then given by

$$z-h_s = - \frac{1}{dn/dz+2/a}(\ln(\cos(\sqrt{C_1 \cdot C_2} \cdot x) - \sqrt{C_1/C_2} \cdot tg\alpha_s \cdot \sin(\sqrt{C_1 \cdot C_2} \cdot x))) \tag{7}$$

h_s height of the transmitter; α_s angle of radiation of the transmitting antenna.

The parameters we are interested in a view to experiment A follow directly from equation (7):
<u>height of the centre of the scattering volume</u>

$$h_m = h_s - \frac{1}{C_1} \ln(\cos(\sqrt{C_1 \cdot C_2} \cdot D/2) - \sqrt{C_1/C_2} \cdot tg\alpha_s \cdot \sin(\sqrt{C_1 \cdot C_2} \cdot D/2)) \tag{8}$$

D distance between transmitter and receiver

<u>Scattering angle</u>

$$\Theta = \operatorname{arctg}\left(\frac{tg\,\psi_1 + tg\,\psi_2}{1 - tg\,\psi_1 \cdot tg\,\psi_2}\right) \tag{9}$$

ψ_1, ψ_2 angles between the rays within the centre of the scattering volume and the x-axis.

$$\Delta x = D/2 - x_1 = D/2 - \frac{1}{\sqrt{C_1 \cdot C_2}} \cdot \operatorname{arctg}\left(\frac{1 + a_1 b_3 - b_4}{b_3 + a_1 b_4 + a_2}\right)$$

$$z_1 = h_s - \frac{1}{C_1} \ln\left(\cos(\sqrt{C_1 \cdot C_2} \cdot x_1) - a_2 \cdot \sin(\sqrt{C_1 \cdot C_2} \cdot x_1)\right) \qquad (10)$$

Δx: distance between the observation point and the centre of the propagation path.
z_1 height corresponding to x_1.

$$a_1 = \sqrt{C_1/C_2} \cdot \operatorname{tg}\alpha_s; \qquad a_2 = \sqrt{C_1/C_2} \cdot \operatorname{tg}(\alpha_s + \beta_H/2);$$

$$b_3 = \sin(\sqrt{C_1 \cdot C_2} \cdot D); \quad b_4 = \cos(\sqrt{C_1 \cdot C_2} \cdot D);$$

β_H half beam width of the transmitting antenna.

I don't intend to present the detailed tables of the geometrical parameters, but only give you a feeling about the spatial transpositions of the scattering volume. The maximum vertical shifts of the volume during experiment A amounted to 2.5 km; the horizontal shifts during experiment B ranged from 10 to 25 km with respect to the centre of the path, the vertical shifts in this case amounted to 300 m. The scattering angles ranged from 4 mrad to 110 mrad.

E. EXPERIMENTAL LAYOUT AND EVALUATION OF THE DATA

The frequencies used were 7 and 16 GHz, the propagation path (fig. 5) was situated between the island of Heligoland and Bremerhaven (length 77.2 km).

The height of both antennas amounted to approximately 30 m, the half beam widths of the antennas were 1.8 degrees for both frequencies.

In addition to the radio measurements radiosonde ascents were simultaneously carried out in the neighbourhood of the receiving station, to measure the vertical refractive index stratification and the vertical wind stratification (the latter by radar).

The refractive index data were needed to determine the vertical gradients of refractive index in formula 8 till 10. Only quasi-linear refractive index profiles were elected.

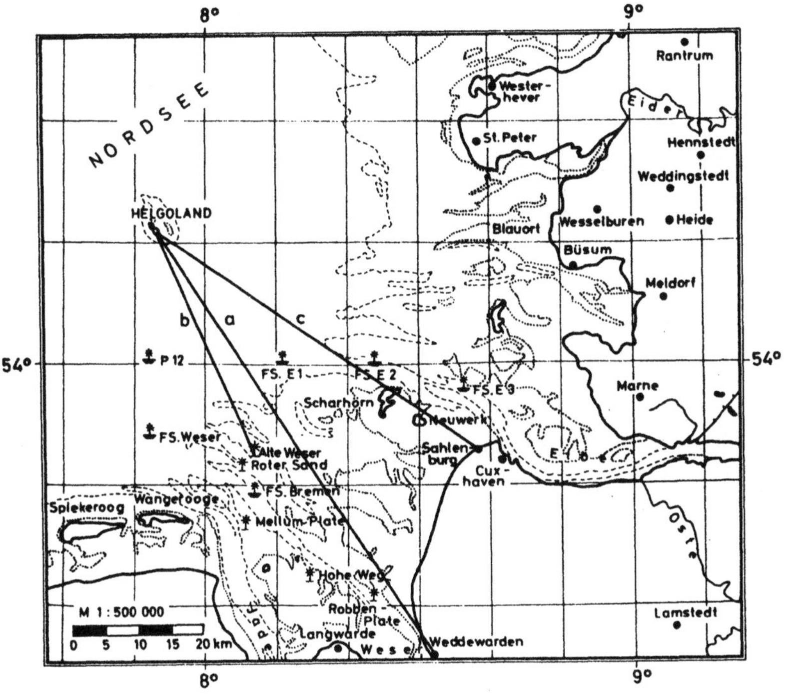

Fig. 5: General Map of the Measuring Area

The set of data consisted in a large number of records (length between 2 and 3 minutes).

The fieldstrength fluctuation intensity was characterized by

1. the number of positive crossings through a given level (i.e. the fading frequency),
2. the 1/e-width of the autocorrelation function.

Fig. 6 shows an example of the scintillation fading to be analyzed. In the case of such a fading both parameters ought to be proportional to oneanother, provided that the fading is stationary.

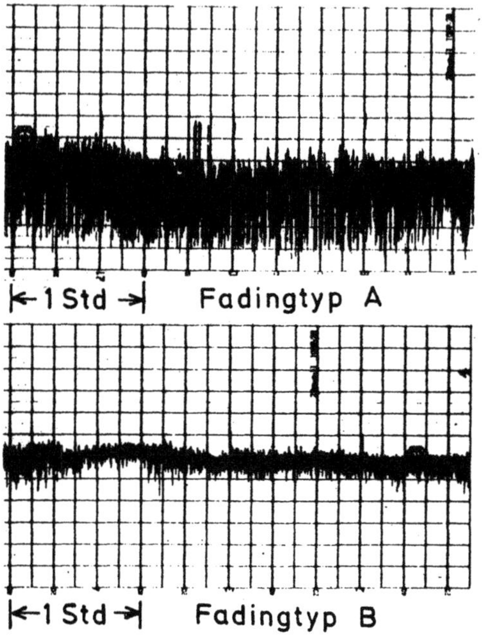

Fig. 6: Record of the 7 GHz scintillation fading

F. SOME RESULTS

According to formula (4) fading frequency and 1/e-width first of all are proportional to three unknown parameters, wind speed, scattering angle, and the angle between the vector $\vec{K}$ and the wind vector $\vec{V}$. The experimental tests were carried out in such a way that the direction of $\vec{K}$ remained constant. The slight variation of $\vec{K}$ during experiment B could be shown to be negligible. Hence only the influence of $|\vec{V}|$ and scattering angle Θ on the fading intensity has to be studied.

First of all an almost linear relationship between scattering angle and fading intensity was observed. The results indicate that firstly the fading intensity is only slightly influenced by the scattering angle, secondly there is no marked difference between both frequencies. This means that the variations of the fading intensity measured during a horizontal or vertical shift of the scattering volume are almost exclusively caused by variations of the wind speed within the scattering volume.

Fig. 7 shows the measured fading frequencies and 1/e-widths as a function of the wind component normal to the path. On the left side the fading frequencies are connected with arbitrary wind directions and contain measuring periods the stationarity of which is not guaranted in each case. In what follows that the fading frequencies are not always correct.

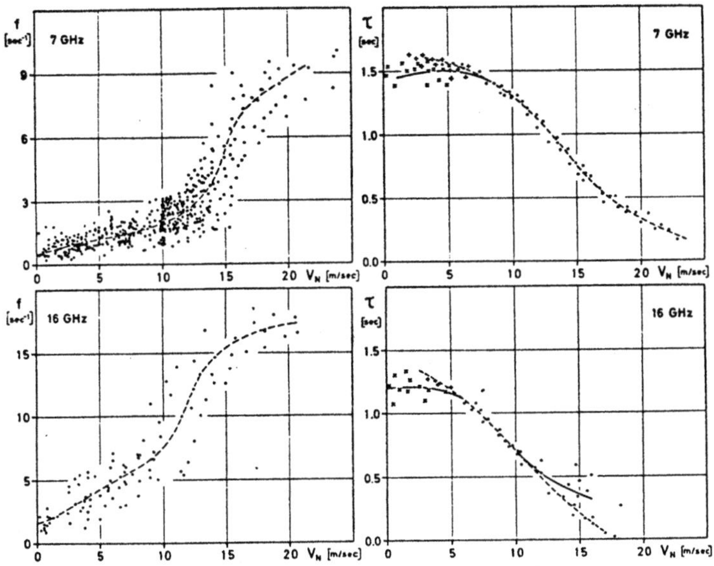

Fig. 7: Fading rate and 1/e-width of the AKF as a function of the normal component of the wind speed

In spite of a strong scatter of the points there is a rather good but non-linear dependence between both parameters. The non-linearity is not in agreement with the theory of single scattering and similar experiments which were carried out in the USA and the Federal Republic of Germany. Nevertheless those experiments don't present truthworthy results because of the few available data. On the right side of fig. 7 there is a strong correlation between 1/e-width and wind. This part of the figure contains only measuring periods with wind directions from the sea (i.e. from SW to NNE). Note that for both frequencies the scatter of points increases in the case of low wind speeds and large τ . Probably it is a question of incorrect direct wind measurements inasmuch as the measuring area near the harbour of Heligoland was situated in the lee-side of the main island. The slanting crosses mean wind measurements corresponding to elevations below 500 m (i.e. eventually influenced by the island). The solid curve indicates the propably incorrect graph, the broken curve the true graph, i.e. in this case the wind measurements are not modified by the island. The rather strong scatter of points at 16 GHz in the upper part of the figure is probably caused by special difficulties in the data evaluation, I don' want to describe this effect here. Very slight sinusodial trends can simulate too high 1/e-widths in what follows an incorrect graph like the solid curve.

The experiment A carried out continously over half a day or a day enables us to determine a temporal and spatial diagram of the isolines of the 1/e-width. In a similar way you obtain a spatial diagram of the 1/e-width by combination of one simultaneous experiment A and B. According to the last sections both diagrams give an insight into the temporal and spatial wind field.

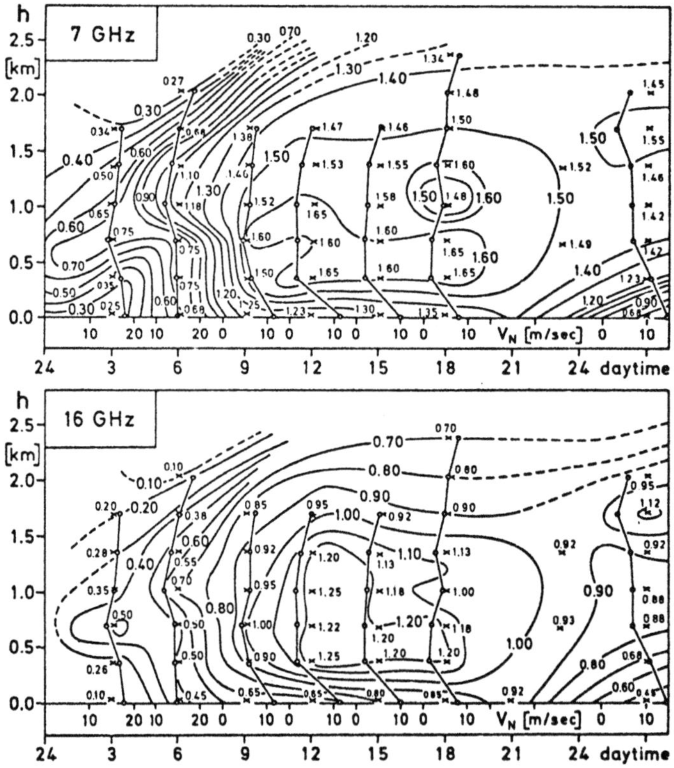

Fig. 8: Isolines of the 1/e-width of the AKF (spatial and temporal wind field)
upper part: 7 GHz; lower part: 16 GHz
upper abscissa: normal component of wind speed, referring to the vertical profiles of wind speed
lower abscissa: daytime
ordinate: elevation

The fig. 8 shows a spatial and temporal diagram of the isolines of the 1/e-width at 7 and 16 GHz. The lower abscissa denotes the daytime and the ordinate the height. The upper abscissa relates to the vertical wind profiles measured for an interval of 3 hours at Heligoland. The gradients of the 1/e-width-isolines are a criterion for the temporal variations of the wind and the vertical wind shear, resp. If there is no agreement between wind field and variation of fading intensity in detail, this may be caused by the fact that the direct measurements were not carried out within the scattering volume but at the terminus of the propagation path at Heligoland. As can clearly be seen strong gradients of the 1/e-width-isolines are in rather good agreement with the directly measured profile.

Especially the zones of strong vertical wind shear between 9^h and 20^h and 23^h and 5^h near the ground as well as the sloping zone of vertical wind shear from 1.5 up to 2.5 km between 3^h and 12^h are evidently discernible. The zone of very slight vertical wind shear for the almost whole observation time and the height interval from 0.5 up to 1.5 km is reflected by the weak gradients of the isolines.

The combination of a simultaneous experimen A and B enables a spatial presentation of the 1/e-width in the vertical plane along the connection line between transmitter and receiver. This way of presentation and its connection with the wind field can only imperfectly be proved by direct wind measurements. This was caused by the fact that the measuring area above Heligoland was already situated outside the region which could be covered by the beam-swinging experiments. Fig. 9 shows two typical examples of spatial fading contours. The broken lines denote only the presumable graph which is not covered by measurements. Nevertheless those contours are distinguished by some degree of reality, as the rather good agreement with the measured wind profile above Heligoland shows. On principle in a view to both examples it is a question of two different meteorological situations. On the one hand the wind field is evidently inhomogeneous according to the first example, whereas the second example contains typical contours of a rather homogeneous wind field. Note the well-marked similarity of the contours for both frequencies.

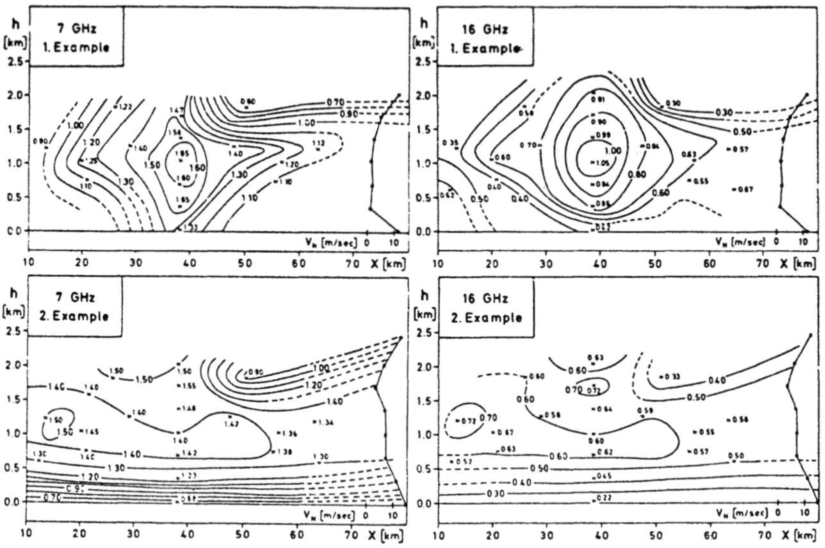

Fig. 9: isolines of the 1/e-width of the AKF (spatial wind field)
left side: 2 periods with 7 GHz
right side: 2 periods with 16 GHz
Abscissa: horizontal coordinate along the connection line between transmitter and receiver
Ordinate: elevation

TURBULENT TRANSPORT AND REFRACTIVE INDEX STRUCTURE IMMEDIATELY ABOVE THE AIR-SEA INTERFACE

R. Bolgiano, Jr.

School of Electrical Engineering
Cornell University

I. INTRODUCTION

In order to study the structure of the radio refractive index n in the maritime atmosphere (since the refractivity $N(\underset{\sim}{x},t) \equiv [n(\underset{\sim}{x},t)-1] \times 10^6$ depends on pressure, temperature, and moisture according to (Bean and Dutton, 1966):

$$N = 77.6\, p/T\,(1 + 4810\, e/T) = 77.6\, p/T\,(1 + 7.73\, q/T),$$

where p is total air pressure in mb, T is temperature in °K, e is partial pressure of water vapor in mb, and q is specific humidity in g/kg) it is necessary to examine the mechanisms that give rise to the distributions of pressure, temperature, and humidity, both in the mean and from instant to instant. Refraction, determined by the gradient of mean refractive index, $\nabla\bar{n}$, depends upon $\nabla\bar{p}$, $\nabla\bar{T}$, and $\nabla\bar{q}$. Therefore, the mean profiles of pressure, temperature, and specific humidity are required. Scintillation and scattering, on the other hand, determined as they are by the spatial and temporal fluctuations in refractive index, n', depend upon the detailed variations p', T', and q'. This fine-scale structure must be known if these radio phenomena are to be described accurately. Moreover, as will be discussed more thoroughly below, the mean profiles and the random irregularities are intimately connected through the mechanism of turbulent transport. Hence, it is essential to look closely at this process.

II. CONSERVED QUANTITIES

As noted above, the radio refractive index depends upon pressure, temperature, and humidity. The atmosphere, being a fluid in motion, convects and mixes these properties. Analysis of the consequences of these motions is simplified considerably if it is possible to express the desired relations in terms of quantities that are conserved for the flow conditions of primary interest. For this reason potential temperature θ is here introduced and moisture vapor will henceforth be accounted for solely in terms of specific humidity q. Justification for this choice of variables will follow shortly.

First, however, consider pressure. Its mean variation is essentially exponential: $p(\underset{\sim}{x},t) = \bar{p}(z) = p_0 \exp(-z/H)$. Coefficient p_0 and scale-height H vary slightly over hundreds of kilometers and tens of hours. For the purposes needed in this study they may be considered constant. Instantaneous fluctuations $p'(\underset{\sim}{x},t)$ occur only in response to velocity fluctuations (Bernoulli effect; presuming no explosions). At the small Mach numbers characteristic of the boundary layer flow being investigated these fluctuations are imperceptibly small. That is, the atmosphere acts as an incompressible, stratified medium. Consequently pressure contributes to $\nabla \bar{n}$, and thus to refraction, through the term $-\bar{p}/H$. Its effect on scintillation and scattering is negligible.

In order to examine temperature an appeal is made to the conservation of heat. The First Law of Thermodynamics (for the case of no condensation or chemical reaction, and assuming viscous dissipation and radiation contributions are negligible) may be expressed in terms of potential temperature thus:

$$\partial\theta/\partial t + \underset{\sim}{u}\cdot\nabla\theta - k\nabla^2\theta = 0.$$

Molecular conductivity k = thermal conductivity/(density x specific heat at constant pressure) = $k/(\rho c_p)$. Potential temperature is related to absolute temperature in such a way that, in the surface boundary layer, $\nabla\bar{T} = \nabla\bar{\theta} - g/c_p$ and $T' = \theta'$. (g is acceleration due to gravity.) In the above equation $\underset{\sim}{u}$ stands for the vector velocity. Since the air acts as though it were incompressible, $\nabla\cdot\underset{\sim}{u} = 0$ and the equation may be rewritten:

$$\partial\theta/\partial t + \nabla\cdot(\underset{\sim}{u}\theta - k\nabla\theta) = 0.$$

Thus, under the assumptions stated, changes of potential temperature are accounted for entirely by convection and molecular diffusion of temperature-bearing air.

Since specific humidity describes the amount of moisture vapor content in a given quantity of air, for the same restrictions as were applied to conservation of heat above, the specific humidity at a point can change only as new air flows into the region or as moisture molecules diffuse in or out. Therefore q satisfies the same conservation equation as θ:

$$\partial q/\partial t + \nabla \cdot (\underset{\sim}{u} q - D \nabla q) = 0,$$

where D = molecular diffusivity for moisture vapor.

The similarity of these two conservation equations (and the invariant nature of the pressure contribution) suggests the introduction of a dummy variable ψ to represent either θ or q, subject to the conservation equation:

$$\partial \psi/\partial t + \nabla \cdot (\underset{\sim}{u} \psi - D \nabla \psi) = 0. \tag{1}$$

Examination of ψ, determination of its mean distribution and its fluctuation structure, will then provide needed insight into the variations of refractive index. Note immediately that, in accord with the notion of a conserved property, all transport of ψ is reflected in the divergence term of Equation (1). Inasmuch as mixing (to be discussed below), which is an important mechanism of transport, is ultimately achieved only through the local homogenization of the medium by molecular processes, one may anticipate that the convection summarized in the term $\underset{\sim}{u}\psi$ serves principally to increase molecular diffusion by steepening local gradients of ψ.

III. MEAN PROFILES

If, following Reynolds, the dependent variables in Equation (1) are expressed as the sum of mean and fluctuating parts ($\underset{\sim}{u} = \underset{\sim}{U} + \underset{\sim}{u}'$, $\psi = \overline{\psi} + \psi'$), a mean equation may be deduced as follows:

$$\underset{\sim}{U} \cdot \nabla \overline{\psi} + \nabla \cdot (\overline{\underset{\sim}{u}'\psi'} - D \nabla \overline{\psi}) = 0. \tag{2}$$

The first term describes advection by the mean wind whereas the divergence term accounts for transport by turbulent and molecular fluxes. Here the turbulent term is usually dominant and one must have extensive knowledge of $\overline{\underset{\sim}{u}'\psi'}$, or introduce a scheme by which to parameterize its role, if one is to determine the mean distribution $\overline{\psi}$.

One approach that has been applied successfully is to invoke similarity theory (parameterization). This is discussed extensively by Dr. Coantic in his second lecture. However, where an understanding of the intricate mechanisms of transport is sought, and in order to study the fluctuation structure, a detailed investigation of ψ' must be made.

IV. TURBULENT FLUX

It is instructive initially to examine the scalar fluctuation term ψ' and its mean square, $\overline{(\psi')^2}$. The turbulent flux $\overline{\underset{\sim}{u}'\psi'}$, which appeared in Equation (2) for the conservation of mean scalar property $\bar{\psi}$, also plays a role here. Clearly turbulence is at the heart of the processes responsible both for the mean distribution and for the fluctuations and, in all likelihood, serves to couple the two structures together. Further inspection sheds some light on this situation (see, for example, Lumley and Panofsky (1964), Bolgiano (1968)).

A. Conservation of ψ' and of $\overline{(\psi')^2}$

Subtracting Equation (2) from Equation (1) yields:

$$\partial\psi'/\partial t + \underset{\sim}{u}'\cdot\nabla\bar{\psi} + \underset{\sim}{U}\cdot\nabla\psi' + \nabla\cdot\left(\underset{\sim}{u}'\psi' - \overline{\underset{\sim}{u}'\psi'} - D\nabla\psi'\right) = 0$$

for the conservation of the fluctuations of ψ. Multiplying this relation by $2\psi'$ and averaging leads to:

$$\partial\overline{(\psi')^2}/\partial t = \underset{(i)}{-2\overline{\underset{\sim}{u}'\psi'}\cdot\nabla\bar{\psi}} \; \underset{(ii)}{-\underset{\sim}{U}\cdot\nabla\overline{(\psi')^2}} \; \underset{(iii)}{-\nabla\cdot\left[\overline{\underset{\sim}{u}'(\psi')^2} - D\nabla\overline{(\psi')^2}\right]} \; \underset{(iv)}{-2D\overline{(\nabla\psi')^2}}, \quad (3)$$

which expresses contributions to the budget of mean-square fluctuations of ψ as the four terms (i)-(iv). In the first, turbulent flux accounts for the production of fluctuations by interaction of irregular velocity components with the mean distribution of ψ; (ii) describes the advective transport of fluctuation intensity by the mean wind; (iii) summarizes turbulent and molecular transport; and the last term expresses the rate of destruction of deviations of ψ from its mean value, by molecular diffusion. This corresponds to the rate of viscous dissipation of turbulent kinetic energy, ε, and will be represented by the symbol χ_ψ.

Although detailed knowledge of the entire mean velocity field is seldom available, it is frequently of interest to consider the influence of shear in term (ii) above. This may be realized by writing the mean wind as the first two terms in a Maclaurin series thus:

$$U_i = U_{i_c} + x_k \, \partial U_i/\partial x_k .$$

(Dyadic tensor notation has been introduced, and will be used hereafter, for the ease it affords in writing some rather complicated expressions.) In a coordinate system moving with the mean velocity U_{i_c}, Equation (3) becomes:

$$\partial\overline{(\psi')^2}/\partial t = -2\,\overline{u_j'\psi'}\;\partial\bar{\psi}/\partial x_j - \partial U_i/\partial x_k \cdot x_k \cdot \partial\overline{(\psi')^2}/\partial x_i$$
$$-\partial\left[\overline{u_\ell'(\psi')^2} - D\,\partial\overline{(\psi')^2}/\partial x_\ell\right]/\partial x_\ell - 2D\,\overline{\partial\psi'/\partial x_r \cdot \partial\psi'/\partial x_r} . \quad (4)$$

The influence of shear is now depicted in the second term on the right. It is still difficult however to visualize, or to comprehend fully, the transport mechanisms represented by terms (ii) and (iii) of Equation (3).

B. Spectral representation

As a further aid to visualization of the exchange processes arising in the non-uniform motions it is helpful to make spectral analyses, in wave number space, of the fluctuating quantities. The manner in which different scale components of the velocity fluctuations and of ψ' interact in contributing to the budget of mean-square fluctuations of ψ is made evident and some insight may thus be gained into the details of the transport processes.

Therefore a velocity fluctuation spectrum $Z_i(\underset{\sim}{k},t)$ and a ψ-fluctuation spectrum $A_\psi(\underset{\sim}{k},t)$ are introduced by:

$$u_i'(\underset{\sim}{x},t) \equiv \int_{\underset{\sim}{k}\text{-space}} e^{i\underset{\sim}{k}\cdot\underset{\sim}{x}}\, Z_i(\underset{\sim}{k},t)\, d\underset{\sim}{k} ; \quad \psi'(\underset{\sim}{x},t) \equiv \int_{\underset{\sim}{k}\text{-space}} e^{i\underset{\sim}{k}\cdot\underset{\sim}{x}}\, A_\psi(\underset{\sim}{k},t)\, d\underset{\sim}{k} .$$

Mean-product spectra are also needed, which are defined by:

$$\Phi_{ij}(\underset{\sim}{k}) \equiv \frac{\overline{Z_i^*(\underset{\sim}{k},t)\,Z_j(\underset{\sim}{k},t)}}{V_o} ; \quad \Phi_\psi(\underset{\sim}{k}) \equiv \frac{\overline{|A_\psi(\underset{\sim}{k},t)|^2}}{V_o} ; \quad \Phi_{\psi i}(\underset{\sim}{k}) \equiv \frac{\overline{A_\psi^*(\underset{\sim}{k},t)\,Z_i(\underset{\sim}{k},t)}}{V_o} ;$$

where * indicates complex conjugate and V_o is a measure of the volume within which the irregular structure occurs. The budget of mean-square fluctuations may now be stated for individual wave number components. Equation (4) transforms to:

$$\partial\Phi_\psi/\partial t = \underset{\text{(i)}}{-2\,\partial\bar{\Psi}/\partial x_j\,\mathcal{R}\{\Phi_{\psi j}\}} + \underset{\text{(ii)}}{\partial U_i/\partial x_k \cdot k_i \cdot \partial\Phi_\psi/\partial k_k}$$
$$+ \underset{\text{(iii)}}{\int_{\underset{\sim}{k}\text{-space}} Q_\psi(\underset{\sim}{k},\underset{\sim}{k}')\,d\underset{\sim}{k}'} - \underset{\text{(iv)}}{2Dk^2\Phi_\psi} . \qquad (5)$$

$\mathcal{R}\{\ \}$ stands for the mathematical real part of the quantity enclosed. Here again the four terms on the right describe respectively production, redistribution by shear, redistribution by turbulent mixing, and dissipation by molecular diffusion. The quantity $2\mathcal{R}\{\Phi_{\psi j}\}$, which is the spectral representation of the turbulent flux $\overline{u_j'\psi'}$, accounts for production of ψ fluctuations in term (i) through mixing of the mean gradient $\partial\bar{\Psi}/\partial x_j$ at relatively large scales. Mean-square local gradient $k^2\Phi_\psi$ is acted upon by molecular diffusion in term (iv) to smooth out, and thus eliminate, ψ fluctuations at very small scales (large k). Redistribution of ψ deviations is effected by term (ii) or (iii). That these terms serve only this redistribution function may be seen by close examination of each.

Consider (ii). In the Fourier analysis

$$x_k \cdot \partial\psi'/\partial x_i = \int_{\underset{\sim}{k}\text{-space}} e^{i\underset{\sim}{k}\cdot\underset{\sim}{x}} \{-\partial[k_i A_\psi(\underset{\sim}{k},t)]/\partial k_k\}\,d\underset{\sim}{k} .$$

It follows that

$$\int_{\underset{\sim}{k}\text{-space}} k_i(\partial\Phi_\psi/\partial k_k)\,d\underset{\sim}{k} = \lim_{\underset{\sim}{r}\to 0} (-r_k)\,\partial[\overline{\psi'(\underset{\sim}{x},t)\psi'(\underset{\sim}{x}+\underset{\sim}{r},t)}]/\partial x_{r_i} \equiv 0$$

and thus that term (ii) makes no net contribution to the total rate of change of $\overline{(\psi')^2}$. It only shifts contribution from one scale to another.

In the same manner, since

$$Q_\psi(\underset{\sim}{k},\underset{\sim}{k}') \equiv i\left[k_\ell\overline{Z_\ell^*(\underset{\sim}{k}-\underset{\sim}{k}',t)A_\psi^*(\underset{\sim}{k}',t)A_\psi(\underset{\sim}{k},t)} - k_m\overline{Z_m(\underset{\sim}{k}-\underset{\sim}{k}',t)A_\psi(\underset{\sim}{k}',t)A_\psi^*(\underset{\sim}{k},t)}\right] ,$$

$$\iint\limits_{\underset{\sim}{k}\text{-space}} Q_\psi(\underset{\sim}{k},\underset{\sim}{k}')\, d\underset{\sim}{k}'\, d\underset{\sim}{k} \equiv 0$$

also and term (iii) is solely an agent for redistribution of contributions to $\overline{(\psi')^2}$.

V. SPECTRAL FORMS

Under equilibrium conditions, such that the mean fluctuation structure is constant, Equation (5) suggests a process in which ψ irregularities are produced at 'large' scales by term (i), transformed to ever smaller scales by term (ii) or (iii), and destroyed at very small scales by term (iv). In a subrange, at scales between the two extremes and at which neither production nor dissipation make significant contributions, predictions as to the form of $\Phi_\psi(\underset{\sim}{k})$ are possible on dimensional grounds alone.

For the case of fully turbulent mixing, the rate χ_ψ of production (and dissipation) of $\overline{(\psi')^2}$, the intensity of turbulence as reflected in the viscous dissipation rate ε, and the wave number k should determine the structure. One finds the usual 'inertial subrange' form:

$$\Phi_\psi(\underset{\sim}{k}) = C_1\, \chi_\psi\, \varepsilon^{-1/3}\, k^{-11/3},$$

where C_1 is a constant of order unity.

For the case of shear straining, in which shear of the mean flow is the primary redistribution mechanism, $\partial U_1/\partial x_3$ replaces ε as a determining parameter and Φ_ψ takes the form:

$$\Phi_\psi(\underset{\sim}{k}) = C_2\, \chi_\psi \left[(\partial U_1/\partial x_3)^2\right]^{-1/2} k_3^{-3},$$

indicating a highly anisotropic structure in the equilibrium range. C_2 is again a constant of order unity.

In the event the velocity field is comprised of a regime of waves, say internal gravity waves or other irrotational motions, the redistribution process embodied in term (iii) of Equation (5) may still operate through its kernel $Q_\psi(\underset{\sim}{k},\underset{\sim}{k}')$ to transfer ψ deviation from one wave number component to another. However the mean product spectrum $\Phi_{\psi j}$ will tend toward zero, if not for each wave number, at least on balance, when summed over all $\underset{\sim}{k}$, since the flux $\overline{u_j'\psi'}$ will be negligible in this case. That is, the waves will not, on the average, produce ψ fluctuations. However, if there is an extensive hierarchy of sufficiently random waves, they may further "mix" (temporarily and reversibly) any irregularities of ψ present from some other cause. This type of structure has not been studied in depth and no particular spectral form is known. Important work remains to be done here.

VI. BUDGET OF TURBULENT FLUCTUATION ENERGY

A relation may be developed similar to Equation (3) that expresses the budget of turbulent kinetic energy in a density-stratified medium. It takes the form (Bolgiano, 1968):

$$\partial(\tfrac{1}{2}\overline{u_i'u_i'})/\partial t = -\overline{u_i'u_j'}\,\partial U_i/\partial x_j + \partial[-\tfrac{1}{2}\overline{u_i'u_i'u_j'} + \nu\partial(\tfrac{1}{2}\overline{u_i'u_i'})/\partial x_j + \nu\partial(\overline{u_i'u_j'})/\partial x_i]/\partial x_j \tag{6}$$

$$- 1/\rho_0\,\partial(\overline{u_i'p'})/\partial x_i + g_i/\rho_0\,\overline{u_i'\rho'} - \varepsilon .$$

ν is kinematic viscosity, p' represents pressure fluctuations, and ρ_0 and ρ' stand for the reference and fluctuation portions respectively of potential density. Because the turbulent mixing results in local inhomogeneities of density, due both to temperature and moisture ($\rho'/\rho_0 = -\theta'/T - 0.61\,q'$), a field of density irregularities will coexist with the turbulent velocity field.

Each such deviation of density from the mean distribution represents an amount of potential energy

$$\phi(\underset{\sim}{x},t) = -\tfrac{1}{2}\,g_i[\rho'(\underset{\sim}{x},t)]^2/(\rho_0\,\partial\bar{\rho}/\partial x_i).$$

A budget of this 'turbulent' potential energy may also be derived. It is:

$$\partial\bar{\phi}/\partial t = -g_i/\rho_0\,\overline{u_i'\rho'} - \partial(\overline{\phi u_j})/\partial x_j - \chi_\phi , \tag{7}$$

in which χ_ϕ is the rate of elimination of density deviations, and hence of turbulent potential energy, by molecular diffusion. Note that the first term on the right of Equation (7), which represents the production of potential energy by work done against the Archimedean force, is exactly the negative of the next to last term on the right of Equation (6). In the latter equation it represents abstraction of kinetic energy by the same buoyancy phenomenon. Thus, the overall effect is that turbulent energy is produced through working of the mean shear against the Reynolds stress (first term on right in Equation (6)) and is ultimately destroyed by molecular smoothing (ε and χ_ϕ).

However, whether or not the turbulent velocities necessary to sustain a given fluctuation energy situation can, in fact, be maintained depends upon the ratio of production of potential energy to production of kinetic energy. This ratio is given by:

$$R_f \equiv \left(-g_i/\rho_0 \, \overline{u_i'\rho'}\right)/\left(-\overline{u_i'u_j'}\,\partial U_i/\partial x_j\right) = 1 - \varepsilon/\left(-\overline{u_i'u_j'}\,\partial U_i/\partial x_j\right), \quad (8)$$

the flux Richardson number. It may be expressed in terms of temperature and humidity deviations according to:

$$R_f = \left(g_i/T \, \overline{u_i'\theta'} + 0.61\, g_i \, \overline{u_i'q'}\right)/\left(-\overline{u_i'u_j'}\,\partial U_i/\partial x_j\right), \quad (9)$$

thus exhibiting the individual contributions of sensible heat and latent heat to the dynamic stability. Immediately over the sea these two factors are of comparable magnitude.

The dynamic role of moisture vapor in the turbulent energy balance has only recently been considered critically (Coantic, 1969; Brutsaert, 1972; Warhaft, 1973). There is still much uncertainty as to the precise manner in which moisture vapor exerts its influence, and this matter is receiving further study.

VII. REFERENCES

Bean, B.R. and E.J. Dutton, Radio Meteorology, Chap. 1, pp. 1-22, National Bureau of Standards, Monograph No. 92, Washington; 1 March 1966.

Bolgiano, R., Jr., Part I. The general theory of turbulence; Part II. Turbulence in the atmosphere: in Winds and Turbulence in Stratosphere, Mesosphere and Ionosphere, K. Rawer (ed.), North-Holland Publishing Co., Amsterdam; 1968.

Brutsaert, W., Radiation, evaporation, and the maintenance of turbulence under stable conditions in the lower atmosphere, Boundary-Layer Meteorology, 2, pp. 309-325; 1972.

Coantic, M., Les interactions atmosphère-océans: les processus physiques et les équations qui les gouvernent, Rapp. Institut de Mécanique Statistique de la Turbulence Cahiers Océanogr., 21, pp. 17-46, 105-143, 223-249; 1969.

Lumley, J.L. and H.A. Panofsky, The Structure of Atmospheric Turbulence, Interscience Publishers, New York, Chap. 2, pp. 59-98; 1964.

Warhaft, Z., The relation between temperature and humidity in the free atmosphere under conditions of stable stratification and strong thermal intermittency--a case study, Quart. J. Roy. Meteorol. Soc., 99, pp. 89-104; 1973.

II GROUP

Air-Sea Interaction and Ducting: Radiopropagation Aspects

THE STRUCTURE OF MOISTURE AND TEMPERATURE PROFILES IN THE NEAR-SURFACE LAYERS, THEIR VARIABILITY AND EFFECTS ON RADAR PROPAGATION AS DEMONSTRATED WITH RAY TRACING TECHNIQUES

CDR G. D. Hamilton, USN and T. Laevastu

Environmental Prediction Research Facility
Monterey, California

1. INTRODUCTION AND REVIEW OF REFRACTIVE INDEX

A knowledge of the detailed structure of water vapor profiles near the surface, and their variability in space and time, is badly needed for both the estimation of evaporation and for the prediction of radar propagation peculiarities, especially with respect to the so-called evaporation duct near the surface. This summary paper reviews the present state of knowledge of water vapor and temperature profiles near the surface, the factors affecting these profiles, and their variability.

The intention of the authors is to answer two basic questions:

(1) How pronounced is the variability of radar propagation near the surface caused by humidity profiles?

(2) Can these (the humidity profile variability and radar propagation) be forecast with sufficient accuracy with available knowledge and techniques?

The length of this paper does not allow detailed references to be made for each statement (indeed voluminous monographs could be written about the subject).

Conventionally, the refractive index is presented in terms of atmospheric pressure, temperature, and water vapor pressure with the following formula:

$$N = 77.6 \frac{p}{T} + 3.73 \cdot 10^5 \frac{e}{T^2}$$

The first part of the formula is called the "dry part" and the second the "wet part." The relative importance of pressure temperature and water vapor pressure on refractive index is demonstrated in Figure 1.

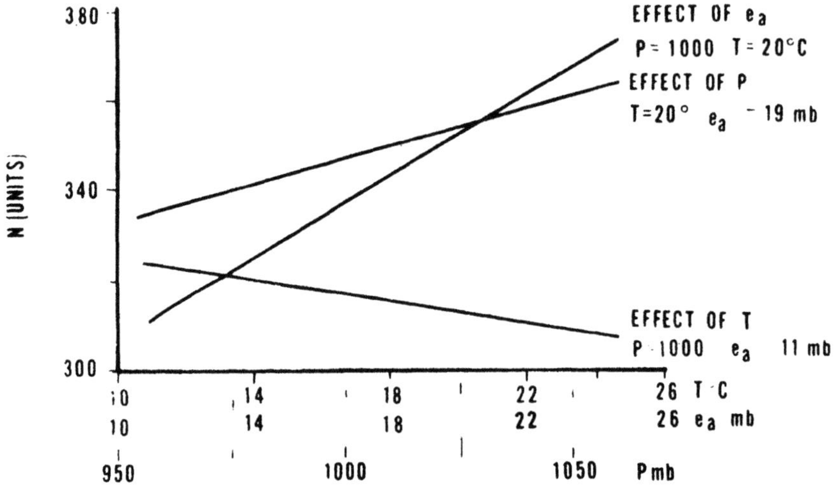

Fig. 1. Effects of water vapor pressure (e_a), temperature (T) and pressure (P) on refractive index (N)

As in all sound and other wave propagation problems, there are critical gradients which cause refraction and ducting. Two different expressions to present these critical gradients are in general use: First the critical gradient is often given as -48.6 N units per 1000 feet of height (Randall, 1964); and secondly the condition where ducting occurs is usually given by:

$$\frac{dn}{dh} \geq 157 \text{ N units/km (Cahoon and Riggs, 1964).}$$

2. FACTORS AFFECTING THE REMOVAL OF WATER VAPOR FROM THE SURFACE

The main factor affecting the removal of water vapor from the sea surface are (1) the "turbulent transfer" caused by wind-generated turbulence (also affected by waves if wind speeds are low); (2) the heating of the lower layers of the atmosphere by the sea (causing convection); and (3) the addition of moisture to the atmosphere also causing convection. The speeds of turbulent upwards transfer depend on the vertical gradient of temperature and humidity and thus, on the vertical stability of the lower layers.

It has been ascertained in technological-physical investigations (especially during the intensive investigations of the steam boiler during the last decennium) that heat and moisture transfers follow the same basic laws (i.e., proceed parallel to each other). The only exception to this rule is in the relatively rare unstable atmospheric condition when the temperature profile is isothermal for some vertical distance above the sea surface, the temperature of the air is that of the sea surface, and when there still exists a water vapor deficit in the lower layers of the atmosphere. In this case there can exist a vapor transport upwards, but little or no heat transport. However, this condition is rare in nature.

The latest studies of the detailed thermal and humidity structure near the sea surface indicate that there is always a so-called "equilibrium deficit" between the sea and the atmosphere with respect to both the temperature and moisture at which heat and vapor transport should decrease. (see Bøyum, 1962) Several explanations have been offered about this equilibrium difference, but one cannot say with certainty which of the explanations is more valid than the other (see Laevastu, Rabe, and Hamilton, 1972).

In most cases, the sea surface is warmer than the atmosphere above and there is continuous evaporation from the sea with moisture being transported into upper layers of the atmosphere by turbulent and convective transfer mechanisms. Thus, a pronounced and thick evaporation duct is unlikely to occur in these conditions. However, there are certain meteorological conditions and areas where the sea surface can be cooler than the air flowing over it. This often occurs in coastal regions and especially along the west coast of continents where warm dry air moves over colder "up-welled" water. Furthermore, the same condition occurs at oceanic fronts with sharp thermal contrasts, and in specific meteorological conditions where there is air flow across the sea surface isotherms towards the colder side.

Inversion conditions are created in coastal regions where warmer air moves over the colder ocean surface (resulting in a cooling of the lower layers of the air) in combination with subsidence of warmer air from aloft. This condition off the coast of California north of San Francisco is illustrated in Figure 2(A). This figure also gives an indication of the formation of California coastal stratus under the inversion, and indicates that with further cooling of the lower layer of the air, the stratus can reach the surface as fog. There can be a breakthrough of the inversion in some synoptic conditions, resulting in a rapid transfer of moisture upwards (as can be seen from Figure 2(B).

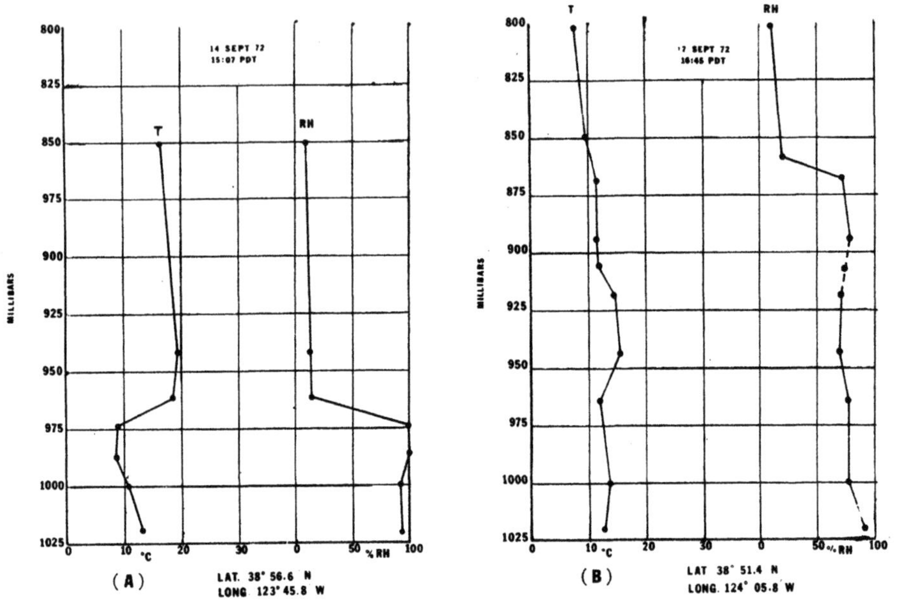

Fig. 2(A) & 2(B). Temperature and relative humidity profiles off Point Arena, September 1972.

3. MEASUREMENT OF MOISTURE PROFILES OVER THE OCEANS AND THEIR VARIABILITY

Accurate measurements of humidity profiles in the lower layers of the atmosphere are relatively difficult and require specific instrumentation. Regular radiosonde observations do not provide the detailed structure near the surface as the lower levels are either missed or are corrupted by the effect of the ships from which they were launched. Only in special cases are 1000 mb observations reported. Thus, special radiosondes are required for the detailed measurement of humidity and temperature structure near the surface.

Experimental methods for the measurement of detailed near-surface temperature and humidity include utilizing masts built into the bottom in shallow water, or the use of floating masts in connection with research vessels. Radio refractometers have also been in extensive use in the last decade. Although these methods have produced very useful observations, none have produced accurate data in the near-surface layers which are needed, e. g., for the computation of evaporation and heat exchange.

Most detailed measurements of the temperature and humidity structure in near surface layers have been recently made by Instructor Captain G. P. Britton, (RN) from Reading University. He has developed a "yo-yo" type radiosonde which upon release rises to a predetermined level using a leaky valve combined with a pressure operated valve. The instrument then descends until a weight touches the surface. The weight is released when an Alka Seltzer tablet melts and the balloon rises again (this time from surface) to a predetermined level at a constant, relatively slow ascent rate (1.2 m sec^{-1}). Data is reported in short intervals. Britton used GRAW sondes M60 (6-sec response time for humidity sensor) or G120 (3-sec response time). With the ascent rate used, the "height error range" does not exceed $\pm$ 5 meters.

The results of Captain Britton's observations are of great importance for an estimation of the variability of temperature and humidity profiles over the sea. Some of his evaluations from a sequence of radiosonde releases at two hourly intervals in NE Atlantic are reproduced below:

"The main interest in the sequence lies in the ascent made at 1530Z (6 June 1970) (Figure 3) which achieved a double "hop" - i.e. four transits of the boundary layer. The following points are of particular interest:

(1) Examination of Figure 3 and the corresponding tephigrams for those ascents show that there was a marked change in the environment curves in a short interval of time (20 minutes). In the second phase there is much greater evidence of the intrusion of warm air ascribed to the developing wave depression.

(2) Two inversions are clearly visible during the second ascent phase. Such "double structure" inversions fairly close together have been noted in many other ascents and merit further study.

(3) A sudden decrease in descent rate occurs between 960 and 970 millibars (Figure 3) during the second descent. The data have been carefully checked and no instrumental error found. Hence it is believed that the falling balloon was arrested in the frontal inversion at a level of 967 millibars. Instability in the lowest layer may even have carried it upwards again for a short distance before the descent was resumed."

The sounding in Figure 4 (Britton, 1971) shows pronounced temperature and humidity variations at lower levels in the vicinity of the Strait of Gibraltar. Of further interest are the ascent and descent profiles (about 30 min apart) through fog in the vicinity of Bouee Laboratoire (about 70 miles south of Toulon),

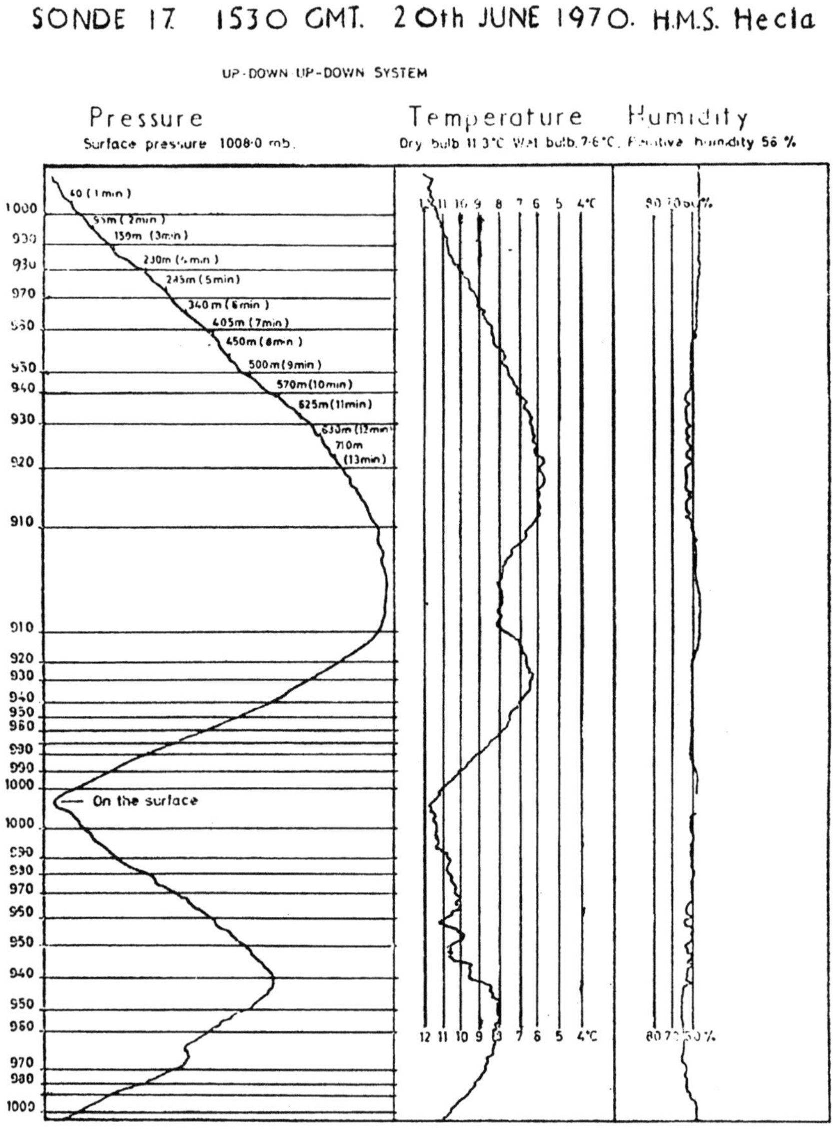

Fig. 3. Radiosondesounding with twice up and down movement 1530Z, 20 June 1970 in NE Atlantic (Britton, 1971)

which are reproduced in Figure 5. These profiles are described by Britton as follows:

"The graphs of the virtual potential temperature indicate that the atmosphere was stable at all levels, particularly in the lowest 80 meters. The graphs of temperature indicate two inversions - one at 80 meters and a second at 210 meters although the latter is not particularly marked on the graphs of temperature, potential temperature and virtual temperature during the descent phase. The two inversion levels are better defined on the graphs of dew point, wet bulb potential temperature, specific humidity and wet bulb temperature, all quantities which involve moisture content. Visual observations at the same time, suggested that the fog bank was confined to the lowest 70 meters of the atmosphere and that there was little or no cloud associated with the second inversion at 200 meters. Clear skies could be seen above the fog between banks.

There is very good agreement between the ascent and descent graphs in almost all parameters. Marked changes in the dew point and specific humidity graphs occur at 996.5 mbs (170 meters) on ascent and 999.0 mbs (145 meters) on descent. Other significant levels are from 75 to 80 meters during ascent and between 55 and 70 meters during descent. These differences in mean height may be due to changes in the inversion levels over the interval of approximately 30 minutes but they are more probably attributable to a small amount of response time in the sensors."

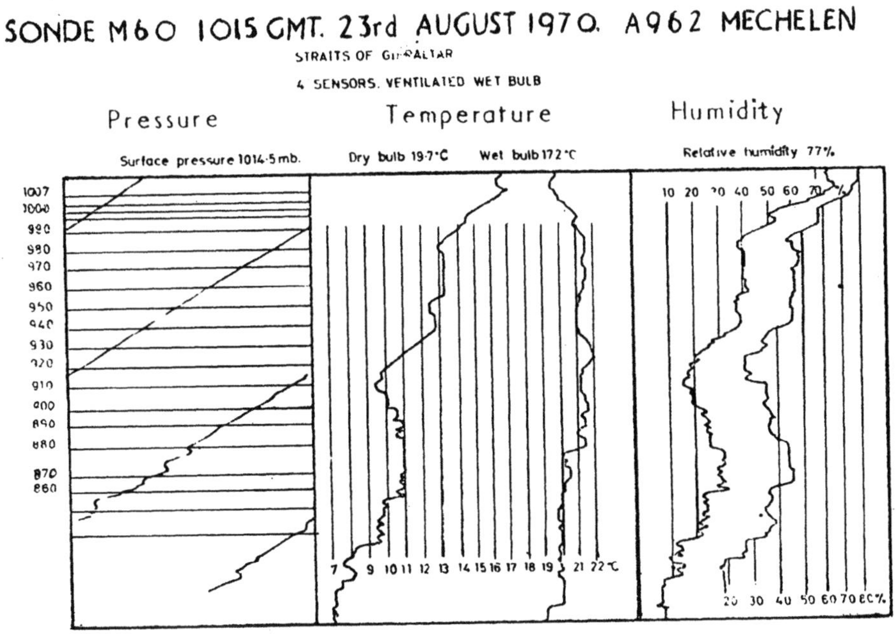

Fig. 4. Radiosonde sounding, 1015Z, 23 August 1970 in the vicinity of Strait of Gibraltar (Britton, 1971)

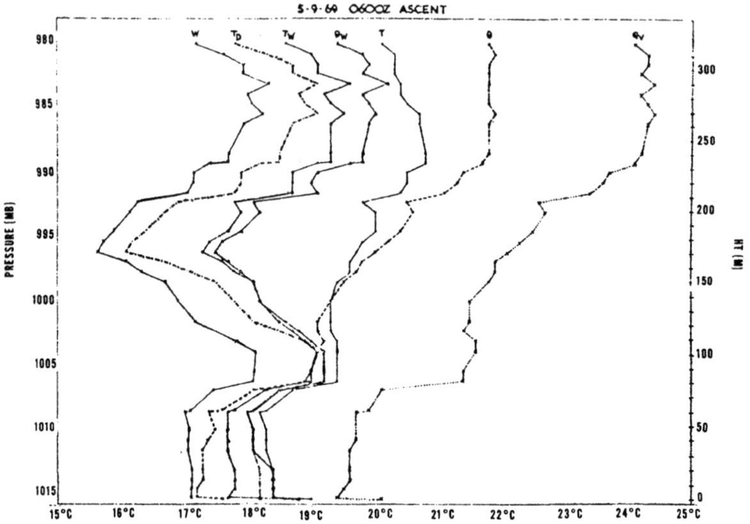

Fig. 5. Radiosond ascent through for on 5 September 1971, ca 70 nm. south of Toulon, (Britton, 1972)

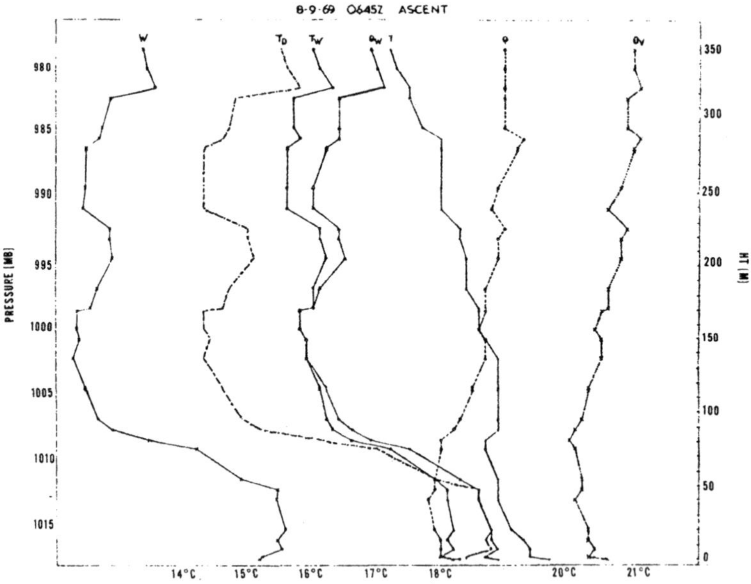

Fig. 6. Radiosonde sounding in a stable atmosphere on 8 September 1971, ca 70 n mi south of Toulon (Britton 1972)

Figure 6 (same location as the sounding given in Figure 5, but about three days later) shows a layer of high humidity below 70 m. The ray traces based on the above mentioned profiles are given in Chapter 4 of this paper.

The above described studies of Britton and other available observations, such as our own observations off the California Coast (see Figure 2) demonstrate that considerable short term and short-distance variations of moisture and temperature profiles exist. These are caused partly by changing synoptic conditions and partly by subsynoptic sea-air interaction processes.

At present no other means exist for the prediction of details of the temperature and humidity profiles near the surface, except measurements with special sondes like the Britton's "yo-yo" sonde or by parameterization and typing (see Chapter 5).

4. RAY TRACING AND PROPAGATION LOSS COMPUTATIONS IN "EVAPORATION DUCT"

The basic numerical ray tracing technique, combined with complex propagation loss computations by numerical means is still the most powerful tool in submarine acoustics and has almost entirely replaced the normal mode approach and its application. This paper is concerned with propagation loss with respect to the near-surface (up to 2000 m. height) radar duct. "Duct" is usually defined as a long propagation channel, whereas "hole" designates the areas from which the electromagnetic waves have been diverted. Ducting normally does not occur with elevation angles above 0.5 deg.

Based on the FNWC numerical models of sound propagation loss predictions in the oceans, Clarke, Hensen and Clark (1973) have devised a computerized method to predict the radar propagation loss in the atmosphere.

The method uses a ray trace technique where the essential input is the refractive index, which is computed with the formulae given in the first chapter of this paper. The essential features of the propagation loss are determined by spreading loss, which is a direct function of density of rays and distance from the source. Of secondary importance in the propagation loss are the absorption loss, which is a function of frequency, temperature, and humidity, the scattering from various surfaces, and leakage from the ducts.

The propagation loss plots, computed by Clarke, et al, and presented in Figures 7 to 10 are based on the temperature and humidity profiles presented in Figures 3 to 6. Considering the

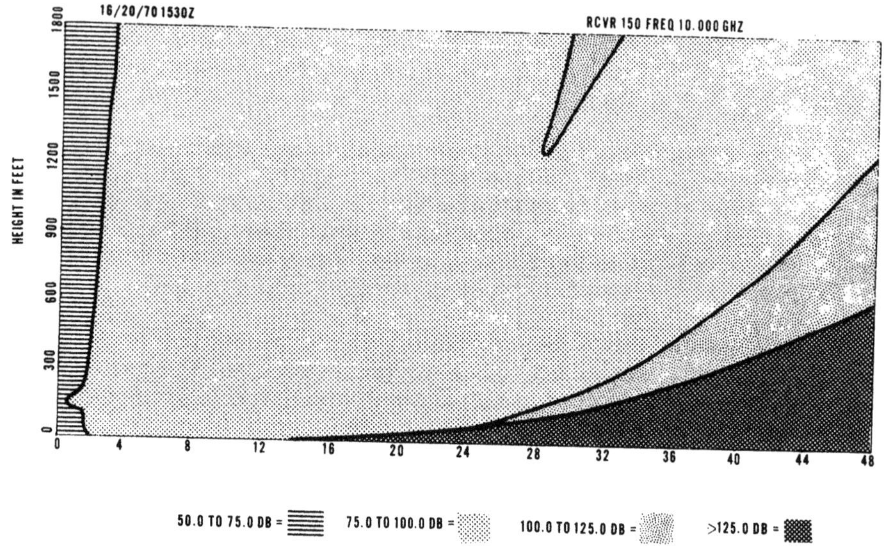

Fig. 7. Propagation loss of radar waves, 10HGZ, receiver 150 feet, temperature and humidity profiles see Figure 3 (Clarke, et al)

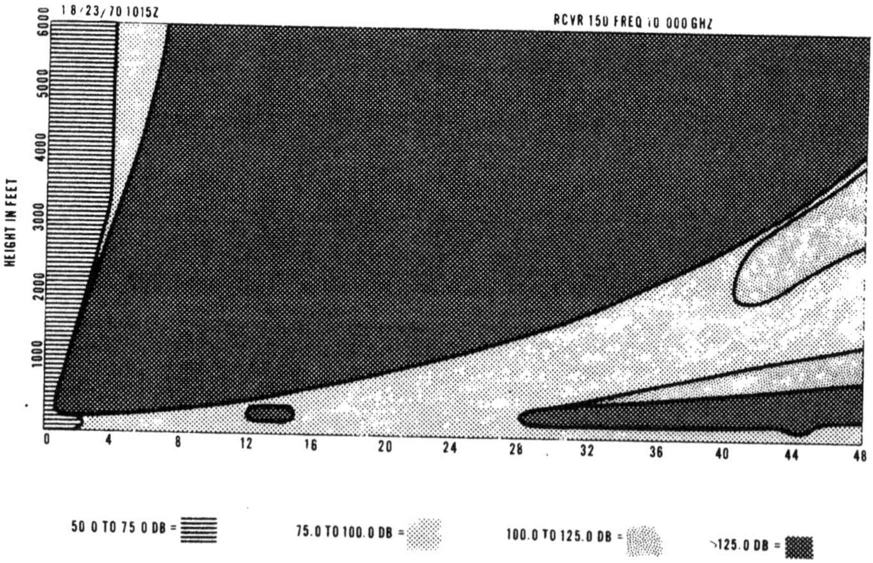

Fig. 8. Propagation loss of radar waves, 10HGZ, receiver 150 feet, temperature and humidity profiles see Figure 4 (Clarke, et al)

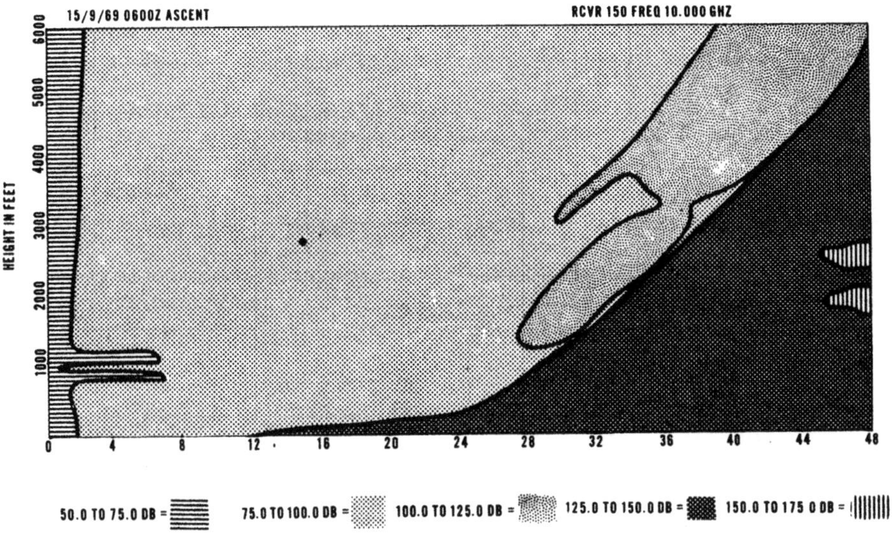

Fig. 9. Propagation loss of radar waves, 10HGZ, receiver 150 feet, temperature and humidity profiles see Figure 5 (Clarke, et al)

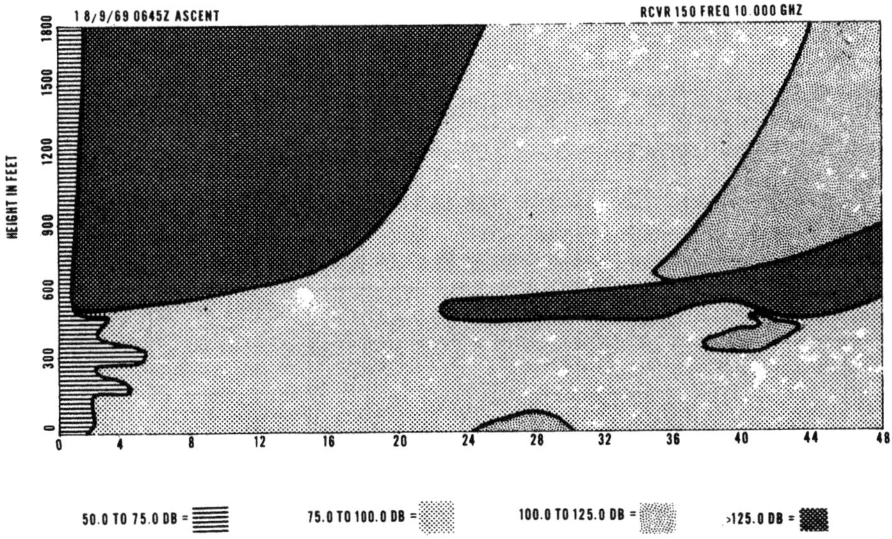

Fig. 10. Propagation loss of radar waves, 10HGZ, receiver 150 feet, temperature and humidity profiles see Figure 6 (Clarke, et al)

relative uniformity of profiles 5 and 6, one can notice the "sensitivity" of the radar propagation even with small changes in these profiles (compare Figures 9 and 10.

Figure 8 presents a fully ducted condition from the area of the Strait of Gibraltar with a very pronounced shadow zone.

Of interest with respect to the sensitivity of the lower part of the humidity profile is an experiment where the surface data was omitted and the refractive index was extrapolated to the surface from a height of 30 m. A considerably different near-surface propagation loss picture resulted, as compared to the run with the input of measured data.

5. RECOMMENDATIONS FOR PARAMETERIZATION AND TYPING OF THE "EVAPORATION DUCT"

As briefly demonstrated in earlier chapters, prediction of the detailed structure of temperature and humidity profiles in the lower layers of the atmosphere is, at present, a quasi-impossible task. Furthermore, if detailed profiles could be obtained with special sondes, their subjective interpretation with respect to radar propagation is not possible without considerable previous computerized propagation loss computations.

A reasonable solution to the above problems could be achieved by the detailed parameterization and typing of the nature and occurrence of various low-level temperature and humidity profiles in defined synoptic meteorological conditions and in specific locations with relation to underlaying sea surface.

Furthermore, applicable typing and characterization of evaporation ducts on the basis of environmental conditions can be achieved only if they are characterized by different types of radar propagation in the duct and propagation loss profiles. Some characterization of conventional air masses with respect to refractive characteristics have been made already by Bean and Dutton (1966).

However, much more detailed typing and characterization is needed than is available at present. In order to accomplish this task, a considerable amount of detailed measurements with Britton-type sounders in a variety of conditions are required together with ray trace and propagation loss computations as presented in this paper.

6. REFERENCES

1. BEAN, B. R. and E. J. DUTTON, "Radio Meteorology" Nat. Bur. of Standards Monogr., 92:435 pp., 1966.

2. BOROVIKOV, A. M., V. V. KOSTAREV, I. P. MAZIN, V. I. SMIRNOV, and A. A. CHERNIKOV, "Radar Measurement of Precipitation Rate" Gidromet, Izd., Leningrad, 112 pp. 1967.

3. BØYUM, G., "A Study of Evaporation and Heat Exchange Between the Sea Surface and the Atmosphere" Geophys, Pbl. (Geophy. Norv) 22(7):15 pp. 1962.

4. BRITTON, G. P., "Marine Meteorology and Oceanography Operations 1970." Univ. of Reading, Dept. of Geophys. Rpt., (mimeo), 1971.

5. BRITTON, G. P., "Sea-Air Interaction Studies, 1971" Univ. of Reading, Dept. of Geophys. Rpt., (mimeo) 1972.

6. CAHOON, B. A. and L. P. RIGGS, "Climatology of Elevated Superrefractive Layers Arising from Atmospheric Subsidence" Proc. 1964 World Conf. on Radio Meteorology, 52-55, 1964.

7. CLARKE, L., J. CLARK, and D. HENSEN, "Operational Propagation Loss Computations for Radar Propagation", (in preparation, 1973.

8. LAEVASTU, T., K. RABE, and G. D. HAMILTON, "The Effects of Oceanic Fronts on Properties of the Atmospheric Boundary Layer" ENVPREDRSCHFAC Techn. Paper, 2-72 1972.

9. RANDALL, D. L., "A Summary of Tropospheric Radio Propagation as Observed on a Trip Around the World Feb. 8 to Mar 15, 1962." Proc. 1964 World Conf. on Radio Meteorology, 62-69, 1964.

STATE AND LIMITS OF PREDICTION METHODS OF RADAR WAVE PROPAGATION CONDITIONS OVER SEA

H. Jeske

University of Hamburg
Meteorologisches Institut

ABSTRACT:

The vertical stratification of the refractive index immediately over sea may be interpreted on the basis of the theory of the atmospheric boundary-layer turbulence. Detailed experimental investigations showed that a parametrization of the vertical gradients (profiles) is possible, i.e. the evaluation by simple meteorological routine parameters (vertical differences air-sea), making up the basis for a radar-range prediction. For this problem, the thickness of the evaporation duct (mostly present over sea) proves to be a decisive parameter for characterizing the propagation conditions in the maritime boundary layer. This fact is verified by a great number of transmission measurements on transhorizon and line-of-sight radio links in the wave-length range of 1.8 cm to 1.8 m.

For coastal areas horizontal gradients in the refractive-index field and with this, horizontal differences of the duct thickness may be important. By means of the diffusion theory, apt models can be given which have to be taken into account especially during summerly advection of air masses from land.

If elevated refractive-index layers are effective, in general, aerological soundings are necessary to determine propagation conditions in such cases. For "large" distances, scatter propagation is predominant.

But all meteorological models of the refractive-index distribution are only a first step towards a detailed prediction of the field strength (amplitude and phase) or radar echo. The second step is the including of realistic models within the propagation theory. A certain meteorological stratification may yield different effects according to path geometry, wave length, and polarization of the radio/radar-links.

1. INTRODUCTION

With the conception of the boundary layer theory a description and interpretation of the structure of the meteorological field near the sea surface, including the distribution of the refractive index, may be given. Great progress in the field of the atmospheric boundary-layer research - in theoretical as well as in experimental regard - allows a parametrization of the physical processes, i.e. a representation of the laws by means of simple meteorological routine measurements. This step was necessary for practical application of the theoretical results, especially also for the determination or prediction of E.M. Wave-propagation conditions (sometimes shortly called radar weather).

All dielectrical propagation phenomena are determined by the time-space structure of the radio refractive-index, which is a well-known function of the meteorological parameters atmospheric pressure p, temperature T, and water-vapor pressure e. Being acquainted with realistic refractive-index models, we can make an attempt to interpret the propagation phenomena by means of Maxwell's electrodynamics.

2. REFRACTIVE-INDEX MODELS OF THE MARITIME BOUNDARY LAYER

2.1. Vertical gradients

The vertical structure of the meteorological parameters immediately above the sea surface is controlled by turbulent transport processes. The flux-gradient relationship for a conservative property i is given by:

$$S_i = - cK_i \frac{di}{dz} \quad , \tag{1}$$

in which S_i is the vertical flux, K_i the transport coefficient, z the vertical coordinate, and c a known constant. According to the similarity theory we write for K_i:

$$K_i = i^* \ (z+z_o) \quad , \tag{2}$$

here i^* is a scaling property (proportional to the ratio of the vertical flux of i and the friction velocity u^*), and z_o the roughness length. Thus for the vertical gradient we obtain:

$$\frac{di}{dz} = - \frac{S_i}{ci^* \, (z+z_o)} \quad . \tag{3}$$

This formula (logarithmic vertical profile) is valid for neutral stratification. For diabatic conditions eq.(3) is extended by a stability function Φ_i.

The gradient of the refractivity N(=(n-1) · 10^6; n refractive index) may be given by the formula:

$$\frac{dN}{dz} = c_1 \frac{dp}{dz} + c_2 \frac{de}{dz} - c_3 \frac{dT}{dz} \quad , \tag{4}$$

where dp/dz is the pressure gradient [mb/m] , de/dz the gradient of water vapor [mb/m] , dT/dz the gradient of air temperature [°K/m] , and c_1 , c_2 , c_3 are known coefficients. In temperate climates we have near the ground:

$$c_1 \simeq 0.3, \quad c_2 \simeq 4.5, \quad \text{and} \quad c_3 \simeq 1.3$$

As we are looking for a routine procedure for the determination of the propagation conditions (radar weather) we must try to approximate the gradients by vertical differences. For this purpose we pursue the following course. By integration of eq. (3) between the surface and a reference height z (in our case mostly 6 m) under the assumption of constant transports in this layer we get a relation for the vertical difference:

$$i_{z_1} - i_o = \Delta i_1 \quad = - \frac{S_i}{ci^*} \ln \frac{z_1 + z_o}{z_o} \quad . \tag{5}$$

Then we replace S_i/ci^* in eq. (3) by means of eq. (5), receiving for the gradient:

$$\frac{di}{dz} = - \frac{\Delta i_1}{(z+z_o)} \cdot \frac{1}{\ln(z_1+z_o)/z_o} = - \frac{\Gamma \Delta i_1}{(z+z_o)} \quad . \tag{6}$$

The factor

$$\Gamma = \frac{1}{\ln(z+z_o)/z_o} = \frac{1}{\Delta i_1} \frac{di}{d \ln z} \tag{7}$$

is called the profile coefficient and may be obtained by measured vertical profiles (it is the slope of the profile in a semilogarithmic system divided by Δi_1). The values of Γ or z are known from extensive measuring series. For neutral conditions: z=0.015 cm; $\Gamma \simeq 0.1$ (reference height 6 m) [Brocks and Krügermeyer 1970] .

Even for not too large deviations from the neutral case, the logarithmic profile shape given in eq. (6) is a useful approximation. As our measuring series showed, Γ is a definite function of ΔT (vertical difference between temperatures of the air in 6 m and the water) and u (wind velocity), i.e. a function of the

Ri-bulk-number ($Ri_b = 2\,\Delta T/u^2$) (see fig.1).

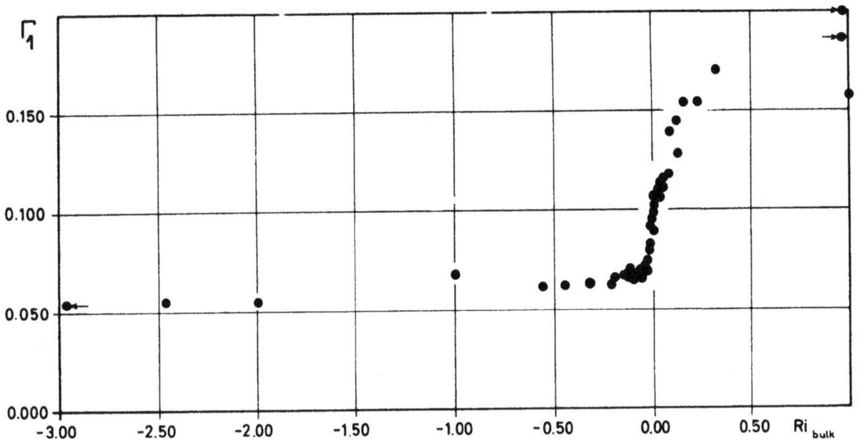

Fig. 1: Mean value of the profile coefficient Γ_1 (see equ. 6 and 7) as a function of the Bulk-Richardson-number ($Ri_b = 2\,\Delta T_1/u_1^{\,2}$) .

But generally, we have to introduce a stability function. Many statements of this function may be combined in the equation:

$$\phi\,(Ri) = \left(1 + \frac{\alpha Ri}{n}\right)^n \quad , \qquad (8)$$

e.g. n=-1 results in the log.-lin.profile, n=1/2 in the Rossby-Montgomery profile, n=-1/4 in the so-called "Keyps"-profile. Here Ri is the Richardson number.

In analogy to the calculations given above, we may derive for the diabatic case:

$$\frac{di}{dz} = - \frac{i}{z+z_o}\,\phi\,(Ri) \Big/ \int_o^z \frac{\phi(Ri)}{(z+z_o)}\,dz \qquad (9)$$

Based on our propagation experience, we can recommend the Keyps profile for instable situations ($\alpha = 4.5$) and the log.-lin. model for stable cases ($\alpha = 5.2$). For extreme stability conditions, where turbulence is gradually replaced by inertia oscillation and internal waves, there are no generally accepted models. This is a real limit for the gradient-determination procedure given here.

Our formula system permits the determination of the gradients of refractivity at any height within the lower boundary layer, of the whole vertical profile, or of mean gradients for certain height intervals, if we know T_{6m}, T_w (water temperature), e_{6m}, E_w (saturation pressure at T_w), and u. These values we may get from simple meteorological measurements.

2.2. The thickness of evaporation duct

The thickness of the so-called evaporation duct, a phenomenon mostly present over sea, proved to be a characteristic parameter for the comprehension of the refractive-index field in the maritime boundary layer. The evaporation duct is that layer, where the ray curvature produced by a strong refractive-index gradient is larger than the earth curvature. In the case of horizontal radiation we find for the critical gradient:

$$\frac{dN}{dz} = - \frac{10^6}{r} = -0.157/\text{m} \quad \text{(r: earth radius).} \tag{10}$$

Inserting this gradient into eq. (4) we get the height belonging to this gradient. This is the upper boundary of the duct. In the simplest case of a logarithmic shape we have with equations analogic to eq. (6):

$$z_* = \frac{\Gamma_1 [c_2(e_1 - E_w) - c_3(T_1 - T_w)]}{-0.157 - c_1 dp/dz} \tag{11}$$

$$= \frac{\Gamma_1 [c_2(e_1 - E_w) - c_3(T_1 - T_w)}{-\ 0.125}$$

For practical purposes we may write:

$$z_* \doteq \frac{\Gamma_1 \Delta N_1}{-0.125} \tag{12}$$

(Strictly speaking ΔN_1 is the difference of the potenzial refractivity [Jeske, 1965]).

Analogous terms may be gained for diabatic profile forms (instable and stable conditions), starting with eq. (9). For the Keyps profile the z -value must be gained by numerical methods [Jeske, 1965] . The real height dependence of the Richardson number is given by:

$$\frac{z}{L'} = \text{Ri}\,\phi(\text{Ri}) \quad \text{(see text books of boundary-layer theory)} \tag{13}$$

L' is Monin-Obukhov's scaling length in the gradient form (independent of height!) and can be approximated by:

$$L' = \frac{T_1 u_1^2 \Gamma_1}{g\ \Delta T_1}\ , \tag{14}$$

the index 1 refers to the reference height z ; Γ_1 you may find in fig. 1.

It may be pointed out that the duct thicknesses z_* thus derived are defined on the basis of geometric optics. It is especially dependent on the wave length used, whether such a duct becomes really effective for radar-wave propagation; a certain critical duct thickness has to be exceeded. Here, z_* stands as a characteristic feature of the refractive-index stratification. With z_* we may find a very practical relationship:

$$\frac{dN}{dz} = - 0.033 - 0.125 \frac{z_*}{z} . \tag{15}$$

This ground-based duct is called evaporation duct, as the water-vapor gradient is predominant in its formation; the negative temperature gradient mostly present in temperate latitudes even weakens the refractive-index gradient.

The derived continuous profiles with modified logarithmic shape have been fully verified, if m e a n v a l u e s of several minutes were taken; the "momentary" profile shows strong variations and makes the logarithmic profile form hardly recognizable.

To illustrate this chapter, some figures may be given. <u>Fig. 2</u> gives refractive-index profiles for different duct thicknesses. It is to be seen that upper boundary does not demonstrate a N-jump; we have to be cautions with too simple comparisons to a technical waveguide.

<u>Fig. 3</u> shows a diagram of the duct thickness as a function of 1/L' (gained from eq. (14) with ΔN_1 as the parameter for the Keyps profile.

<u>Fig. 4</u> compares duct thicknesses evaluated according to different relationship.

By means of an annual variation, the lower part of <u>fig. 8</u> offers an impression of the occurring evaporation-duct thickness (50%-values). Additionally some statistical information may be given: Over ten years, the a n n u a l 50%-values at light-ship Elbe I in the German Bight vary between 4.5 and 6.5 m, the 10%-values between 10.7 and 13.5 m. That means, the influence of the evaporation duct in radio propagation is restricted to cm-waves only.

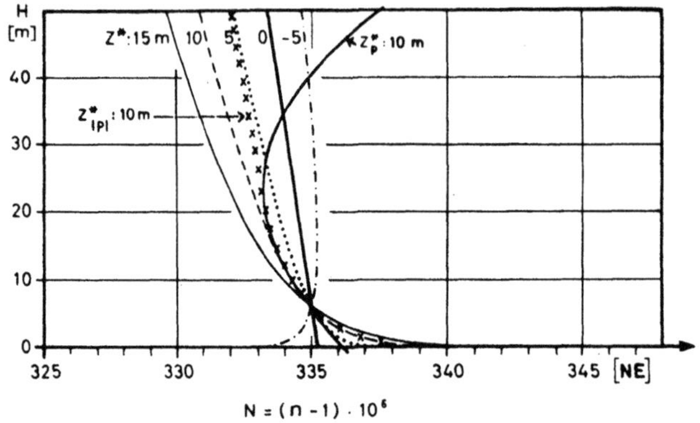

Fig. 2: Refractivity profiles for different duct thicknesses using a logarithmic profile shape.
For z_* =10 m profiles for a parabolic and a parabolic-linear height distribution of N are given too.
[s. Früchtenicht, 1971] .

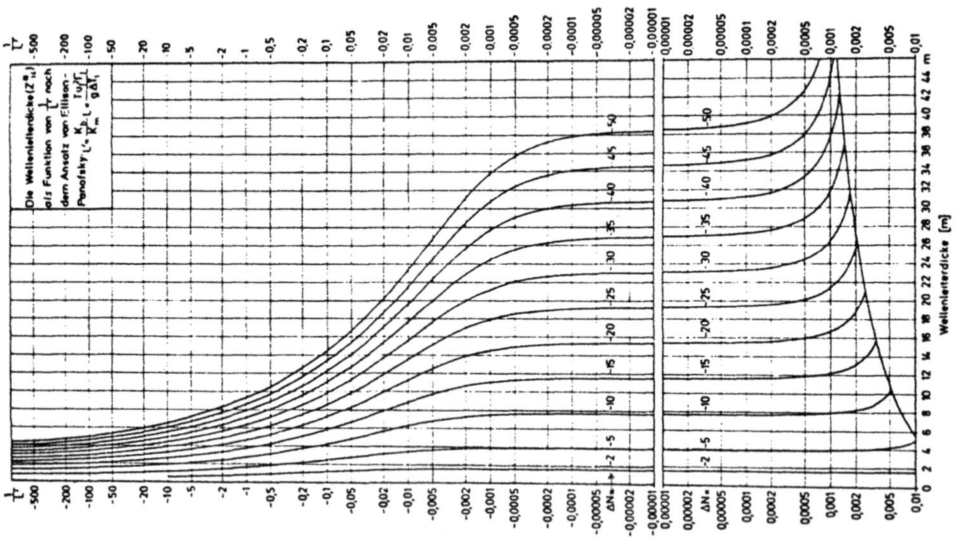

Fig. 3 : The thickness of evaporation duct from the Keyps-model as a function of the stability parameter 1/L' for different ΔN-values.

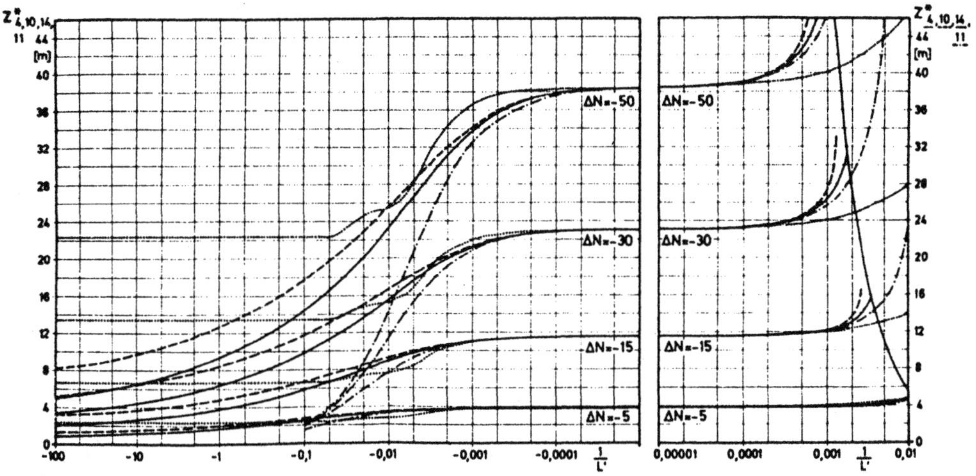

Fig. 4: Comparison of the thickness of the evaporation duct determined with the log. model (z_{*4}, see eq.12), the log.-lin. model (z_{*11}), and the Keyps-profile (z_{*14}). The z_{*10} curves are also due to the Keyps model, but with the assumption of a linear height dependence of the Richardson number (instead of eq.13).

2.3. Verification of the meteorological models by transmission measurements

For many years, transmission measurements form a part of the research work of the Meteorological Institute and the Institute for Radiometeorology and Maritime Meteorology at Hamburg under the leadership of the late Prof. Brocks. We frequently reported on the facts. Some examples may be quickly called back to your mind.

a) T r a n s h o r i z o n p a t h s

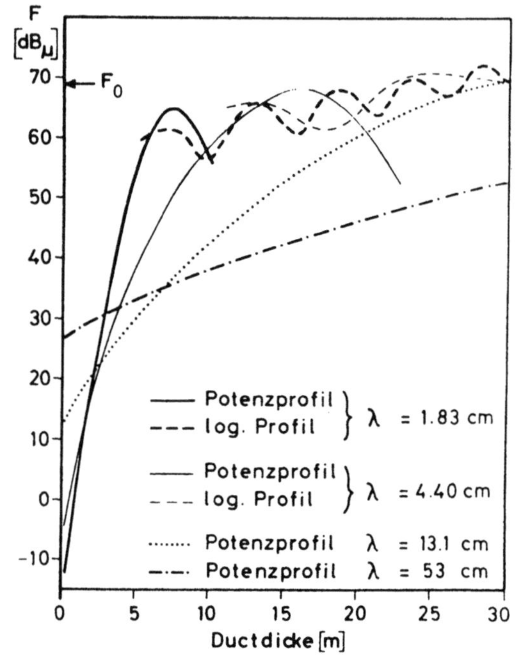

Fig. 5: Fieldstrength as function of duct thickness for various wave lengths and two different profile forms (s. Jeske 1971)

Fig. 5 gives an insight into the theoretical relationship between field strength and thickness of the evaporation duct (s. Jeske 1971). In the selected coordinates we get a quasi-linear dependence of the field strength (in dB) on the duct thickness up to the critical duct thickness (corresponding to the first maximum in the relation 'field strength - duct').

The figures 6 and 7 show the course of field strength parallel to the duct thickness in a single case (fig. 6) and in an annual variation (fig. 7, upper part). The influence of the duct thickness decreases with growing wave length and disappears totally for the 1.8 m-waves.

Fig. 8 shows the correlation 'field strength versus duct thickness' in the form of scatter diagrams for the 7GHz- and the 16.5GHz-links.

Fig. 9 is very interesting, showing that this correlation may be improved by the consideration of more complex boundary-layer models. But even with an exact duct thickness, a certain scatter will remain. The propagation theory reveals certain differences for one and the same duct thickness according to the respective profile model (as in the duct different gradients are present for different models), cf. paper of Mr. Rotheram.

Summarizing we may say that the predominant influence of the evaporation duct is to be distinguished plainly. The propagation over sea can be regarded as a prototype of duct propagation. This, however, is only valid for cm-waves (at best for dm-waves), if the duct thicknesses really present are considered. It may be seen (consult fig. 5) that the interdependence of field strength and duct thickness for transhorizon propagation is roughly in correspondence

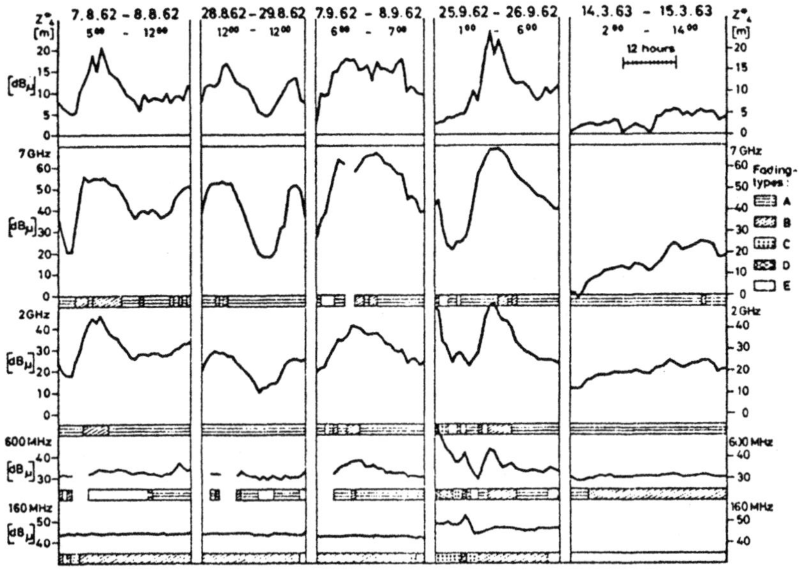

<u>Fig. 6</u>: Simultaneous variations of duct thickness z_* and field strength on wave-lengths of 4.4 cm (7 GHz), 13 cm (2 GHz), 53 cm (600 MHz) and 183 cm (160 MHz) - individual cases -. (Fadingtypes: A,B correspond to a scintillation fading (typical for evaporation propagation); C,D correspond to a slower interference fading. E is a mixed type)

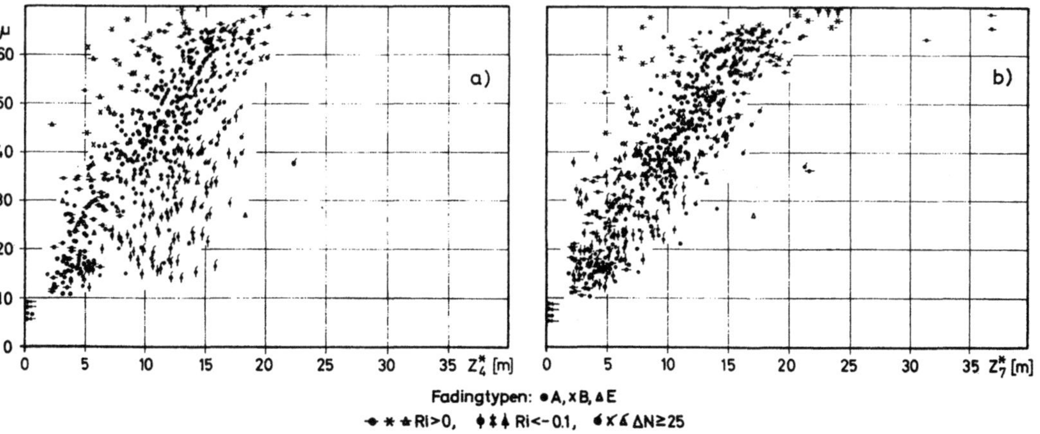

<u>Fig. 9</u>: Correlation between field strength (7 GHz) and the duct thickness z_{*4} (logarithmic model) and z_{*7} (model of Monin and Obukhov), September 1962.

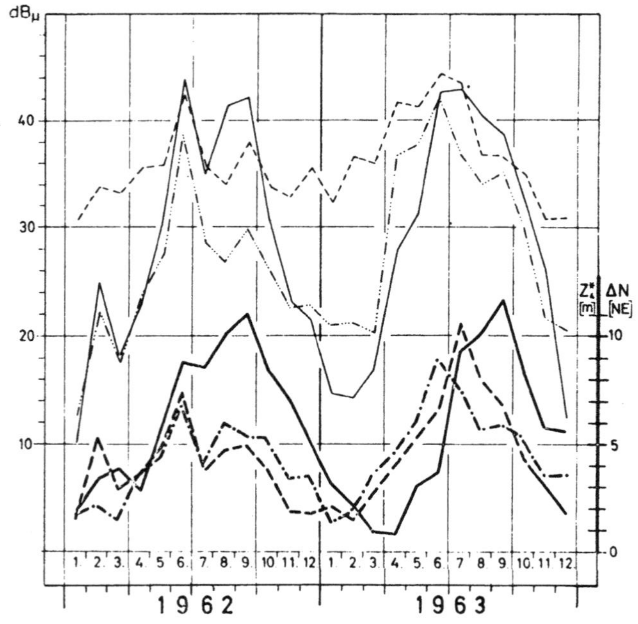

Fig. 7: Upper part: Annual variation of field strength on radio links of 7 GHz, 2 GHz, and 600 MHz between Bremerhaven and Heligoland (distance 77.2 km, antenna height about 30 m above NN). Lower part: Annual variation of the duct thickness z_*, the horizontal refractivity difference between Schleswig and lightship Elbe 1, and the refractivity depression ($N_{Sätt} - N_{500m}$). (This is the difference between the refractivity at the given temperature in 500 m and that of the air at saturation temperature).

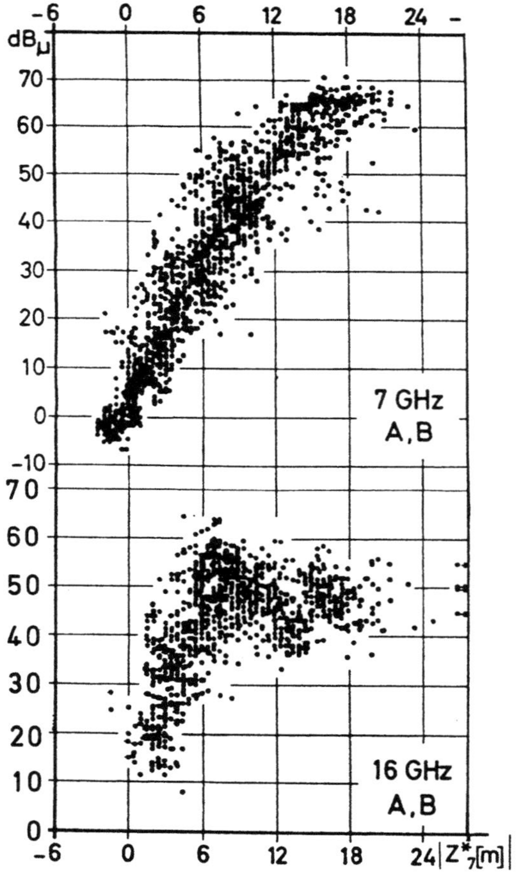

Fig. 8: Correlation between field strength and duct thickness ("scintillation" fading types A, B only)

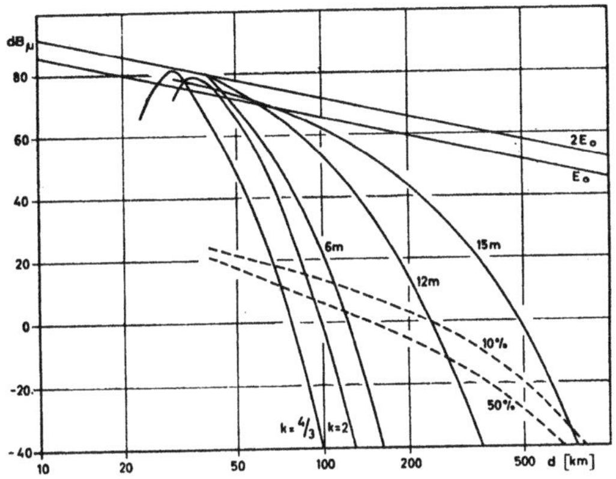

Fig. 10: The dependence of field strength (7 GHz) on distance for different refraction conditions (effective earth radius (4/3) R and 2 R) and various duct thicknesses. The dashed curves correspond to CCIR values.

with the propagation theory [Jeske 1965, 1971] . For large distances scatter propagation will surpass the duct influence.

Fig. 10 demonstrates this fact if the CCIR-curves are above the duct curves. For larger distances, well-known prediction methods are at hand [e.g. Boithias, 1970].

b) Line-of-sight paths

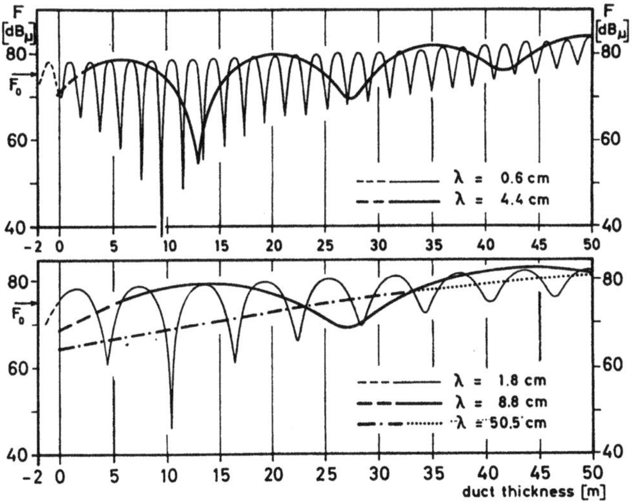

Fig. 11 : 38,2 km line-of-sight path over sea, theoretical field-strength curves for logarithmic refractive-index distribution

Also for line-of-sight propagation the Hamburg group is provided with extensive material within the wave-range of 6 mm up to 1.8 m. This type of propagation is generally marked by a system of interference lobes due to the superposition of direct and ground-reflected field. This characteristic feature is preserved also on ducting conditions with some important modifications (cf. paper of Dr. Früchtenicht). We have to take some pains in tracing the duct effects from the experiments, as they may not be discerned as clearly as for transhorizon propagation.

This is even not to be expected, if we consider the interference-link dependencies of the field strength on duct thickness (fig. 11) and moreover take into account that the duct thickness used during the experiments is affected with a certain error. Additionally, in the German Bight there are tidal variations of ± 3 m which develop their own interference field. The situation is the more difficult, the shorter the wave length is, as may be seen from the 6 mm-propagation curves in fig. 11 [Jeske, 1971] .

c) "A n t i - d u c t"

I may shortly point out to the fact that the method given even permits also the determination of the thickness of subrefraction layers immediately above the sea surface. The negative z_*-values shown in some figures present the thickness of the layer where $\delta N/\delta z > +0.125/m$ is valid ("anti-duct"). These cases, in the German Bight especially occurring in January and February, may lead to longer black-outs of the radio links (not due to "short-lived" interference effects).

3. REFRACTIVE-INDEX MODELS FOR THE COMPREHENSION OF HORIZONTAL REFRACTIVE-INDEX GRADIENTS

The present studies assume - for differentiated vertical structure - horizontal homogeneity. This is obviously a largely realistic statement, for the evaluation of the duct thickness used above was based on measurements from only one point in the measuring area. The horizontal homogeneity was also confirmed directly fengler, 1966 by comparison of meteorological measurements - in any case for north-west or west winds which are predominant in the German Bight. Under these conditions even immediately near the coast homogeneity is given. But if during off-land advection there are larger contrasts between the air over land and sea (summerly high-pressure conditions), we have to reckon with stronger horizontal inhomogeneities. This has been verified by direct meteorological measurements [Jeske, Kruspe, 1967] .

Theoretically this problem might be approached by means of the conception of the diffusion theory. From a meteorological point of view this is a modification of the air masses by boundary turbulence. The cases interesting for us occur in summer, when well-mixed air is drifting over the sea. In a first step we did model calculations for steady-state conditions; calculations for the unstationary case are just being done. Mathematically, this problem is described by the following differential equation and the boundary conditions:

$$u(z)\frac{\delta i}{\delta x} = \frac{\delta}{\delta z}\left(K(z)\frac{\delta i}{\delta z}\right) \quad \text{or} \quad \frac{\delta i}{\delta x} = \frac{K}{u}\frac{\delta^2 i}{\delta z^2} + \frac{K'}{u}\frac{\delta i}{\delta z} \tag{16}$$

$$i(z,o) = i_{\underline{l}and} \text{ and } i(o,x) = i_{\underline{w}ater} \tag{17}$$

(vertical homogeneous air over land, constant water temperature). Here $K' = \delta K/\delta z$, the horizontal coordinate x shows into the main wind direction. Essential for the solution of this differential equation are the used statements for K and u. - For an illustration of the procedure we will comment the simple, but meteorologically unrealistic situation of a constant K/u relation. In this case, the solution of the differential equation is given by the error function

$$i(z,x) - i_w = (i_l - i_w) \ \text{erf} \ [z/2\sqrt{Kx/u}] \ . \tag{18}$$

For the purpose of propagation of electro-magnetic waves we need especially the vertical gradients and the duct thickness as a function of x, respectively. In our simple case we get

$$\frac{\delta i}{\delta z}(z,x) = \frac{(i_l - i_w)}{\sqrt{\pi Kx/u}} \ e^{-z^2/4kx/u} \tag{19}$$

and from this - if i is the potential refractive index - by applying the critical gradient of the potential refractive index and by solving eq. (19) for z, we get the duct thickness z_{**}

$$z_{**} = \left[-\ln\frac{0.125\sqrt{\pi Kx/u}}{(N_l - N_w)} \cdot 4Kx/u\right]^{1/2} \ m] \ . \tag{20}$$

Analytic solutions are also available for power law statements, $u \sim z^m$ and $K \sim z^{m-1}$ [see Sutton, 1949] . But I will directly pass over to more realistic models. During stable conditions - as they are given in the case of summerly advection - the log.-lin. profile seems to be the most realistic model [Webb, 1970] . For u and K, it yields the equations

$$u(z) = \frac{u_1}{[\ln(z_1/z_0) + \alpha z_1/L']} \left(\ln\frac{z}{z_0} + \frac{\alpha z}{L'}\right) , \tag{21}$$

$$K(z) = \frac{k^2 u_1 z}{[\ln(z_1/z_o) + \alpha z_1/L']} \frac{1}{(1+\alpha z/L')} \tag{22}$$

Eq.(22) results from $K = k^2 z^2 \delta u/\delta z \Big/ (1+\alpha z/L')^2$.

The correspondent differential equation is integrated numerically by a difference-approximation method. The computer provides fields of values for $i(x,z)$ or $\delta i/\delta z\ (x,z)$. Let us regard some results. The change of the meteorological parameters at a fixed height with growing distance is given in fig. 12 (for 4 fixed heights).

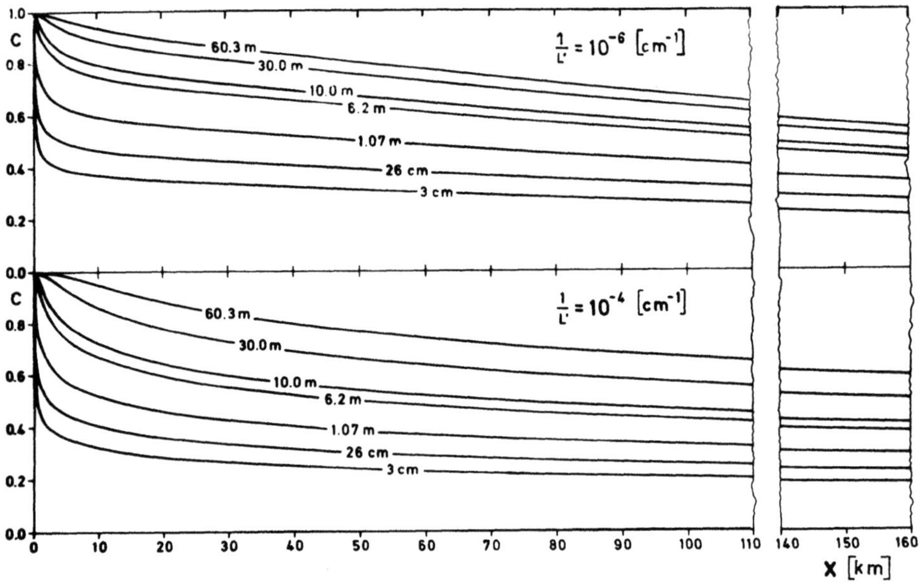

Fig. 12: $C=(i-i_w)/(i_1-i_w)$ as function of distance for fixed heights and two stabilities.

In the first decakilometers we have very strong horizontal gradients. For a first approximation we find a logarithmic decrease with increasing distance. At very short distances from the coast the theory even yields a "horizontal duct". We hope to be able soon to verify these strong gradients by means of theodolite measurements (lateral refraction) or geodimeter measurements on paths normal to the coastal line. The attempt of a rough verification by light-ship observations during advection situations was successful.

Especially interesting for radiometeorological research is the dependence of the "advection duct" on the off-shore distance. Fig. 13 informs on this problem for different stabilities. We generally perceive a strong increase with x up to a maximum height and then a slow decrease towards the z_*-value of the log.-lin. profile in very large distances. In neutral stratification the increase is more rapid than in stable stratification.

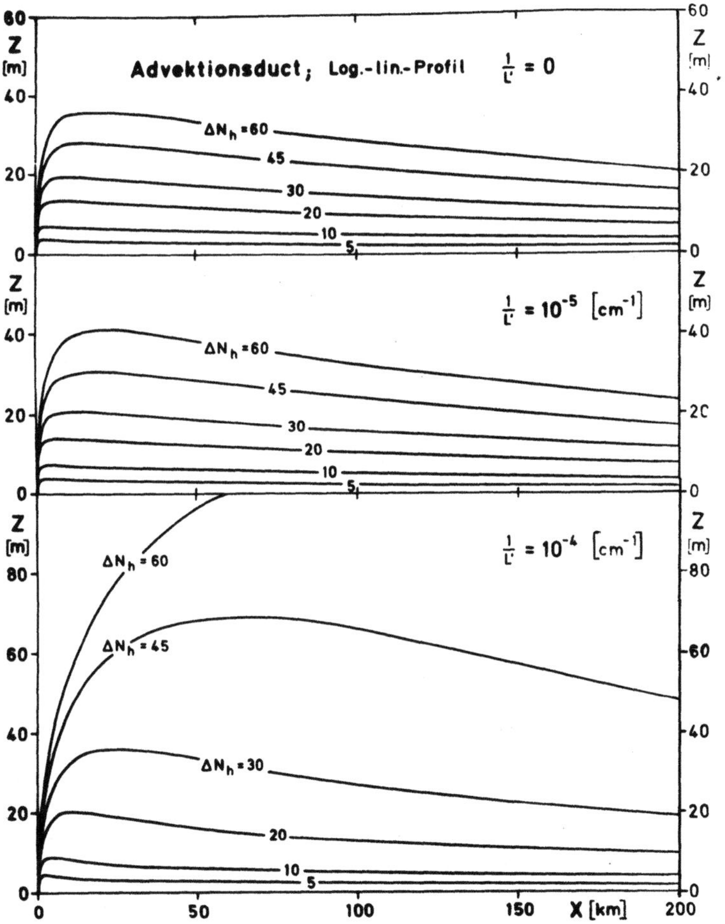

Fig. 13: Advection duct thickness as function of distance and stability (log.-lin. profile shape).

In stable stratification the duct thickness may reach considerable values. In these cases even the propagation in the m-bands is influenced. In the scope of the assumptions the model permits a determination of the advection duct thickness as a function of the horizontal refractivity-difference and a parameter dependent on the stability.

We have to undertake great efforts, if instationarity and the effect of land-sea breeze are to be taken into account. The inhomogeneity during advection situations described here will above all take considerable influence on electro-magnetic distance-measuring procedures, where an accuracy of 10^{-6} is aspired.

4. INFLUENCE OF ELEVATED LAYERS

Elevated layer structures in the refractive-index field may cause effects which exceed the influence of the ground duct. Reflections or ducting are possible (situations with fading types C and D in our material, given in 10-20% of the time of the year). On principle, these effects cannot be determined by means of observations at the ground. For their evaluation or prediction radio sonde measurements are necessary. Because of their large time intervals and their errors in getting the moisture, the routine radiosonde ascents are not very useful for this purpose. But some promising attemps, nevertheless, have been made in this direction, as is shown, for example, by the procedure of the Norwegian radio-meteorologists [see Gjessing et al., 1969] . The continuous control of the atmosphere by means of remote sensing methods (vertical radar, lidar, sodar) might lead to better progress.

5. CONCLUSIONS FROM THE PROPAGATION THEORY

The exclusive knowledge of the refractive-index field is not sufficient for a complete interpretation of the propagation phenomena. According to the geometrical parameters of the respective radio-link (distance, antenna heights, wave length), different effects may occur in the same meteorological situation. The meteorological model is only one part of the problem. The consideration of the propagation theory completes the picture. The influence of a ground duct may be given as an example. For transhorizon propagation we must generally assume superrefraction conditions near the ground. On line-of-sight paths an interference minimum may cause a total black-out (short-time radar hole) [cf. paper of Dr. Früchtenicht] . But the intense dependence of the receiving field on the geometrical and technical parameters offers some advantage. Thus a radar hole, e.g., may be avoided by application of the technique of polarization diversity, frequency diversity, or space diversity. In order to get a reliable registration of the evaporation effects on radar facilities or distance measuring equipments, we must have duct statistics of the respective area. For the occurring cases, the field strength contours or distance errors should be evaluated for different antenna heights and be edited in the form of a field strength atlas. We probably must invest these efforts for areas of special interest.

6. PROSPECT

As in the case of synoptic weather prediction, we may only step by step approach a complete solution of the prediction problems of the propagation conditions. For apart from the topics mentioned in this paper, we have to consider further effects, as there are influences of mist, clouds, precipitation, fronts ground roughness, etc. As far as the homogeneous maritime boundary, layer is concerned, we may say that the achieved results are sufficient for practical application. The horizontally inhomogeneous advection models, however, need further enlargement, e.g. the inclusion of instationarity and land-see-breeze effects. The prediction of influences of elevated layers is least guaranteed. The conditions become complicated by the frequent overlapping of different propagation effects. Of the propagation theorists we must require above all solutions of the problems of elevated duct, "anti-duct" (subnormal layer), and generally the properties near caustics as well as a complete consideration of the rough sea.

REFERENCES

BOITHIAS, L. and P. MISME, Seasonal Predictions of Tropospheric Propagation Conditions, AGARD Conference Proc. n. 70, 1970.

BROCKS, K. and L. KRÜGERMEYER, The Hydrodynamic Roughness of the Sea surface, Studies in Physical Oceanography, Vol. 1, Science Publishers Inc., New York, 1970.

GJESSING, D.T., H. JESKE, N. KLINT-HANSEN, An investigation of the tropospheric fine scale properties using radio, radar and direct methods, Journ. of Atm. and Terr. Phys., Vol. 31, 1157-1182, 1969.

FENGLER, G., Die Homogenität des meteorologischen Feldes über See, Dtsch. Hydr. Zeitschr., 19/3, 1966.

FRÜCHTENICHT, H.W., Atmosphärische Einflüsse auf die Ausbreitung elektromagnetischer Wellen (cm-Band) über See innerhalb des radiooptischen Horizonts - Atmospheric Effects on SHF-Over-Sea-Propagation within the Horizon. Hamb. Geophys. Einzelschr., H. 14, Verlag W. de Gruyter u. Co., Berlin, 1971.

JESKE, H., Weleenleitereinflüsse über See auf die Ausbreitung im mm- und cm-Band bei Vor- und Überhorizontstrecken, Kleinheubacher Berichte, Bd. 14, FTZ Darmstadt, 1971.

JESKE, H., Die Ausbreitung elektromagnatischer Wellen in cm- bis m-Band über dem Meer unter besonderer Berücksichtigung der meteorologischen Bedingungen in der maritimen Grenzschicht, Hamburger Geophys. Einzelschr., H. 6, Cram, de Gruyter u. Co., Hamburg, 1965.

JESKE, H., and G. KRUSPE, Time-space Structure of Atmospheric Refractive Index especially obtained by Refractometer Measurements, Sonderheft 25 der Österreichischen Zeitschr. f. Vermessungswesen, Proc. of the International Sympos. Figure of the Earth and Refraction, Wien 1967.

SUTTON, O.G., Atmospheric Turbulence. London, 1949.

WEBB, E.K., Profile relationships: the log.-lineare range, and extension to strong stability, Quart.J.R.Met.Soc., 96, 67-90, 1970.

CHARACTERISTICS AND APPLICATIONS OF LINE-OF-SIGHT DUCT PROPAGATION

H.W. Früchtenicht

Institut fur Radiometeorologie und Maritime
Meteorologie Hamburg, Germany

SUMMARY

Duct propagation in the frequency bands between 7 GHz and 16 GHz on line-of-sight paths can be treated by ray tracing. Over sea the evaporation duct causes a certain receiving height where the receiving field may black-out. The phenomenon is dependent on the duct thickness, the path geometry, and on the polarization of the radiation. From this it follows that the sea-evaporation duct produces a sort of radar holes. Correspondent to that possible black-out of the receiving signal there is a discontinuity in the variation of the phase of the electromagnetic field amounting to integer multiples of 2π. As phase measuring equipments are not able to detect such discontinuities, it follows that phase measurements over sea are many-valued. If frequency modulation is used, as it is done in geodesy, the results will not be many-valued but, in principle, be affected with an error.

1. INTRODUCTION

Experience shows that line-of-sight propagation may be treated by ray tracing. The receiving field is considered to be a part of an interference field arising in front of the transmitter. One component of the receiving field passes directly (direct ray) from the transmitter to the receiver. The second one (reflected ray) reaches the receiving antenna after reflection at the earth's surface. As known, the field strength may be written as

$$E = A_R e^{i\varphi_R}, \qquad (1)$$

where A_R is the amplitude and φ_R is the phase of the resultant field. However, A and φ_R are dependent on the refractive-index field between transmitter and receiver. The present paper deals with propagation effects due to the influences of sea evaporation ducts on the amplitude and the phase of the receiving field.

2. THEORETICAL FORMULATION OF THE PROBLEM

2.1. The propagation formulas

The magnitude and the phase of a resultant field which consists of two components, are given by

$$|E|^2 = A_R^2 = E_o^2 \, (A_1^2 + |R|^2 A_2^2 + 2|R| A_1 A_2 \cos(\varphi_2 - \varphi_1)) \tag{2}$$

$$\varphi_R = \operatorname{arctg} \frac{\sin\varphi_1 + A \sin\varphi_2}{\cos\varphi_1 + A \cos\varphi_2} ; \quad A = \frac{|R| A_2}{A_1} \tag{3}$$

E_o is the free-space field-strength. $|R|$ considers the power loss due to ground reflection. In the following, Fresnel's coefficients are used for R. φ_1 and φ_2 are the phases of the direct and the reflected ray. A_1 and A_2 denote focussing factors for the electromagnetic energy along the path of the direct and the reflected ray in relation to free-space propagation. The phases and the focussing factors are dependent on the refractive-index profile. For their determination the solution of the differential equation of the ray path

$$\frac{d^2z}{dx_2} = 10^{-6} \frac{dM}{dz} \tag{4}$$

must be known. z and x are the vertical and horizontal coordinates and M stands for the refractive modulus.

The phase belonging to a certain ray path may be derived as [1]

$$\varphi = \frac{2\pi}{\lambda} \int m ds = \frac{2\pi}{\lambda} \left[r + (10^{-6} M(h_1) - \frac{tg^2\alpha(h_1)}{2} r + \int_{h_1}^{h_2} tg\alpha(z) dz \right] , \tag{5}$$

where m is the modified refractive index and λ is used for the free-space wave-length of the radiation. r gives the path length while h_1 and h_2 are the transmitter- and the receiver-height. The quantities m and M, the refractive index n, and the refractivity N are connected by

$$M = 10^6(m-1) = 10^6(n(1+\frac{z}{a})-1) \simeq 10^6(n-1+\frac{z}{a}) = N + \frac{z}{a} 10^6 , \tag{6}$$

where a denotes the earth radius. α is the angle of inclination of the ray. It is given by

$$tg^2\alpha(z) = 2\cdot 10^{-6}(M(z)-M(h_1))+tg^2\alpha(h_1). \qquad (7)$$

The focussing factors may be derived from the cross section of a ray family. They amount to [1]

$$A^2_{RAY} = \left|\frac{r d\alpha(h_1)}{dH_2}\right| \quad , \qquad (8)$$

where $d\alpha(h_1)$ and dh_2 are the differentials of the radiation angle and the receiver height.

As long as a solution of the differential equation of the ray is available, analytic relations for the phases and the focussing factors may be found. Otherwise the integral and differential equations may be solved by numerical treatment. In general, the method of ray tracing will be valid, if the condition $\varphi_2-\varphi_1\geq 2$ is not violated. Moreover, this paper deals only with ray paths without turning points.

If the refractive index is only a function of z, the vertical profile of M may be written as

$$M(z) = \int_{z_1}^{z} f(z)dz+N(z_1)+10^6 \frac{z_1}{a} \quad . \qquad (9)$$

In connection with the equations (5) and (7), it follows that for a fixed path geometry the differential of the phase may be written as

$$d\varphi = \frac{\delta\varphi}{\delta N} dN_1 + \frac{\delta\varphi}{\delta g} dg; \qquad N_1 = N(z_1) \ , \qquad (10)$$

where N_1 is the refractivity in a fixed but arbitrary reference level. g denotes a parameter which considers the gradients of the profile.

For $\frac{\delta\varphi}{\delta N_1}$ there exists the linear relation

$$\frac{\delta\varphi}{\delta N_1} = \frac{2\pi}{\lambda} r \ 10^{-6} . \qquad (11)$$

Therefore, the phase varies proportional to the ratio of path- and wave-length, if the gradients of the refractive index are constant. For the phase variation dependent on the gradients a similar relation cannot be found.

2.2. The radiometecrological conditions

Propagation experiments have indicated that the propagation conditions in the southern part of the North Sea are mainly given by low level evaporation ducts and by large ground-based ducts due to advection or subsidence [2] . The evaporation duct is a characteristic property of the undisturbed sea atmosphere, where horizontal gradients of the refractive index are vanishing. The corresponding vertical profile of the refractive index can be derived in the following way.

From turbulence theory it is known that the vertical flux and the vertical gradient of a property i of the air are related by

$$\frac{\delta i}{\delta z} = -\frac{Si/\varrho}{ku_*}\,\frac{f(Ri(z))}{z+z_o}\ . \tag{12}$$

The quantities Si, ϱ , u_*, and z_o denote the flux, the density, the friction velocity, and the roughness-parameter. k is the v.Karman-constant and f(Ri) is a stability function of the air. If the quantities ϱ , u_* , and Si are constant, the vertical profile of i is given by

$$\Delta i_1 = i(z_1)-i(o) = -\frac{Si/\varrho}{ku_*}\int_o^{z_1}\frac{f(Ri(z))}{z+z_o}\,dz. \tag{13}$$

Combining the relations (12) and (13) it follows

$$\frac{di}{dz} = \frac{f(Ri(z))}{z+z_o}\;\frac{\Delta i_1}{\int_o^{z_1}\frac{f(Ri(z))}{z+z_o}\,dz}\ . \tag{14}$$

Using the well known relations

$$\frac{dN}{dz} = c_1\frac{dP}{dz} + c_2\frac{de}{dz} + c_3\frac{dT}{dz} \tag{15}$$

and

$$\Delta N_1=N(z_1)-N_W=c_1(P(z_1)-P_W)+c_2(e(z_1)-E_W)+c_3(T(z_1)-T_W)\ , \tag{16}$$

where P,e,T,P_W,E_W, and T_W stand for pressure, water-vapour pressure, and temperature of the air at the height z_1 and at the sea-surface, it follows from equation (14) that the vertical profile of the refractive modulus is given by

$$\frac{dM}{dz} = c_v\left[1 - \frac{d+z_o}{f(Ri(d))}\;\frac{f(Ri(z))}{z+z_o}\right]\ . \tag{17}$$

In this relation c_v is a constant and d stands for the duct thickness. The integration of equation (17) results in

$$M(z) = c_v\left[z-z_1 - \frac{d+z_o}{f(Ri(d))}\int_{z_1}^{z}\frac{f(Ri(z))}{z+z_o}\right]+ M(z_1). \tag{18}$$

If the stability of the air is independent of z, the relation reduces to a logarithmic profile of the form

$$M(z) = c_v \left[z-z_1-(d+z_0)\ln \frac{z+z_0}{z+z_1} \right] + M(z_1) \quad . \tag{19}$$

The profile given by equation (19) is an admissible simplification of the evaporation duct in the undisturbed sea atmosphere. Using this approximation, the duct thickness can be calculated from simple meteorological measurements [3] . If the duct thickness becomes zero, the logarithmic profile virually turns into a linear one.

$$M(z) = c_v(z-z_1) + M(z_1) \tag{20}$$

3. THEORETICAL AND EXPERIMENTAL RESULTS

3.1. The amplitude of the receiving field

The propagation formulas of section 2.1. have been solved for the linear and the logarithmic (duct) profile of the refractive modulus given by the equations (19) and (20).

The receiving field strength maximally and minimally possible is essentially governed by the focussing factors. In fig. 1 the dependence of A_1 and A_2 on the receiver height is shown. For the linear profile, the focussing factor of the direct ray always remains unity. That one of the reflected ray is always less than unity, except immediately near the ground it increases with the receiver height. On logarithmic stratification conditions, A_1 increases with height, becoming asymptotically unit-. A_1 has a maxim on a relatively low level. In this range it is greater than A_1 and may even reach values greater than unity.

The height dependence of the receiving field strength is presented by fig. 2. It may be seen that with increasing receiver height the field strength passes periodically through maxima and minima. With increasing path length the system of interference lobes is shifted upwards, while the fine structure of the lobes decreases

The linear profile always causes the most intensive elevation of the lowest interference maximum above ground.

At a logarithmic refractive-index stratification this maximum approaches the ground the further, the greater the duct thickness is.

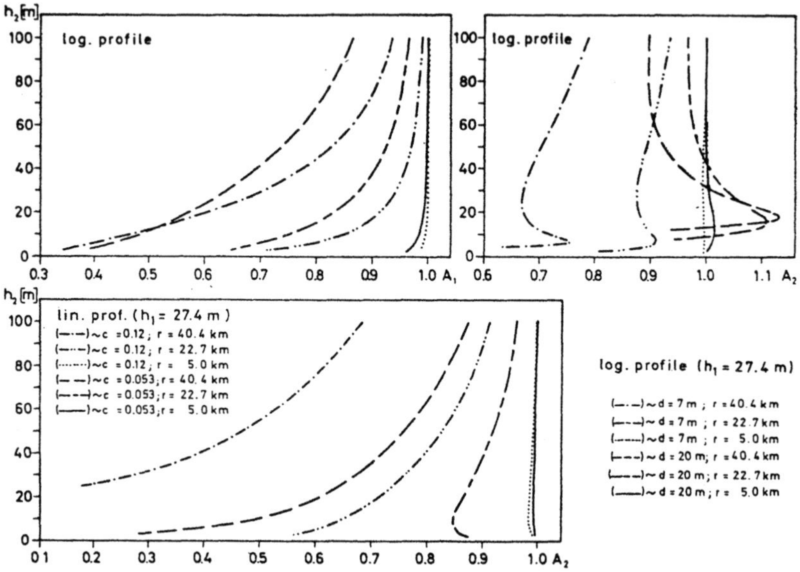

Fig. 1: Focussing factors as a function of the receiver height.

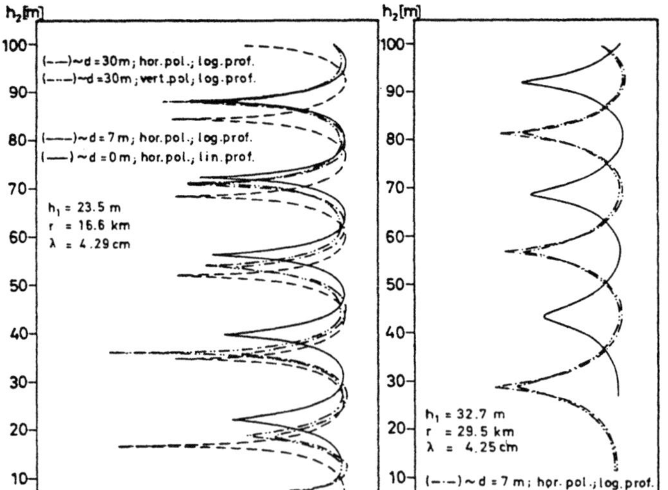

Fig. 2: Receiving field-strength as a function of the receiver height.

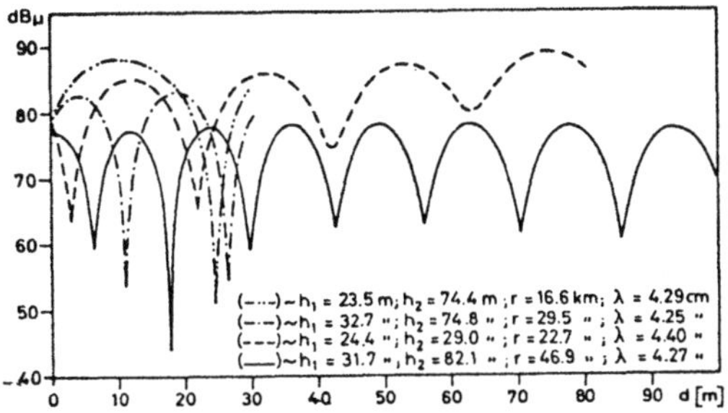

Fig. 3: Receiving field-strength as a function of the duct thickness.

Fig. 3 presents the receiving field as a function of duct thickness. The graph shows that - similar to the effect of growing receiver height - with increasing d a certain number of interference lobes is passed through. The positions and spaces of the minima in the diagram are dependent on the geometry of the paths. Generally, the interference lobes are the more frequent, the longer the path length is. If the duct thickness vanishes, the logarithmic profile virtually turns into a linear one. Thus, if d equals zero, both the profiles result in the same field strength.

A comparison between theory and measurements carried out in the German Bight of the North Sea is presented in fig. 4. The field-strength recordings were made on two 7 GHz-links and one 16 GHz-link, which had been installed over the same path with 22.7 km in length. One of the 7 GHz-links was a transhorizon path (low antennas) and is called TL-RL in the following. The second 7 GHz-link, called TU-RU, and the 16 GHz-link were line-of-sight paths. The part c of fig. 4 shows the correlation between the receiving field strength of the TL-RL link and the simultaneously observed duct thickness. In the figures 4a,b the receiving field of the TL-RL link is plotted against that one of the line-of-sight paths TU-RU (fig. 4a) and 16 GHz (fig. 4b). Part 4d gives theoretical field-strength curves corresponding to the links 7 GHz (TU-RU) and 16 GHz, and to different water levels. The observed duct thickness are determined in the way already mentioned in connection with equation (19). The method has been derived by Brocks [4,5] in 1955.

Following meteorological conditions are valid for the points: $-0.02 \leq Ri_{\Gamma_{1m}} \leq 0.002$, $u \geq 5$m/sec., $270° \leq DD \leq 360°$, $0 \leq P \leq 1.5$ m above sea level ($Ri_{\Gamma_{1m}}$: approximate value of Richardson's number at the height of 1m, u: wind velocity at the height of 6m, DD: wind direction, p: water level at gauge "Hole Weg").

As the receiver of the TL-RL path is situated beyond the horizon, the corresponding field strength increases continuously with the duct thickness. On the line-of-sight paths, however, the receiving field passes through an interference pattern. The scatter of the points in fig. 4a,b,c is partly caused by tidal effects, although only such values have been used which belong to water levels between 0m and 1.5m above sea level (see fig. 4d).

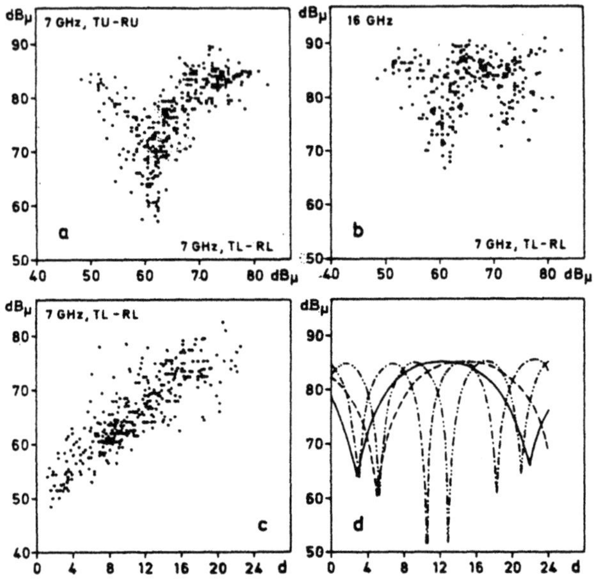

Fig. 4: a-b: Correlations of field strengths measured simultaneously
c: Correlation of field strength and duct thickness measured simultaneously
d: Theoretical field-strength curves.
7 GHz(TU-RU):(-)~p=0m, (--)~p=1.5m,
16 GHz:(-.-)~p=0m, (-•••-)~p=1.5m
(p=water level above sea level).

Another reason, especially in fig. 4c, is the fact that the real refractive-index profile is approximated by the assumed logarithmic stratification. In the main, the theoretical curves conform with measured correlations. They are only shifted to smaller duct thicknesses.

From the facts given above it follows that the field strength level in the frequency bands between 7 GHz and 16 GHz is governed by the maritime evaporation duct and that duct propagation within the horizon may be treated by ray tracing. Marked differences between the propagation conditions of a duct profile and a linear profile are revealed by the maximum and minimum field strengths. They are given by the envelopes of the interference patterns treated above. Fig. 5 gives the maximum and minimum field strength for certain path geometries and duct thicknesses as a function of the receiver height. The areas bordered by the envelopes are hatched in different ways for both the polarizations. The examples show that the depth of the interference minimum virtually varies only as a function of the minimum field strength. In the case of the logarithmic profile there always exists a certain receiving height where the field strength may vanish totally. A black-out of the receiving field occurs, if the height of an interference minimum the receiver height, and the height of the minimum of the "lower" envelope coincide. The phenomenon of "vanishing field strength" is caused by interference, but it originates in the fact that the focussing factors become equal at a certain height.

With increasing duct thickness the point, where the focussing factors are equal, is shifted upwards. Moreover, for vertical polarization the receiving height of "vanishing field strength" is situated on a lower level and, with increasing duct, is shifted upwards more slowly than for horizontal polarization. If the linear profile is realized, A_2 is always smaller than A_1. Now the minimum field strength will never become zero and the minima of the interference lobes will always be deeper for horizontal polarization than for vertical polarization. From these facts it may be drawn the conclusion that over sea the evaporation duct produces a sort of radar holes.

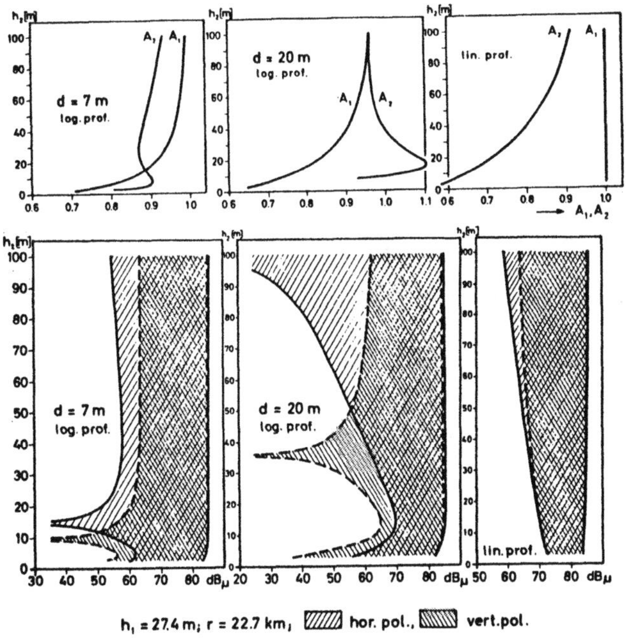

Fig. 5: Upper part: Focussing factors as a function of the receiver height
Lower part: Maximum and minimum field strength as a function of the receiver height

Fig. 6 shows lines of constant maximum-, minimum-, and resultant field strength for a duct profile given by equation (21)

$$M(z) = \frac{c_v}{2d}\left((z-d)^2-(z_2-d)^2\right)+M(z_2); \quad z \leq 2d, \text{ par.part.} \qquad (21a)$$

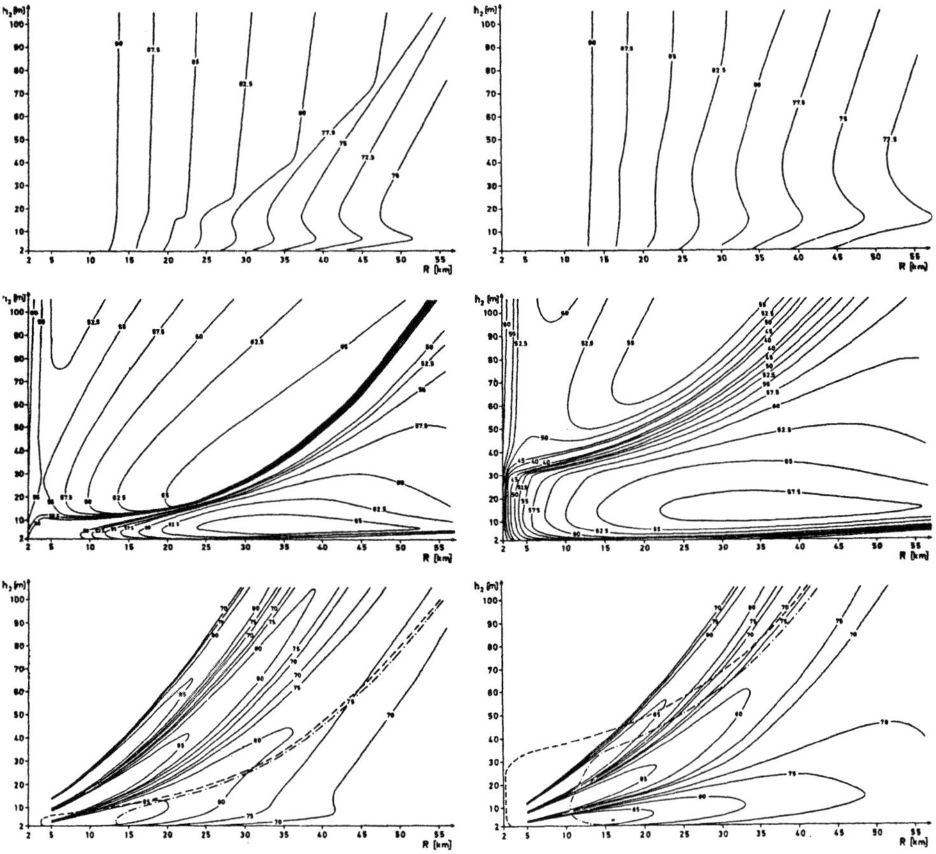

Fig. 6: Lines of constant field strength (dBμ) for a par.-lin. profile (h_1= 27.4 m, hor.pol.)
left hand parts: d = 7m, right hand parts: d = 15m,
upper parts: max. field strength, central parts:
min. field strength, lower parts: resultant field
strength

$$M(z) = c_v\,(z-z_3) + M(z_3) \quad ; \quad z \geq 2d, \text{ lin. part.} \qquad (21b)$$

z_2 and z_3 are again reference levels. As shown in [6] this profile reveals also the main features of the radar holes mentioned above. The graph (central part) indicates, that the receiving height, at which the field strength may vanish, increases with increasing path length. Near this curve the minimum field strength has strong gradients. Above the curve of vanishing field strength and near the ground there are areas where the minimum field strength reaches a maximum. The gradients of the minimum field strength decrease

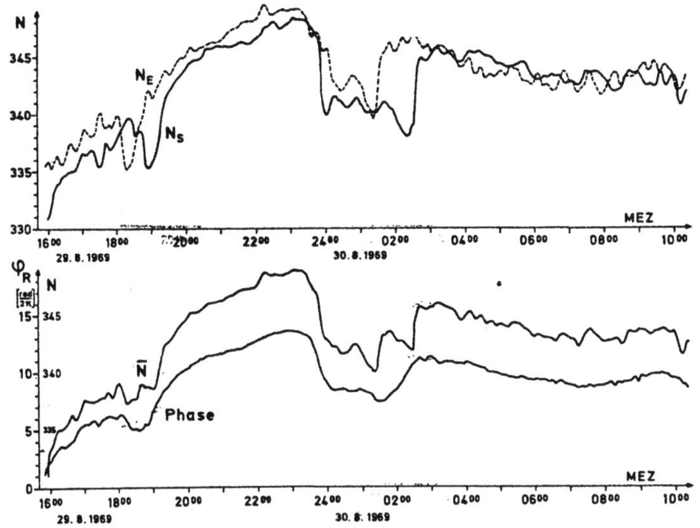

Fig. 7: Upper part: Time variation of the refractivity at transmitter (N_S) and receiver (N_E).
Lower part: Time variation of the phase and of the mean ($\bar{N}$) of N_S and N_E.

resp., become equal, if the phase difference $\varphi_2 - \varphi_1$ amounts to integer multiples of π . The characteristics of the resultant phase are caused by the concurrence of the two focussing factors (cf. fig. 8 left side, lower part). At small duct thickness A_1 is greater than $|R|A_2$, leading to the effect that the resultant phase may differ from the direct phase by maximally $\pm\pi/2$. At greater ducts A_1 is less than $|R|A_2$. Therefore, now a difference of maximally $\pm\pi/2$ to the reflected phase may occur.

Phase discontinuities amounting to integer multiples of 2π cannot be observed by any phase-measuring technique. Therefore, the evaporation duct gives rise to many-valued results of phase measurements near sea surface. This effect may cause errors, which - converted into units of length - are at least equal to the wave length of the radiation.

In geodesy phase-measuring techniques working with frequency modulation are used for distance measurements. So many-valued rerults are avoided. From the observed phase the optical length of the corresponding ray path can be calculated as indicated in equation (5). If then the relation between the optical length and

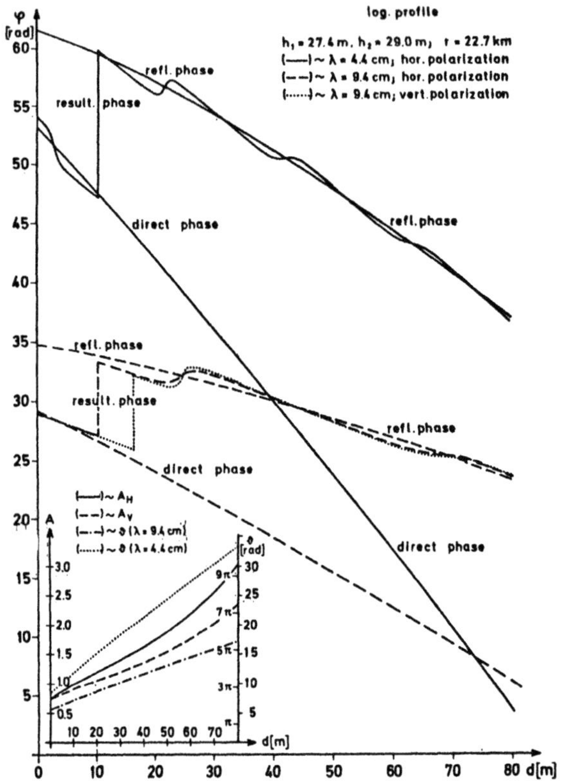

Fig. 8: Variation of the direct, the reflected, and the resultant phase with duct thickness.
Left side, lower part: A and ϑ as a function of the duct thickness
(A_H = A for hor.pol., A_V = A for vert.pol., $\vartheta = \varphi_2 - \varphi_1$).

the path distance is known, r can be determined. For a linear height dependence of the refractive index the reduction formula can be derived as

$$\int m ds = r + 10^{-6} M(h_M) r + \frac{(h_1 - h_2)^2}{2r} - \frac{10^{-12} c_V^2}{24} r^2 , \qquad (22)$$

where $M(h_M)$ is the arithmetic mean of the antenna heights. In order to get a feeling of the corresponding formulas belonging to duct profiles, the parabolic profile given by equation (21a) was considered. Using the propagation formulas (4,5,7) the following approximation of the optical length can be derived

$$\int mds = r+10^{-6}M(h_M)r+\frac{(h_1-h_2)^2}{2r}+\frac{c_v r10^{-6}}{12d}(h_1-d)(h_2-d). \quad (23)$$

Now $M(h_M)$ is given by

$$M(h_M) = \frac{1}{3}(M(h_1) + M(h_2) + M(d)). \quad (23a)$$

On the assumption that the form of equation (23) is typical for ground based ducts, the optical length may be written as

$$\int mds = r+10^{-6}M(h_M)r + \frac{(h_1-h_2)}{2r} + K_r \quad , \quad (24)$$

where K_r stands for the remaining difference. Theoretical calculations for a log. duct-profile and a path length of 20 km proved that K_r exceeds the quantity of 2 cm only if the difference between the duct thickness and the antenna heights is very large. So, neglecting K_r in equation (24), one gets a suitable approximation for geodetical purposes, where an accuracy of 1 mm/km is required.

If a tellurometer is used for the phase measurements mentioned above, there may arise some difficulties, because of the reflected wave. According to Poder [7] the modulation phase is affected with an error, which for a linear vertical profile of the refractive index is given by

$$\Omega = \operatorname{arctg}\frac{\sin\gamma\,(A^2+A\cos\vartheta)}{1+A\cos\vartheta+\cos\gamma\,(A^2+A\cos\vartheta)} \quad . \quad (25)$$

ϑ is the phase difference of the carrier (λ_T) between the direct and the reflected ray. γ gives the corresponding difference for the modulation (λ_m), but without the phase shift associated with the reflection coefficient. It may be demonstrated that equation (25) is also valid for any other vertical profile of the refractive index.

Fig. 9 shows the errors given by equation (25) for a fixed path geometry as a function of the duct thickness.

The modulation phase is converted into units of length. It is to be seen that the error alternates between two envelopes. The envelopes themselves are essentially governed by focussing factors. Whereas the nearly rectilinear "inner" envelope increases continuously with increasing A, the "outer" envelope shows a discontinuity similar to that one of the phase without frequency modulation. If A is less than unity, the outer envelope belongs to the positive error range. At A=1 it skips from $+0.25\lambda_m+c$ to $-0.25\lambda_m+c$ and for $A>1$ it remains in the negative range (c is a constant). The correction necessary is independent of the fre-

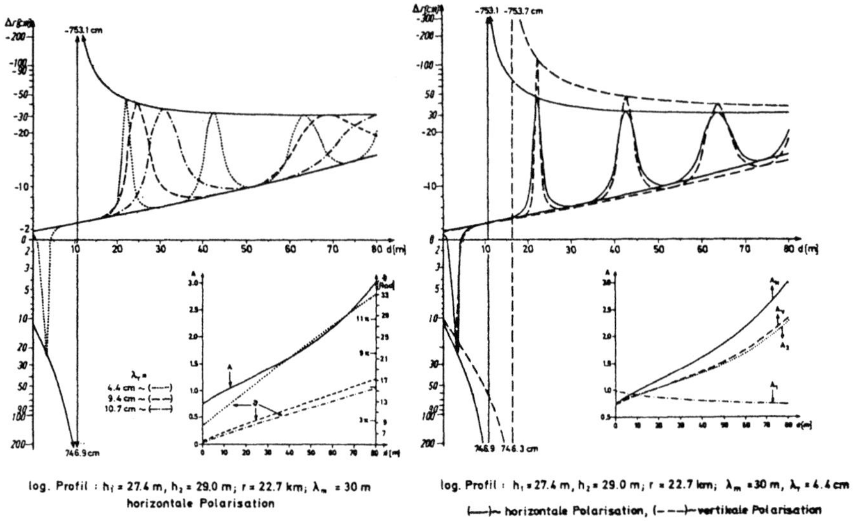

Fig. 9: Distance error as a function of duct thickness. Right sides, lower part: A and ϑ as a function of duct thickness (A_H=A for hor.pol., A_V=A for vert.pol.)

quency for that duct thickness, where A is equal to unity. Then it is given by

$$-\Delta r = \frac{\lambda_m}{2\pi} \text{ arctg } \frac{\sin\gamma}{1+\cos\gamma} \quad . \tag{26}$$

At this point, however, the correction changes its direction. For smaller duct thicknesses it is smaller and for greater duct thicknesses it is greater than the corresponding value of the inner envelope. The spacing of the peaks of each curve is governed by the phase difference of the carrier. For vertical polarization the discontinuity is found at a greater duct thickness than for horizontal polarization. But besides this, the extrema occur rather simultaneously for both polarizations.

REFERENCES

[1] L. Fehlhaber, "Beitrag zur Ausbreitung der dm-und cm-Wellen über See", Techn.Bericht Nr. 5501 d.Fernmeldetechnischen Zentralamtes der Deutschen Bundespost, Darmstadt, 1960.

[2] K. Brocks and H. Jeske, "The Meteorological Conditions of Electromagnetic Wave Propagation above the Sea". Publ. in Electromagnetic Distance Measurements, Hilger a. Watts LTD, London, 1965.

[3] H. Jeske, "Die Ausbreitung elektromagnetischer Wellen im cm-bis m-Band über dem Meer unter besonderer Berücksichtigung der meteorologischen Bedingungen in der maritimen Grenzschicht". Hamburger Geophys. Einzelschriften, H.6, Verlag Cram, de Gruyter & Co., Hamburg 1965.

[4] K. Brocks, "Der Brechungsindex für elektromagnetische Wellen (cm-bis m-Band) in der maritimen Grenzschicht der Atmosphäre". Dt. Hydr.Zeitschr. nr. 8, 1955.

[5] K. Brocks and L. Hasse, "Der Brechungsindex für elektromagnetische Wellen (cm-bis m-Band) in der maritimen Grenzschicht der Atmosphäre". 2. Beitrag, Dt. Hydr. Zeitschr., Nr. 9, 1956.

[6] H.W. Früchtenicht, "Atmosphärische Einflüsse auf die Ausbreitung elektromagnetischer Wellen über See innerhalb des radiooptischen Horizonts". Hamburger Geophys. Einzelschriften, H. 14, 1971.

[7] K. Poder and O. Bedstedt Andersen, "Probleme der Reflexion von Mikrowellen". Nachrichten aus dem Karten- und Vermessungswesen, Reihe I, H. Nr. 35, 1967.

PROPAGATION THEORY FOR THE EVAPORATION DUCT

S. Rotheram

GEC-Marconi Electronics Ltd.,
Great Baddow Research Laboratories, Great Baddow,
Chelmsford, Essex, England.

1. INTRODUCTION

Propagation in radio ducts first attracted attention during the 1940's and was the subject of some notable theoretical studies(1),(2),(3). Recent developments such as the use of the Monin-Obukhov similarity theory(4) to give realistic refractive index models, improvements in theoretical and numerical methods and the need to incorporate other phenomena such as scattering from the sea surface and atmospheric turbulence allow some extensions of the theory. Section 2 of this paper gives a descriptive introduction to duct propagation. Section 3 gives an integral representation which is the theoretical starting point. Section 4 gives the superdiffraction theory and section 5 a sketch of a superrefraction theory. Section 6 applies this to the evaporation duct and gives numerical results. Section 7 assesses the state of the theory.

2. DESCRIPTIVE ASPECTS OF DUCT PROPAGATION

2.1. Characteristics of surface ducts

Generally the refractive index, n, of the troposphere decreases with height, h. This decrease bends rays towards the earth but not sufficiently to overcome the earth's curvature. Instead of using n over a curved earth of radius r_0 it is permissible to use the modified refractive index $m = n + h/r_0$ over a flat earth. Generally m increases with h so that rays bend away from this flattened earth. A radio duct exists where $dm/dh < 0$ within which rays are bent towards the flattened earth. A simple surface duct exists when

$$\frac{dm}{dh} \lesseqgtr 0 \quad \text{for} \quad h \lesseqgtr h_d \qquad (1)$$

where h_d is the duct thickness. Figure 17 shows some models.

2.2. Ray paths in a surface duct

Used with caution ray optics (zero wavelength limit of wave optics) can give useful insights into duct propagation. Figures 1 and 2 show ray paths for a transmitter, height h_T, which is above and below h_d. Figure 1 shows rays A and B that enter the duct, are reflected by the earth and then leave the duct. By symmetry no direct ray paths can lead to a trapped ray for $h_T > h_d$. The ray E never enters the duct. There is a limiting ray C between those that enter the duct and those that do not which is horizontal beyond D. Rays initially above and below C will diverge beyond D indicating a caustic at the height h_d beyond D in the vicinity of which ray optics breaks down. In figure 2 a ray A is reflected from the earth and leaves the duct and a ray E leaves the duct directly. There are limiting rays C and H which become horizontal at D and I beyond which there are caustics. All rays between C and H, e.g. B, are trapped in the duct.

2.3. Wave aspects of propagation in surface ducts

Within the trapped rays there are preferred directions for which the phase at similar points F and G will differ by multiples of 2π (see eqns.(33) to (35) for a more exact statement) giving constructive interference with maximum energy flow. Figure 3 shows that there are two angles giving rise to similar rays. The energy flowing into the preferred directions forms a mode which form the basis of long distance propagation in ducts. According to figure 3 it is impossible to excite a mode unless $h_T < h_a$. This is not so. Consider, by Huyghen's principle, secondary

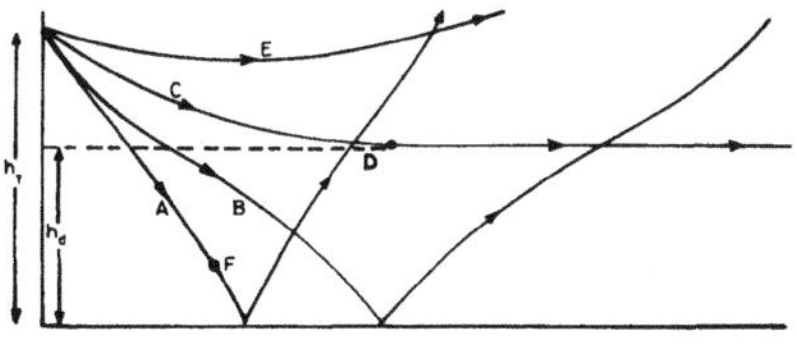

Figure 1.

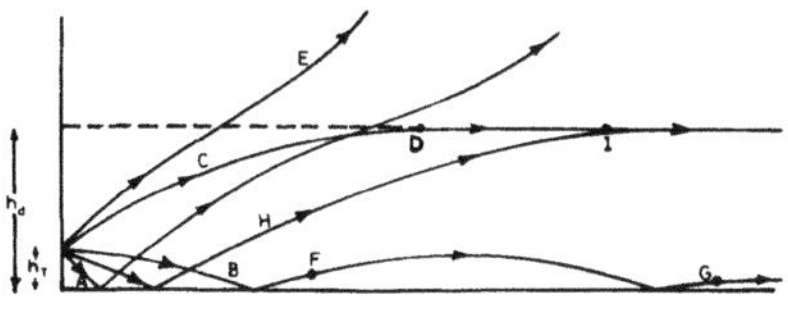

Figure 2.

Ray paths in a surface duct.

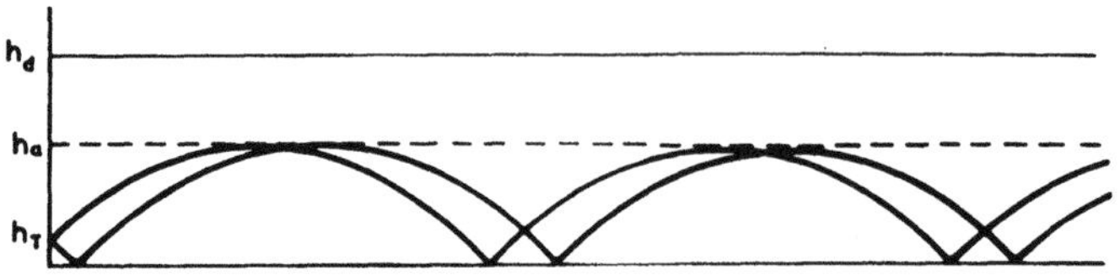

Figure 3. Ray paths for a waveguide mode.

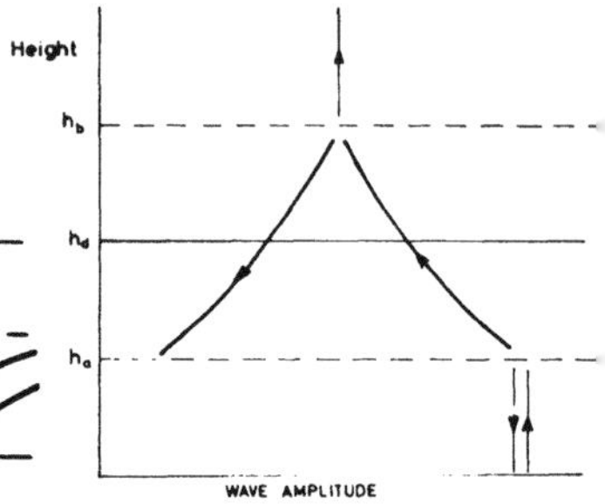

Figure 4. Vertical waves in a trapped mode.

wavelets to emanate from F in figure 1. Some of this energy will excite the modes. Thus energy leaks into a duct. The transmitter height merely affects the excitation. If we consider secondary wavelets from F in figure 2 then some energy will leave the duct giving horizontal attenuation., If all the excitations of a mode are superposed then the field strength-range variation is not periodic as figure 3 might imply but smoothly varying of the form $D^{-\frac{1}{2}} \exp(-i\beta D)$ where D is range and β is a complex constant. $D^{-\frac{1}{2}}$ gives the cylindrical energy spreading whilst Im β gives the range attenuation, i.e. leakage.

Figure 4 shows the vertical variations of a well trapped mode. Below h_a there are upgoing and downcoming waves forming a standing wave. Between h_a and h_b there is an attenuated upgoing wave. This is reflected at h_b producing an attenuated downcoming wave and a transmitted upgoing wave above h_b. The duct acts like a potential barrier through which energy must tunnel. As the frequency or h_d decrease the leakage increases and the attenuation between h_a and h_b decreases. Eventually h_a loses its significance, becoming negative. This is shown in figure 5. Below h_b there are upgoing and downcoming waves and above h_b an upgoing wave.

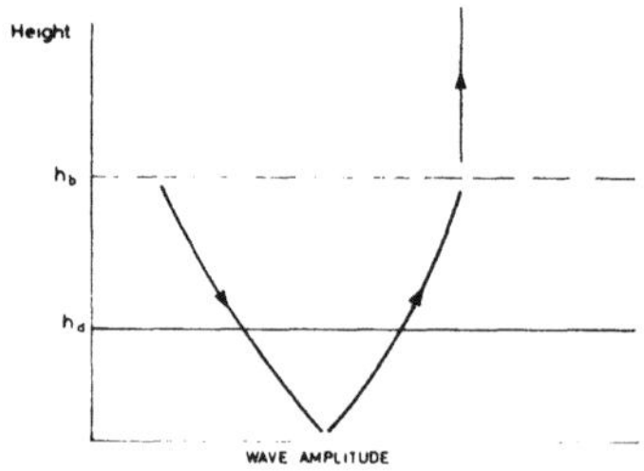

Figure 5. Vertical waves in a leaky mode.

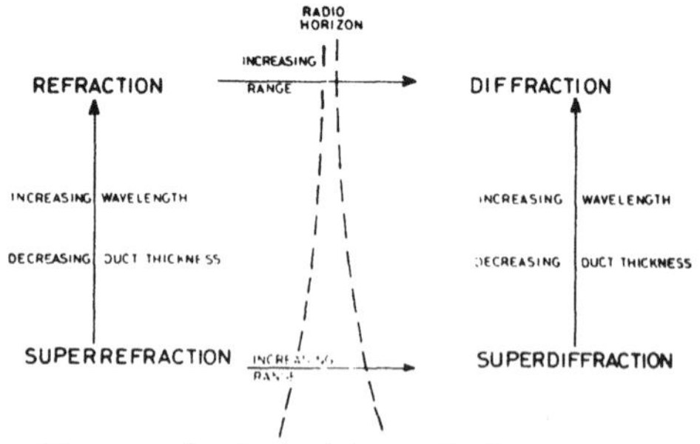

Figure 6. Relations between propagation mechanisms.

The variations below h_b are caused by internal reflections in the inhomogeneous medium, the two waves feeding on each other. The contributions of these reflections decay with height so that the downcoming wave becomes negligible and h_b should be understood as a notional level where this occurs.

With a standard atmosphere there are domains of primary and secondary radiation within and beyond the radio horizon in which refraction and diffraction are the dominant mechanisms respectively. With a surface duct there is a domain of primary radiation near to the transmitter where the mechanism is properly called superrefraction and a domain far from the transmitter dominated by the leaky modes with their properties of excitation and leakage by secondary radiation and the mechanism is properly called superdiffraction. These merge into each other in a less clear way (Fock[3] has considered this problem). Figure 6 shows relations between these mechanisms.

3. THE INTEGRAL REPRESENTATION

If $\underline{E}$ and $\underline{H}$ are the electric and magnetic space vectors (time factor $e^{i\omega t}$), $\underline{r}$ is the distance vector measured from the earth's centre and k and $Z_s \approx 120\pi$ ohms are the free space wavenumber and impedance then the relations

$$\underline{E} = iZ_s \nabla \times V\underline{r} - \frac{1}{n^2 k} \nabla \times \nabla \times U\underline{r} \quad \underline{H} = \frac{1}{iZ_s} \nabla \times U\underline{r} - \frac{1}{k} \nabla \times \nabla \times V\underline{r} \tag{2}$$

are such that U or V alone give the fields of a radial Hertzian dipole of electric or magnetic type respectively. U and V have the source singularities e^x/x as $x = -i\, n_T\, k\, R \rightarrow 0$ where $n_T = n(h_T)$ and R is the distance from the source. U and V can be cast in the form (see Bremmer[5])

$$U,V = \frac{e^{-i\, n_o k\, D}}{(-2i\pi kD)^{\frac{1}{2}}} \int_{-\infty}^{\infty} d\alpha\, e^{ikD\alpha} \Big[g_\alpha^{(1)}(h_L) g_\alpha^{(2)}(h_H) - R_\alpha^{E,M} g_\alpha^{(2)}(h_L) g_\alpha^{(2)}(h_H) \Big] \tag{3}$$

where $n_o = n(0)$, h_T and h_R are the transmitter and receiver heights, $h_L = \min(h_T, h_R)$, $h_H = \max(h_T, h_R)$ and the $R_\alpha^{E,M}$ are reflection coefficients. The $g_\alpha^{(1)}(h)$ and $g_\alpha^{(2)}(h)$ are incoming and outgoing wave solutions for large h, with Wronskian 2ik, of

$$\frac{d^2 g}{dh^2} = - k^2\, p\, g = - 2\, k^2 (m - m_o + \alpha)\, g \tag{4}$$

where $m_o = m(0)$. Eqn.(3) is the integral representation, valid both within and beyond the horizon for any form of $m(h)$.

4. SUPERDIFFRACTION THEORY

4.1. The residue series

If the contour in eqn.(3) is closed by a semicircle in the upper half plane and residues are formed there results

$$U,V = \left(\frac{8i\pi}{k\ D}\right)^{\frac{1}{2}} e^{-in_o kD} \sum_{\alpha=\alpha_\ell} e^{ikD\alpha} \frac{g_\alpha^{(2)}(h_T) g_\alpha^{(2)}(h_R)}{g_\alpha^{(2)}(0) g_\alpha^{(2)}(0)} \frac{\partial\alpha}{\partial\varphi} \tag{5}$$

The α_ℓ $(\ell = 1,2,\ldots,\infty)$ are fixed by the boundary condition

$$\varphi = \left[\frac{d}{dh} \ell n\ g_\alpha^{(2)}(h)\right]_{h=o} = i\ k\ \zeta_{E,M} \tag{6}$$

$$\zeta_M = n_e^2\ \zeta_E = (n_e^2 - 1)^{\frac{1}{2}} \tag{7}$$

where n_e is the earth's refractive index. The α_ℓ are complex propagation constants for which we must have $0 < \arg \alpha_\ell < \pi$. The residue series in eqn.(5), where it converges, is valid for diffraction and superdiffraction, each term representing a wave-guide (residue, creeping etc.) mode. Each mode has an energy spreading term $D^{-\frac{1}{2}}$, a range attenuation $e^{ikD\alpha}$, normalised height

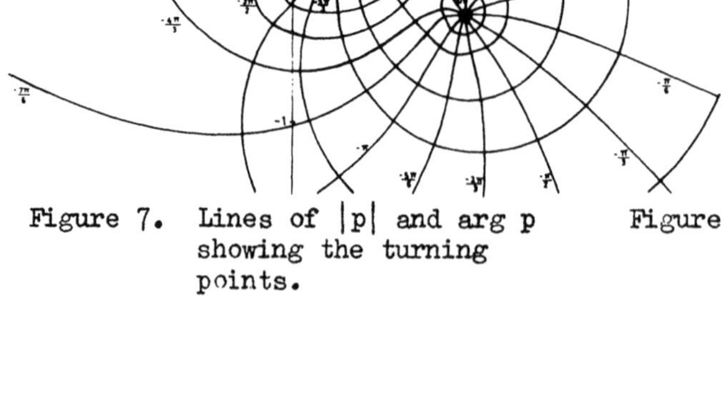

Figure 7. Lines of $|p|$ and arg p showing the turning points.

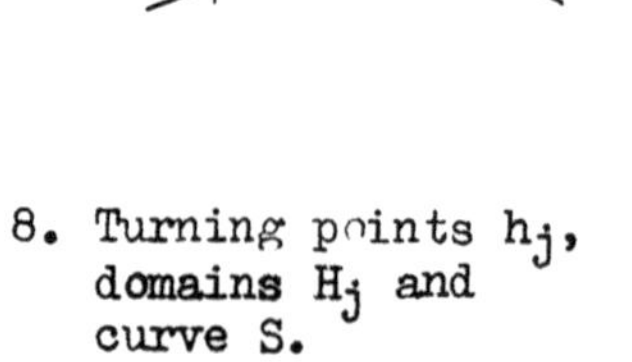

Figure 8. Turning points h_j, domains H_j and curve S.

gain functions $g_\alpha^{(2)}(h)/g_\alpha^{(2)}(0)$ for both terminals and an excitation factor $\partial\alpha/\partial\varphi$. The attenuation is fixed by Im α_ℓ and we order the modes by increasing Im α_ℓ.

4.2. The turning points

$g_\alpha^{(2)}(h)$ and α for each mode are found from eqns.(4) and (6). The solution is conditioned by the positions of zeros (turning points) and singularities of $p = 2(m - m_0 + \alpha)$ in the complex h-plane. From the minimum at h_d, p must have two principal turning points at $h = h_j (j = 1,2)$. Figure 7 shows the map of p in the h-plane for a log-linear m-variation and a moderately trapped mode showing the h_j. Figure 8 shows typical variations of the h_j as a mode goes from leaky to trapped. A formula for the h_j can be found from the Taylor expansion and inverse

$$p = p_d + \sum_{m=2}^{\infty} \frac{(h-h_d)^m}{m!} p_d^{(m)} \tag{8}$$

$$h = h_d + \sum_{m=1}^{\infty} c_m \left[\frac{2(p + p_d e^{-i\pi})\, e^{4i\pi - 2ji\pi}}{p_d^{(2)}}\right]^{\frac{m}{2}} \qquad h \in H_j \tag{9}$$

where $p_d = p(h_d)$, $p_d^{(m)} = d^m p/dh^m$ at h_d, $c_1 = 1$, $c_2 = -p_d^{(3)}/6p_d^{(2)}$etc. and the domains H_j are shown in figure 8 separated by S which is the h-map of $\arg(p - p_d) = \pi$. Setting $p = 0$ in eqn.(9) gives

$$h_j = h_d + \sum_{m=1}^{\infty} c_m \left[\frac{2\, p_d\, e^{3i\pi - 2ji\pi}}{p_d^{(2)}}\right]^{\frac{m}{2}} \tag{10}$$

4.3. Solution by the comparison equation method

This allows approximate solutions to be found in terms of solutions of a similar equation. Given $f(\eta)$ we transform eqn.(4) by

$$\int^{\eta} f^{\frac{1}{2}}\, d\eta = ik \int^{h} p^{\frac{1}{2}}\, dh \qquad g_\alpha^{(2)}(h) = \left(\frac{dh}{d\eta}\right)^{\frac{1}{2}} F(\eta) \tag{11}$$

$$\frac{d^2F}{d\eta^2} = \left[f + \left(\frac{d\eta}{dh}\right)^{-\frac{1}{2}} \frac{d^2}{d\eta^2} \left(\frac{d\eta}{dh}\right)^{\frac{1}{2}}\right] F = \left[f + G(\eta)\right] F \tag{12}$$

If G can be neglected in some η-domain then $F \sim F_o$ where

$$\frac{d^2F_o}{d\eta^2} = f\,F_o \tag{13}$$

which is the comparison equation. As eqn.(4) has two turning points we choose as comparison equations

$$\frac{d^2 v_o}{dt^2} = \left(\frac{t^2}{4} + a\right) v_o \qquad d^2 w_{jo} = z_j\, w_{jo} \tag{14}$$

$$\frac{d^2 u_o}{d\,\zeta_j^2} = u_o \tag{15}$$

which are Weber's, Stokes' and the wave equations with 2, 1 and 0 turning points. The collected transformations giving eqns.(14) and (15) are

$$\zeta_j = \frac{2}{3} z_j^{\frac{3}{2}} = \int_{t_j}^{t} \left(\frac{t^2}{4} + a\right)^{\frac{1}{2}} dt = ik\int_{h_j}^{h} p^{\frac{1}{2}}\, dh \tag{16}$$

$$g_\alpha^{(2)}(h) = \left(\frac{dh}{d\zeta_j}\right)^{\frac{1}{2}} u(\zeta_j) = \left(\frac{dh}{dz_j}\right)^{\frac{1}{2}} w_j(z_j) = \left(\frac{dh}{dt}\right)^{\frac{1}{2}} v(t) \tag{17}$$

$$i\pi a = \int_{t_2}^{t_1} \left(\frac{t^2}{4} + a\right)^{\frac{1}{2}} dt = ik\int_{h_2}^{h_1} p^{\frac{1}{2}}\, dh \tag{18}$$

where $t_j = 2i^{3-2j} a^{\frac{1}{2}}$. v_o is valid at h_1 and h_2, w_{10} at h_1 but not h_2, w_{20} at h_2 but not h_1 and u_o is invalid at h_1 and h_2. None are valid near other turning points and singularities. Other restrictions are given below. For these the h-maps of $\arg \zeta_j = n\pi$ (here and below n is any integer) known as Stokes lines (Heading(6)) are required. Figures 9 and 10 show them for a log-linear m-variation. The h-plane is divided into 5 domains W_1 to W_5. Also shown are branch cuts to make eqn.(16) single valued where necessary and the levels h_a and h_b of section 2.3.

v_o is given by Weber parabolic cylinder functions, denoted $U(ae^{ni\pi}, te^{ni\pi/2})$ (Whittaker(7) or Miller(8)) and used by Fock(3) to

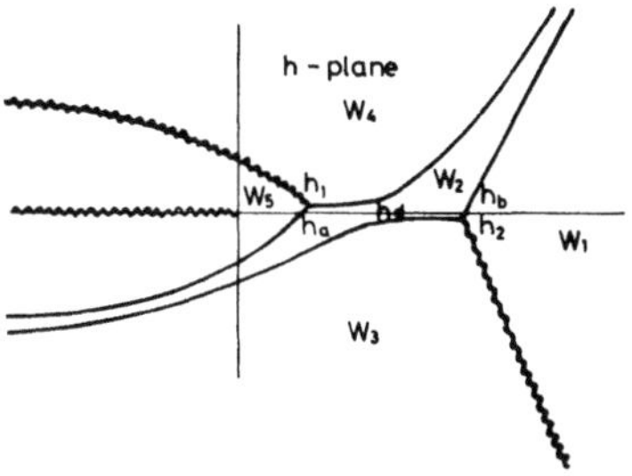

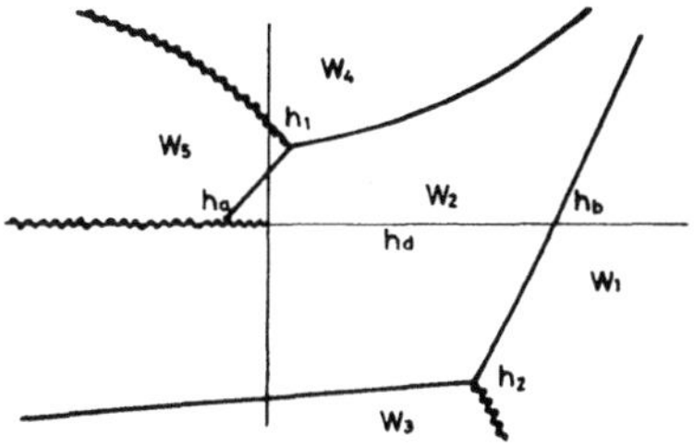

Figure 9. Trapped mode Figure 10. Leaky mode.
Stokes lines (—), branch cuts (〰) and W-domains.

study ducts. Their asymptotic approximations are(7)

$$U(a,t) \sim t^{-a-\frac{1}{2}}\, e^{-t^2/4} \qquad |\arg t| < \frac{\pi}{2} \quad (19)$$

$$\sim t^{-a-\frac{1}{2}}\, e^{-t^2/4} + \frac{(2\pi)^{\frac{1}{2}}}{\Gamma(\frac{1}{2}+a)} \left(t e^{\pm i\pi}\right)^{a-\frac{1}{2}} e^{t^2/4} \quad |\arg t\, e^{\pm 3i\pi/4}| < \frac{\pi}{4}$$
$$(20)$$

The w_{jo} are given by Airy Integral functions, denoted $Ai(z_j e^{2ni\pi/3})$ (Miller(9)). This method was developed by Langer(10). We have

$$2\pi^{\frac{1}{2}}\, z_j^{\frac{1}{4}} Ai(z_j) \sim e^{-\zeta_j} \qquad |\arg \zeta_j| < \pi \qquad (21)$$

$$\sim e^{-\zeta_j} \mp i\, e^{\zeta_j} \qquad |\arg \zeta_j\, e^{\pm 3i\pi/2}| < \frac{\pi}{2} \qquad (22)$$

u_o is given in terms of the WKB approximations(6) $e^{\pm\zeta_j}$.

$U(a,t)$ is an outgoing wave as $h \to \infty$ whence

$$v_o = U(a,t) \qquad (23)$$

Excepting the restrictions above it has unrestricted validity in the h-plane. Olver(11) has approximated $U(a,t)$ by Airy functions whence

$$w_{10} = e^{-i\pi a} Ai(z_1) + e^{i\pi a + i\pi/3} Ai(z_1 e^{4i\pi/3}) \quad h \in W_2 \cup W_4 \cup W_5 \quad (24)$$

$$w_{20} = Ai(z_2) \qquad h \in W_1 \cup W_2 \cup W_3 \quad (25)$$

These lose accuracy when $|a|$ is small when the h_j are close together. There are two kinds of u_o solutions based on Airy or Weber functions.

For the Airy function type eqns.(21) and (22) in eqns.(24) and (25) or the Stokes constant method[6] give

$$u_o = e^{-\zeta_2} \qquad h \in W_1 \qquad (26)$$

$$= e^{-\zeta_2} + i\, e^{\zeta_2} \qquad h \in W_2 \qquad (27)$$

$$= i\, e^{\zeta_2} \qquad h \in W_3 \qquad (28)$$

$$= e^{-\zeta_1 - i\pi a} + i(e^{i\pi a} + e^{-i\pi a})e^{\zeta_1} \qquad h \in W_4 \qquad (29)$$

$$= (e^{i\pi a} + e^{-i\pi a})e^{-\zeta_1} + i\, e^{\zeta_1 + i\pi a} \qquad h \in W_5 \qquad (30)$$

which lose accuracy when $|a|$ is small and can be cast in terms of ζ_1 and ζ_2 using $\zeta_2 = \zeta_1 + i\pi a$. They correspond to those in figures 4 and 5. For the Weber function type we have for $|a/t^2| \ll 1$

$$\left(\frac{t^2}{4} + a\right)^{-\frac{1}{4}} e^{\pm\zeta_j} \sim 2^{\frac{1}{2}} a^{\mp a/2}\, t^{\pm a - \frac{1}{2}}\, e^{\pm(2a - 2i\pi a + t^2)/4} \qquad (31)$$

This with eqns.(19) and (20) and a Stokes constant method[6] gives the same as eqns.(26) to (30) but with no solution in W_2 and $(e^{-i\pi a} + e^{i\pi a})$ replaced by $re^{i\pi a}$ where

$$r = \frac{(2\pi)^{\frac{1}{2}}}{\Gamma(\frac{1}{2}+a)}\, e^{-a}\, a^{a}\, e^{-2i\pi a} \qquad (32)$$

These solutions are valid for small $|a|$. For a trapped mode $|a|$ is large, $\arg a \to 3\pi/2$ and $r \sim 1 + e^{-2i\pi a}$.

Any of these systems can be used in eqn.(6) to find α. An iterative method, e.g. Newton's, must be used with a trial value of α. Space forbids elaboration of this subject except that the u_o solutions give the phase integral relations

$$ik\int_{h_1}^{o} p^{\frac{1}{2}}\, dh + i\pi(\ell - \tfrac{1}{4}) + \tfrac{1}{2}\ln(-R) = \tfrac{1}{2}\ln(1 + e^{-2i\pi a}) \quad \text{trapped} \qquad (33)$$

$$= \tfrac{1}{2}\ln r \quad \text{trapped and intermediate} \qquad (34)$$

$$= -i\pi a \quad \text{leaky} \qquad (35)$$

where $R \approx (\alpha^{\frac{1}{2}} - \zeta_{E,M})/(\alpha^{\frac{1}{2}} + \zeta_{E,M})$ is the earth's reflection coefficient and the factors on the right express the leakage.

5. SUPERREFRACTION THEORY

The method sketched here is similar to Fock's[3]. Eqn.(3) is to be integrated directly and both $g_\alpha^{(1)}(h)$ and $g_\alpha^{(2)}(h)$ are required. α takes all values from $-\infty$ to ∞. The turning points may be found from eqn.(10). For $p_d < 0$ the h_j are real with $h_1 < h_d < h_2$, for $p_d = 0$ $h_1 = h_d = h_2$ and for $p_d > 0$ they move along S in figure 8 with $\mathrm{Im}(h_1 - h_2) > 0$. Using Weber's equation as comparison equation we readily find $g_\alpha^{(1)}(h)$ in terms of $U(-a,-it)$ and $g_\alpha^{(2)}(h)$ in terms of $U(a,t)$ which, normalised by the Wronskian, give a formal solution in eqn.(3). One simple method is then to replace these by their exponential approximations of Weber function type as described in section 4.3. Eqn.(3) can then be integrated approximately by the method of steepest descent (Budden[12]). These solutions are the wave analogues of ray optics and are invalid near h_d because of caustics. A number of methods exist to improve the accuracy and domain of validity of these solutions and these are being investigated.

6. APPLICATION TO THE EVAPORATION DUCT

To apply the theory refractive index variations for the evaporation duct are required. The Debye formula for n[13] leads to

$$m - m_o = \frac{h}{K\,r_o} + c(e_p - e_{po}) + d(\theta - \theta_o) = \frac{h}{K\,r_o} + n_p - n_{po} \quad (36)$$

$$c = \frac{.373}{T^2} \qquad d = -\,7.76 \times 10^{-5}\,\frac{P}{T^2} - .746\,\frac{e}{T^3} \quad (37)$$

$$\frac{1}{K\,r_o} = \frac{1}{r_o} + 7.76 \times 10^{-5}\,\frac{g}{RT^2}\,(P - .378e) - c\,\frac{eg}{RT} - d\,\frac{g}{c_p} \quad (38)$$

where e_p, θ and n_p are potential forms of the water vapour pressure, temperature and refractive index and e_{po}, θ_o and n_{po} are their surface values. K is an effective earth's radius factor and P, e, T, R, c_p and g have conventional meanings. From one year (April 1960-

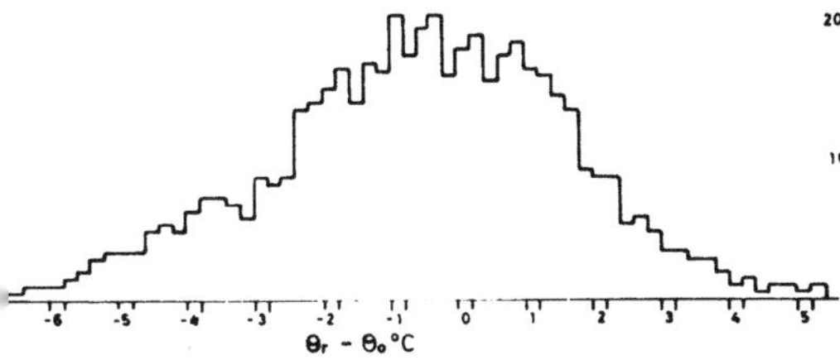

Figure 11. Histogram of $\theta_r - \theta_o$.

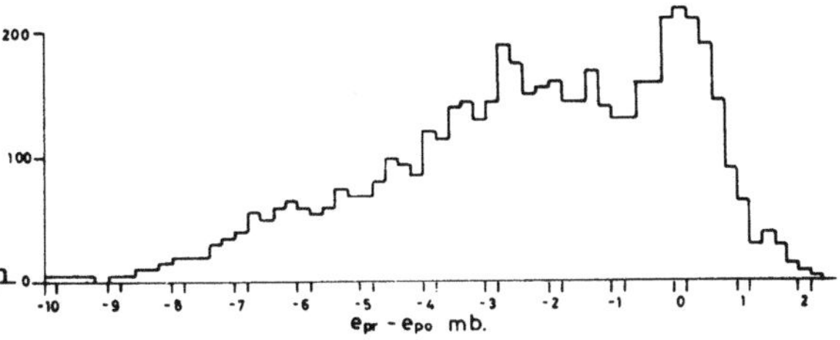

Figure 12. Histogram of $e_{pr} - e_{po}$.

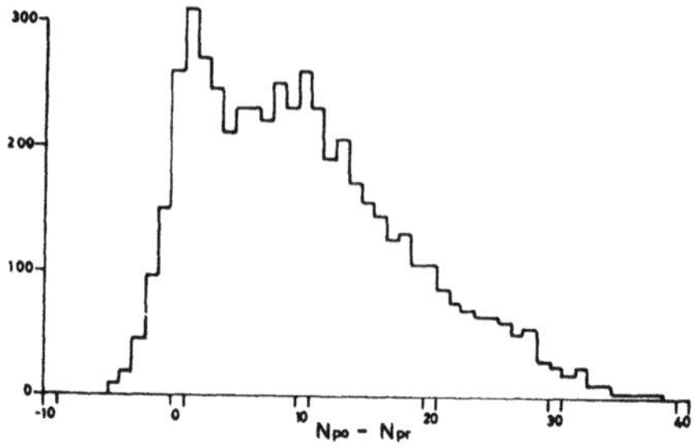

Figure 13. Histogram of $N_{po} - N_{pr}$.

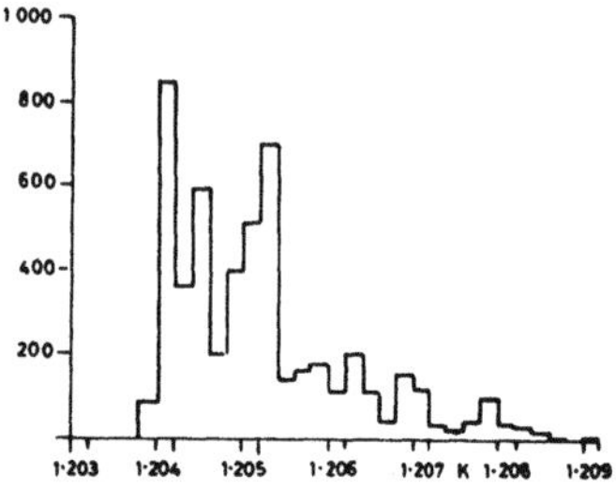

Figure 14. Histogram of K.

March 1961) of hourly means of data from the German Bight(14) figures 11 to 14 show histograms of $\theta_r-\theta_o, e_{pr}-e_{po}, N_{po}-N_{pr} = 10^6 \times (n_{\rho o}-n_{pr})$ and K where θ_r, e_{pr} and n_{pr} are values at $h = h_r = 6$m. K may be taken as 1.205.

From the Monin-Obukhov similarity theory(4) we have

$$s - s_o = \frac{s_*}{k_v}\int_0^h \frac{\varphi_F\left(\frac{h+h_o}{L}\right)}{h + h_o}\,dh \tag{39}$$

$$u_* s_* = -\overline{w's'} \qquad \frac{1}{L} = \frac{k_v\, g}{u_*^2}\left[\frac{\theta_*}{T} + .378\,\frac{e_{p*}}{P}\right] \tag{40}$$

where s is u (horizontal wind speed), θ or e_p, F is M, H or W, $k_v \approx .41$ is von Karman's constant, $h_o \approx 1.5 \times 10^{-4}$ m.(14) is the surface roughness parameter and w' and s' are fluctuations in w (vertical wind speed) and s. The φ_F are Monin-Obukhov functions. With the simplifying assumption $\varphi_H = \varphi_W$ eqn.(36) becomes

$$m - m_o = \frac{1}{K\, r_o}\left[h - \frac{(h_d+h_o)}{\varphi_W\left(\frac{h_d+h_o}{L}\right)}\int_0^h \frac{\varphi_W\left(\frac{h+h_o}{L}\right)}{h + h_o}\,dh\right] \tag{41}$$

$$h_d + h_o = -\frac{(c\, e_{p*} + d\theta_*)}{k_v}\, K\, r_o\, \varphi_W\left(\frac{h_d+h_o}{L}\right) \tag{42}$$

Eqn.(41) is log-linear in neutral conditions. The models

$$\varphi_H(x) = \varphi_W(x) = (1 \pm 16x)^{\pm\frac{1}{2}} \qquad \varphi_M(x) = (1 \pm 16x)^{\pm\frac{1}{4}} \quad x \gtrless 0 \tag{43}$$

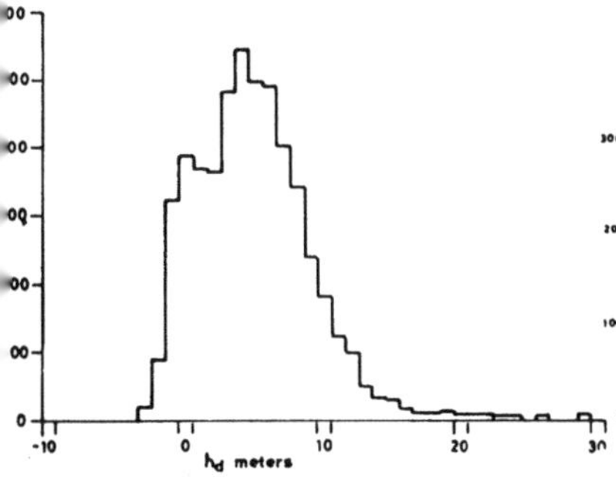

Figure 15. Histogram of h_d.

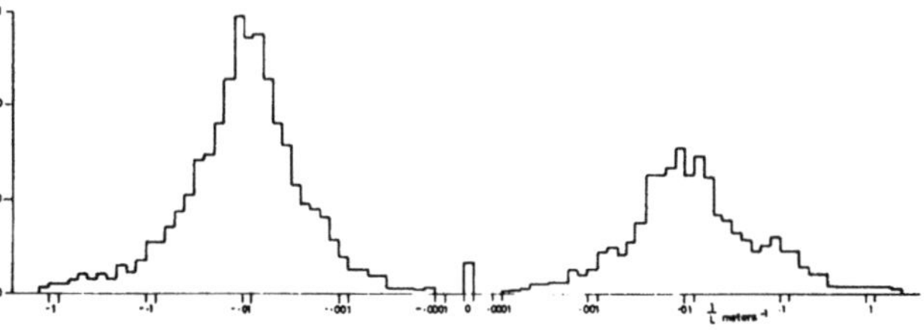

Figure 16. Histogram of 1/L.

are used in our present computer programs though any can be inserted. Figures 15 and 16 show histograms of h_d and L from the German Bight data(14). Figure 17 shows the variations of m with h, h_d, L and K.

With $\mu = 1/(h_d + h_o)$ and $\eta = K r_o/2$ the normalisations $h_n = \mu h$, $L_n = \mu L$, $h_{no} = \mu h_o$, $p_n = \mu\eta p$, $\alpha_n = 2\mu\eta\alpha$, $k_n = \mu^{-3/2}\eta^{-\frac{1}{2}} k$ and $\zeta_{nE,M} = \mu^{\frac{1}{2}}\eta^{\frac{1}{2}}\zeta_{E,M}$ recast eqns.(4) and (6) in the same form so that the theory applies to these equations. These do not contain h_d and K so α_n depends only upon k_n, L_n and mode number ℓ and slightly upon h_{no} and $\zeta_{nE,M}$. The height-gain function depends upon these and h_n.

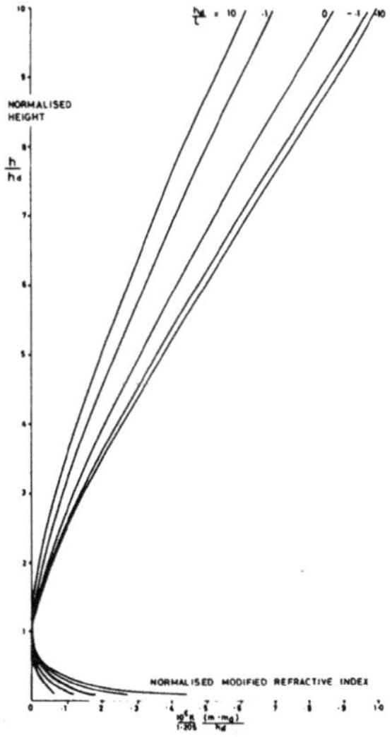

Figure 17. Modified refractive index variations.

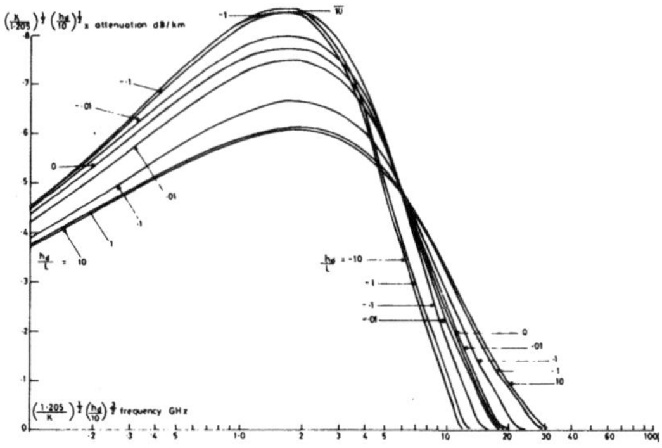

Figure 18. Variation of the attenuation for the 1st mode.

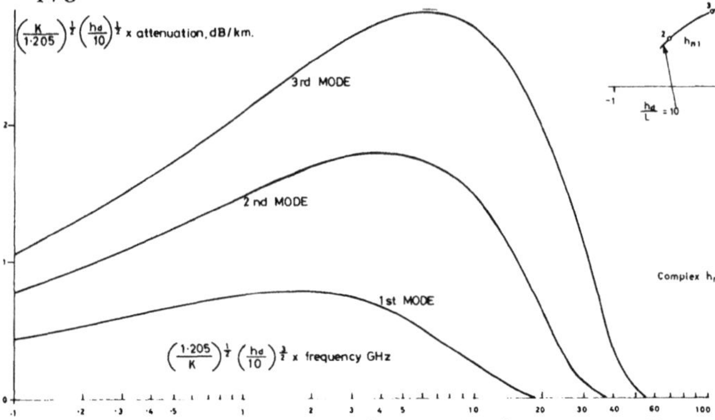

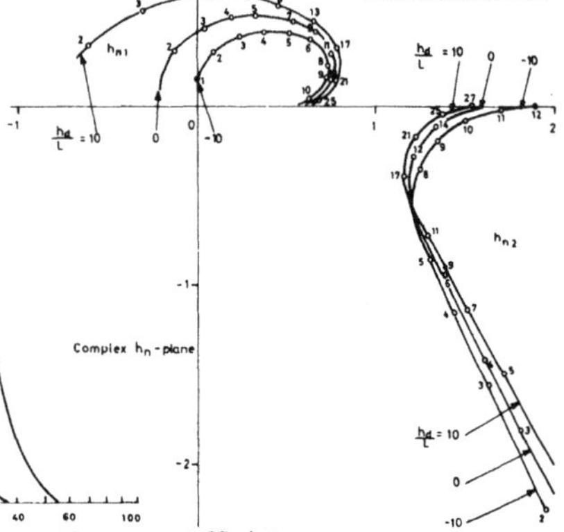

Figure 19. Attenuation in neutral conditions.

Figure 20. Turning points The numbers are 2.96 k_n.

Using the theory of section 4 with eqn. (41) computer programs have been written to compute values of α and $g_\alpha^{(2)}(h)$ and the field strength at a point and field strength-distance, height and duct thickness variations using up to 9 modes of the residue series. Figure 18 gives an attenuation (20 $\log_{10}k$ Im α in dB/km) map for the 1st mode in terms of h_d(meters), L(meters), K and frequency(GHz) and figure 19 for the first 3 modes in neutral conditions. Figure 20 shows the variations of the turning points with k_n. Figures 21 and 22 show the height-gain functions for the first 2 modes against h, h_d, K and frequency in neutral conditions. Figures 23 and 24 show experimental field strength-duct thickness data from the German Bight(14). The numbers are the number of observations (divided by 5 for figure 23 and 7 for figure 24) falling in each small block. Also shown are the theoretical predictions for a series of values of L.

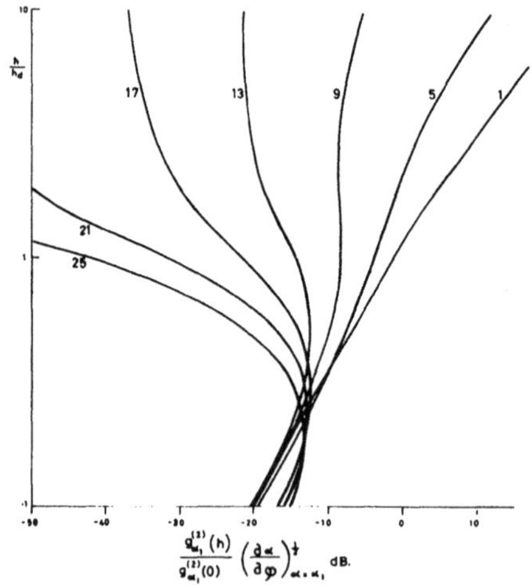

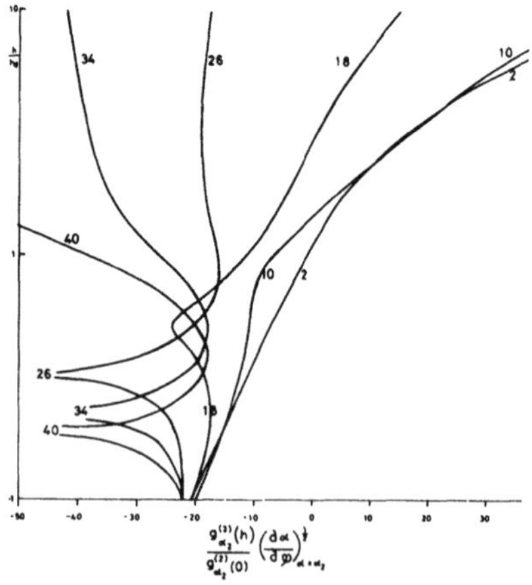

Figure 21. First mode.

Figure 22. Second mode.

Height gain functions. The numbers are $(h_d/10)^{3/2}$ $(1.205/K)^{\frac12}$x frequenc

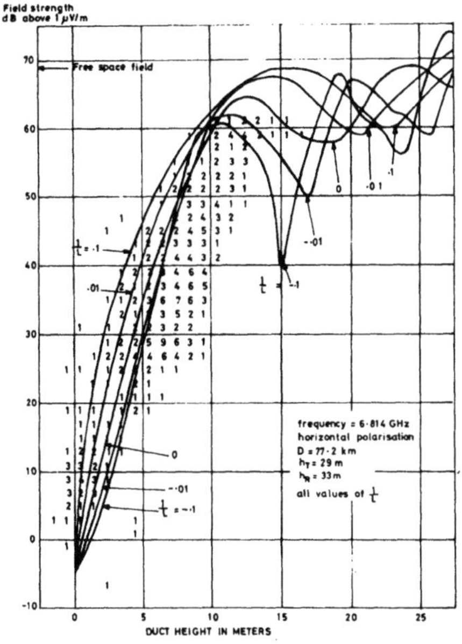

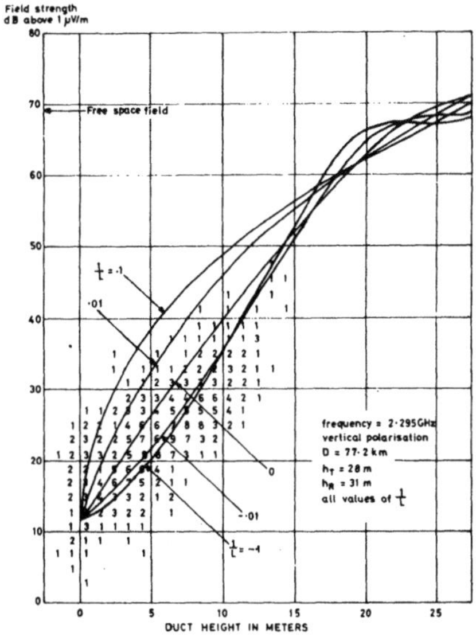

Figure 23. Figure 24.
Experimental and theoretical field strength-duct thickness relations.

7. CONCLUSION

The superdiffraction theory for a smooth m-variation over a smooth sea is well developed. However as figures 23 and 24 show, the theoretical predictions tend to exceed the experimental values. Scattering from atmospheric turbulence and the rough sea seem likely causes for this and are being included in our model. An approach to the rough sea problem can be made by modifying the smooth sea surface impedance $\zeta_{E,M}$. Methods of doing this for a number of roughness models are known[15]. An approach to the turbulence problem and to water vapour absorption can be made by giving the refractive index a lossy component. In addition the superrefraction theory is being developed to the same level.

ACKNOWLEDGEMENT

The author wishes to thank the Director of the Admiralty Surface Weapons Establishment, who sponsored this work, and the Technical Director of GEC-Marconi Electronics Limited, for permission to publish this paper.

REFERENCES

1. Booker, H.G. and Walkinshaw, W. The mode theory of tropospheric refraction and its relation to wave-guides and diffraction. Meteorological Factors in Radiowave Propagation, 1946.
2. Pekeris, C.L. Asymptotic solutions for the normal modes in the theory of microwave propagation. J.Appl.Phys., vol.17, no.12, pp.1108-1124, 1946.
3. Fock, V.A. Electromagnetic diffraction and propagation problems. Pergamon, 1965.
4. Monin, A.C. and Obukhov, A.M. Basic regularity in turbulent mixing in the surface layer of the atmosphere. U.S.S.R. Acad.Sci.Geophys.Inst., no.24, pp.163-187, 1954.
5. Bremmer, H. Terrestrial radio waves. Elsevier, 1949.
6. Heading, J. An introduction to phase-integral methods. Methuen, 1962.
7. Whittaker, E.T. and Watson, G.N. A course of modern analysis. Cambridge, 1953.
8. Miller, J.C.P. Tables of Weber parabolic cylinder functions. HMSO, 1955.
9. Miller, J.C.P. The Airy Integral. Brit.Assoc.Math.Tables, Cambridge, 1946.
10. Langer, R.E. On the asymptotic solutions of ordinary differential equations, with an application to the Bessel functions of large order. Trans.Am.Math.Soc. 33, pp.23-64, 1931.
11. Olver, F.W.J. Uniform asymptotic expansions of Weber parabolic cylinder functions of large orders. J.Res.N.B.S.63B, no.2, pp.131-169, 1959.
12. Budden, K.G. Radio waves in the ionosphere. Cambridge, 1961.
13. Bean, B.R. and Dutton, E.J. Radio meteorology. NBS monograph 92, 1966.
14. Brocks, K. et al. Radiometeorologische Abteilung des Geophysikalischen Instituts der Universitat Hamburg, nos. 3,5,6 and 14, Hamburg, 1960-1970.
15. Beckmann, P. and Spizzichino, A. The scattering of electromagnetic waves from rough surfaces. Pergamon, 1963.

Prof. Straiton: Figure 23 indicates that the signal does not exceed the free space level. Is this the correct interpretation?

Reply: Yes but for links with terminals in the duct or when higher order modes are trapped it is possible to exceed the free space level.

OCEANIC DUCT AND ITS EFFECTS ON MICROWAVE PROPAGATION IN THE MEDITERRANEAN SEA.

M.Anastassiadis and D.Mavrakis

University of Athens
Physical Electronics Laboratory

In 1972 the radiometeorological group of Athens University was invited to collaborate with the Naval Electronics Laboratory Center U.S.A. (principal investigator Dr.J.Richter) in order to investigate oceanic duct and its effects on microwave propagation in Mediterannean Sea, especially in the central part of the Aegean sea.

The program of measurements covered four periods (winter-spring-summer-automn) and each period consisted of about 15 days.

Because the same experiment was carried out in three different parts of the world, for reasons of maximum compatibility the completly oversea transhorizon path was carefully selected to be the same for all parts. The choosen path was between the island of Naxos and the island of Mykonos, a total distance of 19.2 n.m. The transmitting site was organized on Naxos island with the antennas on a height of 4.5 m while the receiving center was organized on the island of Mykonos, with all receiving antennas installed on a mast of 20 m height, at three levels of 4 m, 9 m and 18 m above sea level.

Frequencies from L,S,X, bands used during all four periods of investigations, while for two periods (summer and automn), measurements were taken from Ku-band. Finally during the last period, Ka-band was operated and for the purpose of better understanding of the meteorological conditions, a number of radio sondes were used (two radiosonding per day). Emphasis was also given on the meteorological conditions all over the path. Both sea and air measurements were taken two times every day during the whole period in order to examine the weather conditions all over the path.

From preliminary results which has been presented in this ASI - NATO, it is clear that x-band is affected by ducting conditions while L-band and S-band are not. This is deduced from a number of curves of path loss versus time and path loss versus per cent of total time; the above curves are part of the work which has been already done. Final results relating data from the Mediterannean sea to other parts of the world are now under investigation and will be published in the very near future in colaboration with the Naval Electronics Laboratory Center, U.S.A.

TRANSMISSION OF MILLIMETER RADIO WAVES IN OVER WATER DUCTS

A. W. Straiton

The University of Texas at Austin

INTRODUCTION

The effect of high lapse rates of water vapor on radar ranges over seas was noted in the early shipborne use. The microwave transmissions guided by surface anomalies were reflected by surface targets far beyond the normal line-of-sight.

This paper surveys a series of measurements made at the Gulf of Mexico over near-shore paths. Some of the studies were made as long as twenty-five years ago but they were not widely reported while others were made recently.

This paper is made up of three sections as follows:

I. A description of the geographical features of the shoreline and of the meteorological factors which contribute to trapping conditions which commonly exist over the water. Measurements are reported of the variations of meteorological parameters with height above the water.

II. The amplitude level of three centimeter (10 GHz) radio wave signals as a function of distance was measured over ranges up to eighty kilometers and the result of these observations are reported. From the meteorological data described in Section I, a refractive index model was selected from which the distance and height dependence of the 3 cm signals were calculated. The results of the calculations are compared to the measured values.

III. In the third section, the problems introduced by extending the radio transmission to frequencies of 35 and 70 GHz are considered. The rates of decrease of the signal levels with distance for these frequencies are compared to the 3 cm results. Also the high degree of variability of the transmission levels both on short and long time bases are reported and related to the inhomogeneous refractive index structure and the roughness of the sea surface.

I. GEOGRAPHICAL AND METEOROLOGICAL FEATURES

The upper Texas shoreline from the southwest end of Bolivar Peninsula to the Sabine River is relatively straight with a slight concave shape. Galveston Island lies approximately two miles off the extended shore line. From the northeast tip of the Island or from the southwest end of Bolivar Peninsula, overwater paths may be obtained to points along the beach up to eighty kilometers.

Tides are not significant in the Gulf of Mexico with the usual variation range of only 30 cm. Only one cycle of tide occurs daily.

The refractive index of the atmosphere is a function of the pressure, temperature and water vapor content. The refractive index, n, of air differs from one by less than 0.001 but the changes are important in radio transmissions. It is convenient to define a refractivity, N, which is given by

$$N = (n-1) \times 10^6$$

The value of N for practical purposes may be given by (Bean and Dutton, 1966)

$$N = \frac{77.6}{T} \left(P + \frac{4810e}{T}\right)$$

where P is the atmospheric pressure in m. bars, T is the temperature in degrees Kelvin and e is the partial water vapor pressure in m. bars. Near the water surfaces involved in this study, the refractive index variations are primarily determined by variations in water vapor density. In order to establish water vapor and refractive index profiles above the sea surface, recording psychrometers were moved from heights of one meter up to 20 meters above the surface. These types of measurements were made at various times and at points along the shoreline, from a drilling platform five miles offshore, a lighthouse one-half mile offshore and from a boat cruising offshore.

Although individual soundings at different times and locations varied significantly, there frequently existed surface conditions conducive to microwave trapping. In a horizontally stratified atmosphere, trapping will occur when the downward curvature of a ray launched horizontally exceeds that of the earth. In order to make this effect more easily identified, a modified index of refraction is introduced which relates to the path of a ray to a flat earth. This modified refractive index, M, is given by

$$M = 0.16h + (n-1) \times 10^6$$

when n is the actual index-of-refraction and h is the height in meters above the sea surface. Super refraction or trapping will begin when the rate of change of M with height is negative.

The gradient of the water vapor above the sea surface is dependent on a number of factors. Two important contributors are the wind speed and the relative temperature of the water and the air. Over the radio path along the coast, optimum ducting conditions occurred when the wind was from the sea with a velocity greater than 6 meters per second and with the air warmer than the water. Experiments showed that the decrease in M between 1.5 and 17 meters above the surface correlated well with the wind speed. Under conditions of low wind speed, the water vapor is confined to a shallow layer near the surface. The increased wind speeds tend to move this surface layer of moisture upward. It was observed that the surface changed rather abruptly from a smooth to rough condition at wind speed of 6 or 7 meters per second.

The direction of the wind is also a major factor in the establishment of the trapping conditions along the coast. Uniform wind from the sea tends to establish nearly uniform conditions along the path whereas the wind from the land will be in a transient condition as it moves out over the water. The presence of the shoreline with changes in water depth and water temperature introduces variations in the water vapor distribution. The physical processes involved in establishing the ducting condition have been discussed in detail in other papers in this volume.

The index of refraction at a single point is known to vary rapidly over ranges of several M units in seconds and over even larger variations over longer periods. The psychrometer type measurements tended to smooth out some of the time variation but the individual profiles are quite erratic. The average of a number of soundings at a given site under stable meteorological conditions, however, produced an M curve whose general shape was repeatable. The averaged M curves had the characteristics that they decreased with height up to approximately ten meters above the surface and increased for greater heights. It was felt that this time averaging for similar meteorological conditions at a given location produced an approximation of the space average for these conditions. The data for the median M curve thus obtained were for similar periods of stable wind conditions blowing from offshore.

II. THEORETICAL AND EXPERIMENTAL STUDIES AT 3 cm WAVELENGTH

Measurements of the signal levels of centimeter and millimeter waves were made on a number of different occasions on the overwater paths near Galveston. Perhaps the most comprehensive set of data at 3 cm wavelength were taken in August and September 1948. (Straiton, 1949) These measurements consisted of vertical phase difference and signal strength measurements as functions of transmitter and receiver height for path lengths of 20, 50 and 64 kilometers.

The transmitting equipment including a 76 cm parabolic antenna was mounted on the elevator of a 16 M tower located two meters above sea level. The elevator was raised from 3.3 M to 18 M above sea level in approximately two minutes.

The receiving equipment used two 46 cm parabolas spaced 3.3 meters apart vertically. The amplitude of the signal at each antenna was recorded as well as the phase difference between the two antennas. The receiving antennas were mounted on an elevator of the 20 meter tower. The center point between the two antennas could be raised continuously from 3.3 to 18 meters above average sea level in approximately three minutes with continuous recordings of the phase difference and signal strength.

The general procedure for taking data was to take six sets of phase difference and signal strength changes with height in an interval of approximately twenty minutes. The receiver was moved through its complete height range with the transmitter set successively at elevations of 3.3, 9.9 and 17.5 meters above mean sea level. The transmitter was then moved through its complete height range with the receiver set at 3.6, 9.8 and 17.8 M mean sea level. Fifteen complete sets or ninety height runs were made for the 20 km path and 18 complete sets or 108 runs were made for the 50 km path. With the transmitter at fixed heights, two receiver runs were made for a 64 km path and eight runs for a 75 km path.

All phase difference measurements were taken with reference to that for a plane wave horizontally directed.

For each of the radio paths used, the general characteristic of all of the signal strength and phase difference curves, were approximately the same and only minor variations occurred in the detail of the curves. The homogeneity is attributed to the fact that the meteorological conditions were stable and uniform during the periods of measurements with the breeze consistently from off shore.

In order to represent the results as simply as possible, all of the data for corresponding height runs were averaged to obtain the curves discussed in later sections.

The average signal levels and phase differences for the 20 km path with the receiver moving are shown in Figure 1 and for the same path with the transmitter moving the corresponding curves are shown in Figure 2.

The signal levels increase as the transmitter and receiver are raised and approach a value near the free space level for the greater heights. The curves for the receiver and for the transmitter moving differ somewhat but have roughly similar characteristics. The phase difference which may be interpreted in terms of angle-of-arrival of a plane wave front was such that this wave front was tilted backwards giving an upward component at the receiver.

If an attempt is made to predict the signal condition at the receiver on the basis of a two ray model, a lobe structure would be found with maxima spaced at distance in the order of $h_2 = 28/h$, when h_1 and h_2 are heights of the transmitter and receiver respectively in meters. Such lobe structure is entirely missing in the averaged curves. The height gain curve more nearly resembled that predicted from using the calculations for a wave trapped in the low level duct as will be described later although the path lengths and terminal heights are such that the ducting conditions would not be expected to be established.

The phase data show that the equivalent angle-of-arrival is always below the horizontal. Roughly interpreted, this means that waves seem to be coming from the water surface even when the two terminals are within line-of-sight in the absence of the ducting.

The characteristics of the phase difference and signal amplitude curve for the 50 kilometer path are significantly different for the 20 km path as seen in Figures 3 and 4. For this range, the strongest levels were obtained with both terminals within three meters of the surface. Much weaker signals were experienced when terminals were ten meters or higher. This is obviously the condition for low level trapping since the maximum height would be within line-of-sight in the absence of the atmosphere.

The phase difference (angle-of-arrival) was approximately constant for both paths as the transmitter was moved upward as compared to the condition of decreasing angle with the receiver moving. This indicates that a condition of stability in the duct had been reached so that the launching height changed only the magnitude of the signal and not the phase front inclination.

The analysis of the data taken over the 64 and 67 km paths showed that the signal level remained near the free space value but the limited amount of data did not justify detailed analysis.

In order to establish a meteorological model for comparison of measured and calculated phase difference and amplitude, a single-modified index of refraction profile was selected. The

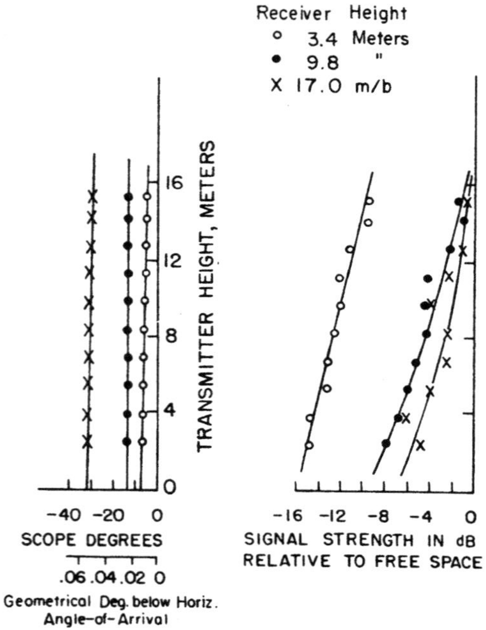

ANGLE-OF-ARRIVAL AND SIGNAL-STRENGTH CURVES FOR 20 km PATH WITH TRANSMITTER MOVING

Fig. 1

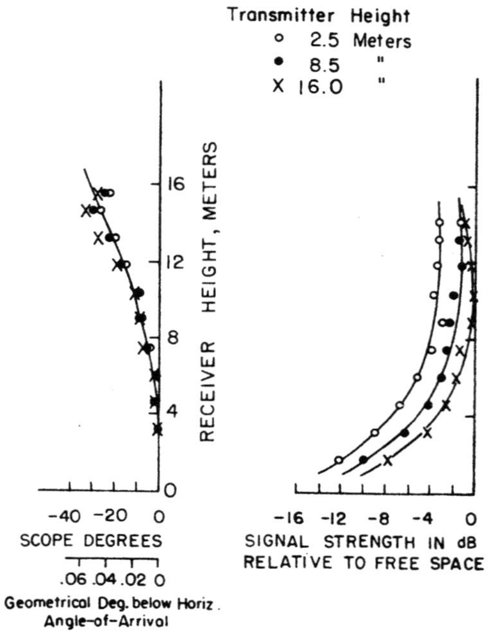

ANGLE-OF-ARRIVAL AND SIGNAL-STRENGTH CURVES FOR 20 km PATH WITH THE RECEIVER MOVING

Fig. 2

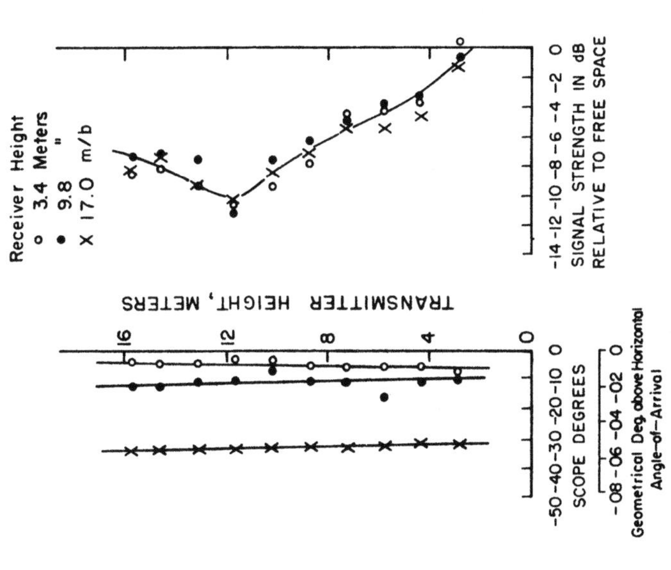

ANGLE-OF-ARRIVAL AND SIGNAL-STRENGTH CURVES
FOR 50 km PATH WITH TRANSMITTER MOVING

Fig. 3

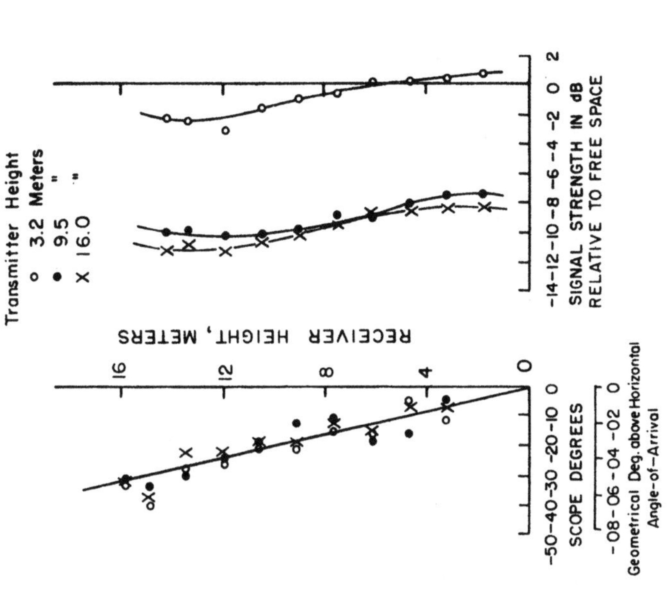

ANGLE-OF-ARRIVAL AND SIGNAL-STRENGTH CURVES
FOR 50 km PATH WITH THE RECEIVER MOVING

Fig. 4

meteorological data taken at both ends of the path over the interval of measurements were averaged. The M curve determined from these averages had the shape of that shown in Figure 7. It should be pointed out that no individual height sounding was very close to the profile taken as a model. However, as indicated earlier, it is assumed that the time average of the soundings provided an approximation of the space average. The curve in Figure 5 is a linear-exponential one which lends itself to theoretical analysis. The equation for the modified index M, is given by

$$M - M_o = 0.16h + 10.5\ e^{-0.43h} - 10.5$$

Thus the terms $10.5\ e^{-0.43h} - 10.5$ give the actual index of refraction difference multiplied by 10^6 and 0.16h gives the modification to account for the earth curvature. For distances well beyond the horizon, a solution may be obtained in terms of a number of normal modes (Furry, 1945) similar to propagation conditions in a waveguide. For the refractive index model chosen to represent the path conditions, the first mode at 3 cm wavelength is found to fall in the transition region between complete trapping in the duct and considerable leakage from the duct. The general shape of a trapped wave is one that increases from near zero at the surface to a maximum at a height slightly less than the height of the minimum modified index of refraction and decreases for greater heights. For complete trapping the only loss would be from the horizontal dispersion of the wave which would reduce the electric field strength as the inverse square root of the distance. The attenuation in excess of this inverse square root of the distance was found to be 0.19 dB/km. The loss of the second mode was found to be 2.6 dB/km below the inverse square root value indicating that it would die out rapidly. For a comparison of the signal level at 20 and 50 kilometers, the relative free space ratio would be 8 dB while the inverse square root loss added to the leakage from the duct would result in about 9.7 dB attenuation. Thus the calculated loss between the two distances would be about the free space value.

The shape of the theoretical height gain curve would be independent of distance for ranges great enough for the single mode condition to be established. As indicated earlier, the level of the curve would decrease 0.19 dB/km above the inverse square root of the distance. The change in level with distance compares well with the measured averages.

The theoretical height gain curve is shown in Figure 6 with the height of four meters taken as reference. This shape should be compared with the shape of the height gain curves obtained with the receiver moving as shown in Figures 2 and 4.

The phase change per meter as calculated for the receiving site is also shown in Figure 6. This curve will be independent of distances great enough so that first mode conditions are

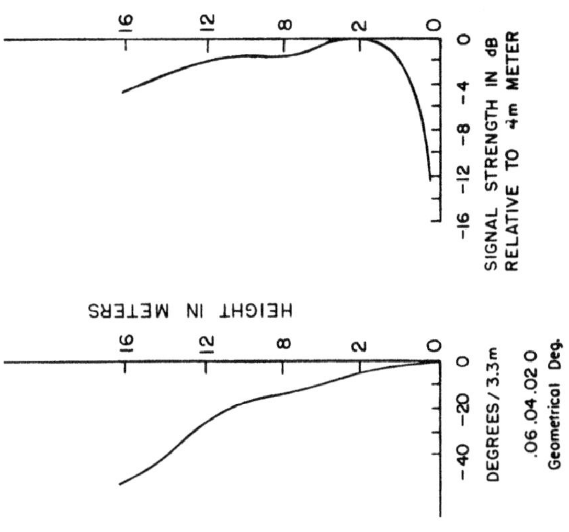

THEORETICAL ANGLE-OF-ARRIVAL AND SIGNAL STRENGTH CURVES

Fig. 6

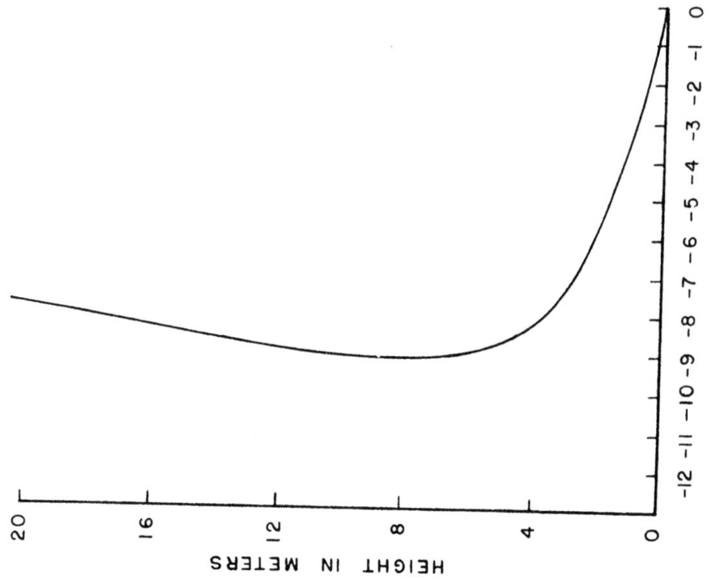

MODIFIED INDEX OF REFRACTION VS. HEIGHT
(Model for Theoretical Studies)

Fig. 5

established. This curve compares well with the phase curves shown in Figures 2 and 4.

III. TRANSMISSION STUDIES AT 8.6 and 4.3 MILLIMETERS

In the theoretical studies conducted using the idealized refractive index profile, the degree of trapping is a function of the wavelength used. It is predicted that as the wavelength is decreased, the degree of trapping will be increased. This neglects the frequency dependence of the effects of the water surface and of the inhomogeneities of the atmosphere.

An experiment was conducted to examine the effect of the water surface water by setting up a radio path between two oil drilling platforms in the Gulf of Mexico. The terminals were approximately 13 km from the shore line and were separated by 1.5 km. Transmitter heights of 5 and 12 meters were used. Simultaneous transmissions were made at wavelengths of 9.0, 5.3, 3.2 and 0.86 cm. Variations in water height were recorded with a step type surf gauge.

Progressive increase with frequency in the peak-to-peak variations was noted from a fraction of a dB at 9 cm to 12 dB at 8.6 mm. The spectra of the water waves compared closely to the 3 cm radio wave spectra while the 8.6 radio waves contained substantially more high frequency components.

The relationship between the increased scattering of radio waves with increased frequency cannot be translated directly to the beyond the horizon trapping of these waves. It indicates, however, that more losses may be expected in the long range transmission because of the incoherent reflection from the rough water surface.

In the same tests on the oil drilling platform microwave refractive index variations were measured with a Crain type refractometer. The rms values of the refractive index changes over a 10 minute period for the sampling cavity 4 meters above the surface were found to be of the order of 3N units for ten minute samples. It was found that the rms value increased for increased sample length. These rms values result from peak-to-peak swings in excess of 10 N-units. The idealized M profile used to calculate the 3 cm transmission has only a difference of 5 M-units from a one meter height to the height of the duct. This shows the difficulty of trying to establish and use a smoothed profile for predicting transmission signal levels.

The results of these refractometer observations show the extreme variability of the refractive index at a single point. In light of this, it is amazing that the 3 cm analysis described in the previous section compared as well as it did with the measurements.

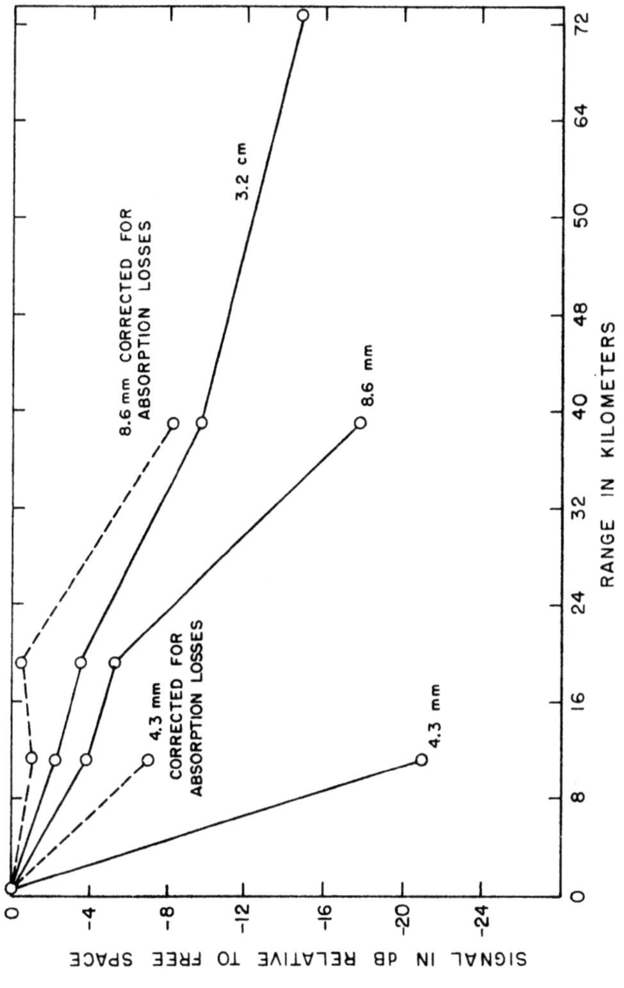

SIGNAL EXCEEDED BY MEDIANS OF 50% OF THE SAMPLES

Fig. 7

For a direct measurement of the frequency dependence of the trapping an experiment was conducted at wavelengths of 3 cm, 8.6 mm and 4.3 mm over the same path previously used for 3 cm tests for a range of distances from 1 to 72 kilometers with the terminals less than 3m above the surface.

Figure 7 shows the average signal level for the three wavelengths. The 4.3 mm level dropped off so rapidly that no data were obtained past 10 kilometers. The maximum range for the 8.6 millimeter transmissions was 39 km. The 3.2 cm waves were easily detected at 72 km. The dotted line in Figure 7 shows the signal levels adjusted to correct for water vapor absorption. This correction is uncertain, however, because of the dependence of the water vapor on height and its variation with time and place.

From the analysis of the curve in Figure 7, it is concluded that the 8.6 mm waves do experience some trapping in the over water duct but that the 4.3 mm waves suffer much greater attenuation. The water vapor absorption in the moist atmosphere does provide additional loss.

The fluctuations of the signal levels of the millimeter waves is also quite large. To study this variability as a function of distance, another field trip was recently made to the same part of the coast line. The transmitter used was a solid state 8.6 mm portable transmitter which was located at various sites along the Bolivar Peninsula beach at a height of about 2 meters above the mean sea level. Receivers on Galveston Island were at two meters and seven meters above mean sea level. Samples of the original recordings at a 16 km range showed a peak-to-peak change in the order of 5 dB in periods of seconds while changes in level of 10 dB or more were experienced in intervals over several minutes. The spectral analysis of these samples showed no significant variations of frequencies greater than a few cycles per second.

REFERENCES

Bean, B. R., and E. J. Dutton (1966), "Radio Meteorology," NBS Mono. 92, U. S. Department of Commerce, National Bureau of Standards.

Straiton, A. W. (1949), "Microwave Phase Front Measurements for 12- and 32-mile Overwater Paths," Proc. of IRE, Vol. 37, No. 7, July 1949, pp. 808-813.

Furry, W. H. (1945), "Theory of Characteristic Functions in Problems of Anomalous Propagation," Report No. 680, Radiation Laboratory, Massachusetts Institute of Technology, Cambridge, Mass.

DIFFRACTION OF RADIO WAVES BY TERRAIN OBSTACLES IN RADIO DUCTS

Dag T. Gjessing

Norwegian Defence Research Establishment

INTRODUCTION

This contribution is intended to serve two purposes:

a) To give a brief theoretical treatment of the general diffraction problem with emphasis on obstacle gain in relation to ducted propagation.

b) To discuss the more practical consequences of obstacles in radio ducts: detection of target behind an obstacle, short wave radio communication in the diffraction zone.

We know that ground-based ducts are very abundant. It is of interest for maritime operations to learn to what extent these ducts open new detection and communication possibilities in areas with many small islands. The current contribution will not give detailed results of numerical calculations. The aim is to establish a set of analytical expressions from which the general properties of a ducted-diffracted propagation mode can be revealed. The potential usefulness of this mode for naval application should then be evident.

Although the analytical work, and indeed the substance of the current paper, is based on a very simple and straightforward approach learned largely from scatter propagation research, it may be pertinent, in view of the tutorial aspect of the contribution, to summarize briefly earlier approaches so as to put the current piece of work in historical perspective. Furthermore, for the benefit of readers who are more attracted by the classical approaches, a method based on Hygen's principle is given in the form of an appendix.

So much about diffraction. The line of sight duct propagation problem is then to be considered. In this contribution we shall lean heavily on results presented in previous contributions to this conference, particularly on the work of Früchtenicht and Rotheram. Specifically, the expression for the vertical profile of field strength as determined by Früchtenicht will be rewritten in a mathematical form suitable for the particular diffraction problem at hand and subsequently used in the expressions for the diffracted field.

THE DIFFRACTION PROBLEM, A SHORT REVIEW

Radio propagation over natural obstructions has been a subject of considerable interest. Classical knife-edge diffraction theory is used as an idealization of the physical problem. However, due to difficulties in obtaining an analytical solution, particularly when ground reflections and ducting are considered, most of the previous work has been restricted to compilations of results from theory and measurements, i.e. magnitude considerations only. These are probably of limited general interest in a presentation such as this. The knife-edge diffraction problem has been formally solved by several techniques. The early work by Sommerfeldt, who based his analysis on the wave equation with appropriate boundary conditions, is perfectly rigorous and not restricted to small diffracting angles, but generalization to the multiple-reflection problem is not readily accomplished.

Fresnel and Kirchhoff based their work on an analytic formulation of Huygen's principle. The Fresnel-Kirchhoff method has the advantage of the simplicity of ray-path concept, and is used almost exclusively in radio propagation literature. However, the method suffers from lack of rigor and from the appearance of the Fresnel integrals which tend to obscure physical interpretations. In order to obtain practical solutions to the problem of diffraction in the presence of multiple reflections, a phasor summation of the Fresnel-Kirchhoff results is performed for the single ray problem.

In a paper by Ratcliffe a very neat and general method for solving complex diffraction problems is sketched. Ratcliffe shows that if, in particular, a one-dimensional diffracting screen is illuminated by a wave (wavelength λ), then the angular power-spectrum $P(\theta)$ produced, expressed in terms of $\sin\theta$, is the Fourier transform of the spatial distribution $E(z)$ of the wave front just as it leaves the screen, provided z is expressed in terms of λ.

BASIC THEORY OF DIFFRACTION

Our problem is the following:

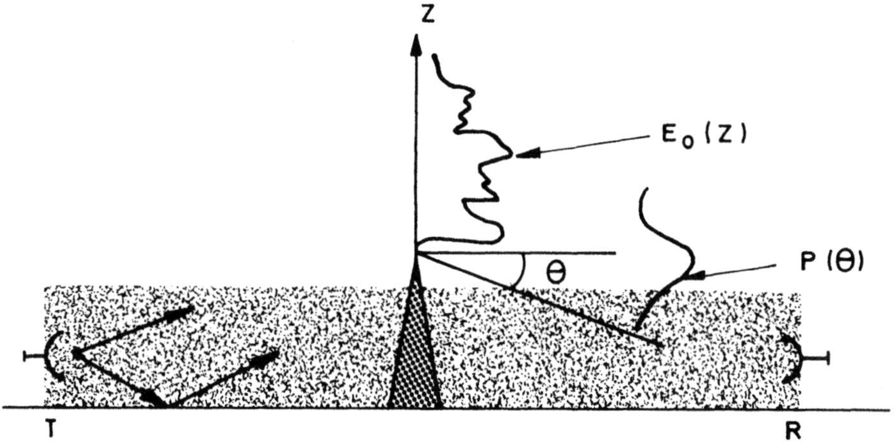

Figure 1 Geometry of the diffraction problem

The transmitter at point T illuminates a region above the obstacle. This gives rise to a distribution of field strength along a vertical direction above the obstacle. This distribution is determined by multipath effects, by the ground reflection coefficient, by the radiation properties of the antenna and by the properties of the radio duct. On the basis of knowledge about E(z), we want to calculate the angular power spectrum P(θ) of the diffracted wave.

We now use exactly the same approach as for scatter propagation (Gjessing 1969) and derive the following expression for the secondary field E_s, (scattered field or diffracted field as the case may be):

$$E_s = \frac{k^2}{4\pi R} \int E_o(z)\ \varepsilon(z)\ e^{-jkz}\ dz \qquad (1)$$

where

R = distance between receiver and diffraction element (distance from obstacle)

$k = \frac{2\pi}{\lambda}$ where λ is the wavelength

$\varepsilon(z)$ = distribution of permittivity above the obstacle

$E_o(z)$ = distribution of field strength above the obstacle

θ = essentially the direction of the diffracted wave under consideration

Thus the secondary field is the Fourier transform of the $E_o(z)\ \varepsilon(z)$ product.

Two limiting cases are of interest:

a) If the field E_o incident on the plane above the obstacle varies far more rapidly with height z than does $\varepsilon(z)$, then $\varepsilon(z)$ does not contribute to the convolution and we can put $\varepsilon(z)$ outside the Fourier integral. We thus have an important relation. The angular distribution of diffracted wave is the Fourier transform of the field strength distribution measured in a plane above the obstacle giving rise to diffraction.

b) Conversely, if $\varepsilon(z)$ varies rapidly in relation to $E_o(z)$, then $E_o(z)$ does not contribute to the convolution and we have an expression for the <u>scattered</u> wave. The angular distribution of the scattered waves is the Fourier transform of the permittivity (refractive index) distribution.

Thus, if we want to calculate the transmission loss resulting from a given obstacle, the procedure is in short the following:

<u>Step 1</u>

Calculate the field strength distribution along a vertical direction over the obstacle, considering the influence of ground reflections, ducting, etc. Note that if the obstacle cannot be considered as a knife edge, then reflection from the obstacle itself may contribute to the field strength distribution over the ridge.

<u>Step 2</u>

Having obtained an expression for $E_o(z)$, we compute the Fourier transform of $E_o(z)$, thus obtaining the angular field strength distribution $E_s(\theta)$ of the diffracted wave. In order to obtain the angular power distribution $P(\theta)$, we shall have to multiply $E_s(\theta)$ with its complex conjugate $E_s^*(\theta)$.

Note that in order for this Fourier transformation to be dimensionally meaningful, the space coordinate will have to be normalized with respect to wavelength. The direction θ is then to be expressed in terms of $\sin\theta$.

Step 3

Having obtained the angular power spectrum $P(\theta)$, we shall essentially have to repeat step number one in order to obtain the desired expression for the power received at a given point in the diffraction zone behind the obstacle.

We shall now consider this procedure in some degree of detail.

LINE OF SIGHT DUCT PROPAGATION, FIELD STRENGTH DISTRIBUTION ABOVE OBSTACLE

In order to obtain simple, comparatively general and in particular physically interpretable expressions for the field strength distribution above the obstacle (first step in the procedure) we shall have to idealize the problem and in doing so make certain approximations.

Inspecting the numerical results of e.g. Früchtenicht, we see that the most striking feature of the vertical field strength distribution is the periodicity. We clearly have a component of field strength which varies in a sinusoidal fashion with height.

From knowledge about the reflection properties of the sea surface under various degrees of roughness, one would expect the influence of multipath on the vertical field strength profile to diminish with height. Hence one would expect the sinusoidal field strength oscillation to be damped in some manner. This too is evident in some of the line of sight ducting results. Lastly, we see that the DC level of the $E_o(z)$ profile is a parameter of some importance.

Our ducting diffraction problem will therefore be based on the following field strength profile:

- a constant term E_1
- a damped sinusoidal term given by $E_2(z) = E_2 e^{-\alpha z \lambda} \sin \frac{2\pi}{L} \lambda z$

Note that the vertical coordinate z is measured in terms of number of wavelengths.

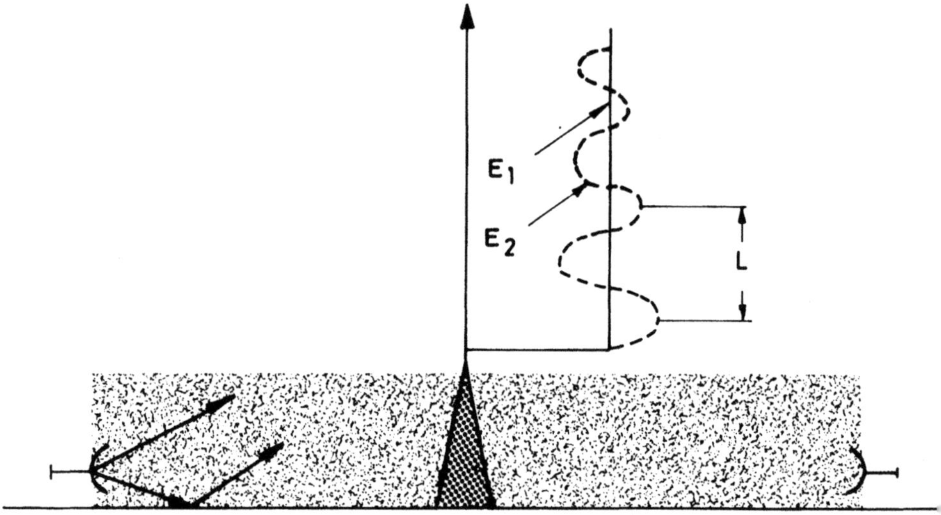

Figure 2 Vertical field strength distribution in line of sight ducting

ANGULAR SPECTRUM OF DIFFRACTED FIELD

The next step in the calculation involves a Fourier transformation of the vertical field strength profile.

a) The first term in the expression for the field is a step function

$$E_1(z) = \begin{cases} 0 & \text{for } z<0 \\ 1 & \text{for } z>0 \end{cases}$$

The Fourier transform of this is the following

$$F_1(\theta) = \tfrac{1}{2}\delta(\theta) + \frac{1}{j\theta} \qquad (2)$$

This assumes small angles θ such that $\theta \approx \sin\theta$. Note that δ is a Dirac delta function.

b) For simplicity we write the second term as follows

$$E_2(z) = E_2(e^{-\alpha z} \sin \beta z) \qquad (3)$$

such that $\beta = \lambda/L$.

This function gives a comparatively simple Fourier transform, namely

$$F_2(\theta) = \frac{\beta}{\alpha^2 + \beta^2 - \theta^2 + 2j\alpha\theta} \tag{4}$$

such that the resultant angular spectrum is given by

$$F(\theta) = F_2(\theta) + F_1(\theta)$$

$$= \frac{\beta E_2}{\alpha^2 + \beta^2 - \theta^2 + 2j\alpha\theta} + \frac{E_1}{j\theta}$$

The angular power spectrum is then obtained from the following

$$P(\theta) = F(\theta) \cdot F^*(\theta)$$

This function has a pronounced maximum for a certain direction, namely the direction corresponding to

$$\theta = \pm \sqrt{\beta^2 + \alpha^2}$$

If we have no damping, i.e. if $\alpha=0$, then we have a Dirac delta function for $F_2(\theta)$ centered at

$$\theta = \beta \quad \text{i.e. at } \theta = \frac{\lambda}{L}$$

This is the same as the results obtained using the Bragg scattering relationship in connection with the scattering from sinusoidal variations with height of the refractive index. From Bragg we have the following.

Maximum scatter in a direction θ given by

$$K = \frac{4\pi}{\lambda} \sin \theta/2$$

where $K=2\pi/L$. As in the diffraction case above, L is the period in space of the sinusoidal variations, i.e.

$$\frac{2\pi}{L} = \frac{4\pi}{\lambda} \theta/2$$

i.e. $\theta=\lambda/L$ (5)

as we obtained directly from equation (4) above.

Disregarding for the moment the radio duct, we know that a single ground reflected component interfering with the direct wave will give a field strength distribution above the obstacle of the form given in equation (3) above. The damping factor α will then be determined by the rate at which the ground reflection coefficient decreases with increasing angle of incidence to the ground. The period of oscillation, L, is given by the geometry. Let R be the distance between transmitter and knife-edge obstacle, h the height of the transmitter above the flat earth and λ the radio wavelength; then the period L is given by

$$L = \frac{\lambda R}{2h} \tag{6}$$

Thus, in this case, we will have maximum energy diffracted in a direction given by

$$\theta = \frac{2h}{R} \tag{7}$$

Thus, by adjusting the height, h, of the transmitting antenna, we can beam the diffracted wave in the desired direction. We see that ground reflection (leading to an oscillating vertical profile of field strength) actually increases the power received at a given point behind the obstacle. This increase is known as obstacle gain. Note that we shall have to transform the coordinate system such that θ is measured relative to the line joining the top of the knife-edge obstacle and the transmitter and not relative to the horizontal line through the obstacle top.

Let us now, as an example, calculate the obstacle gain. As we have seen, the diffracted field, F_1, resulting from the unit step function (the DC term) is given by

$$F_1 = \frac{E_1}{j\theta}$$

The maximum field resulting from the oscillating term (i.e. field in direction θ where $\theta = \sqrt{\alpha^2+\beta^2}$) is, as seen from equation (4), given by

$$F_2 = \frac{E_2\beta}{j2\alpha\sqrt{\alpha^2+\beta^2}} \tag{8}$$

Obstacle gain is then

$$G_o = \left(\frac{F_2+F_1}{F_1}\right)^2 = \left(\left(\frac{E_2}{E_1}\right)\left(\frac{\beta}{2\alpha}\right) + 1\right)^2 \tag{9}$$

Note that this gain refers to the direction $\theta = \sqrt{\alpha^2 + \beta^2} =$

$$= \sqrt{(\frac{\lambda}{L})^2 + \alpha^2}$$

where L is the spatial period of field strength oscillations.

We see from this equation that the diffraction loss relative to that of the no-ducting, no-multipath case can be reduced considerably by proper adjustment of the appropriate parameters, i.e.

- making the field strength oscillations above the obstacle as deep as possible by making the direct wave comparable in magnitude with that of the interfering wave (in equation (9) $E_2 \to E_1$)
- reducing the damping factor, α, by seeking ground reflections from a surface giving a small decrease in reflection coefficient with angle of incidence.

Tables 1 and 2 give some practical examples.

α/β	0.2					
E_2/E_1	1	0.8	0.6	0.4	0.2	0
Obstacle gain G_o db	10.8	9.5	7.9	6	3.5	0

Table 1 Obstacle gain as a function of the depth of the field strength oscillations above the obstacle

E_2/E_1	1				
α/β	0.2	0.4	0.6	0.8	1.0
Obstacle gain G_o db	10.8	7.04	5.26	4.2	3.5

Table 2 Obstacle gain as a function of damping factor α

In the same way as the two tables above give the obstacle gain for various depths of the field strength oscillations and for various damping factors, figures 3 and 4 show the whole spectrum in the diffraction angle range from 0 to some 20 m radians.

We see from figure 3 that the higher the damping factor (for a given period $L = \lambda/\beta$ of the field strength oscillations), the lower is the obstacle gain.

Figure 4 shows the effect of varying the oscillation period, L. We see that changing L results in a change in the direction of maximum diffraction whereas the obstacle gain remains constant since E_2/E_1 and α/β are kept constant.

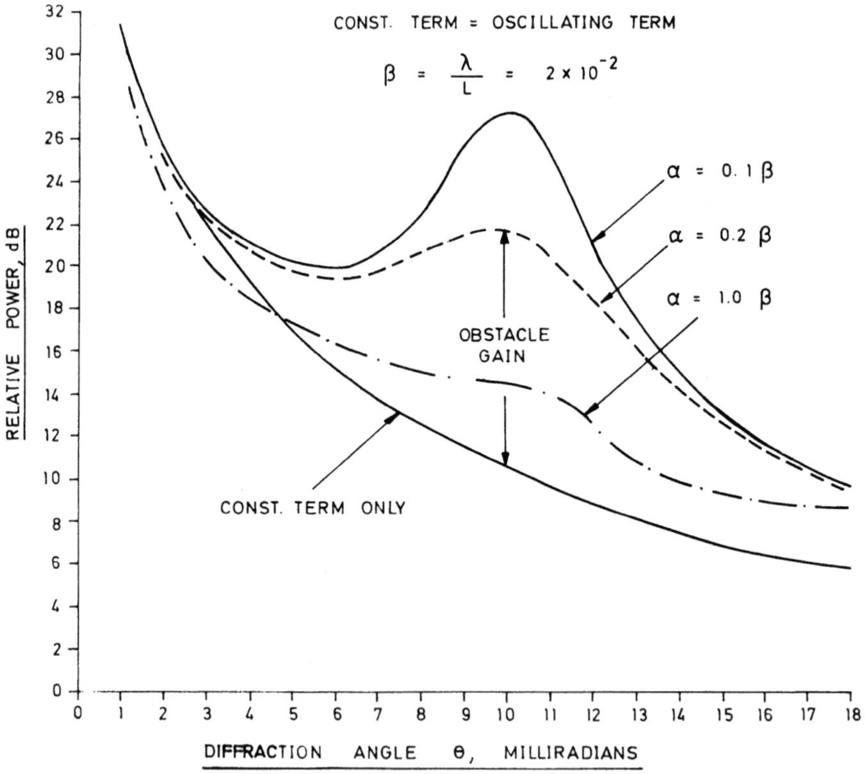

Figure 3 Angular diffraction spectrum for various damping factors α

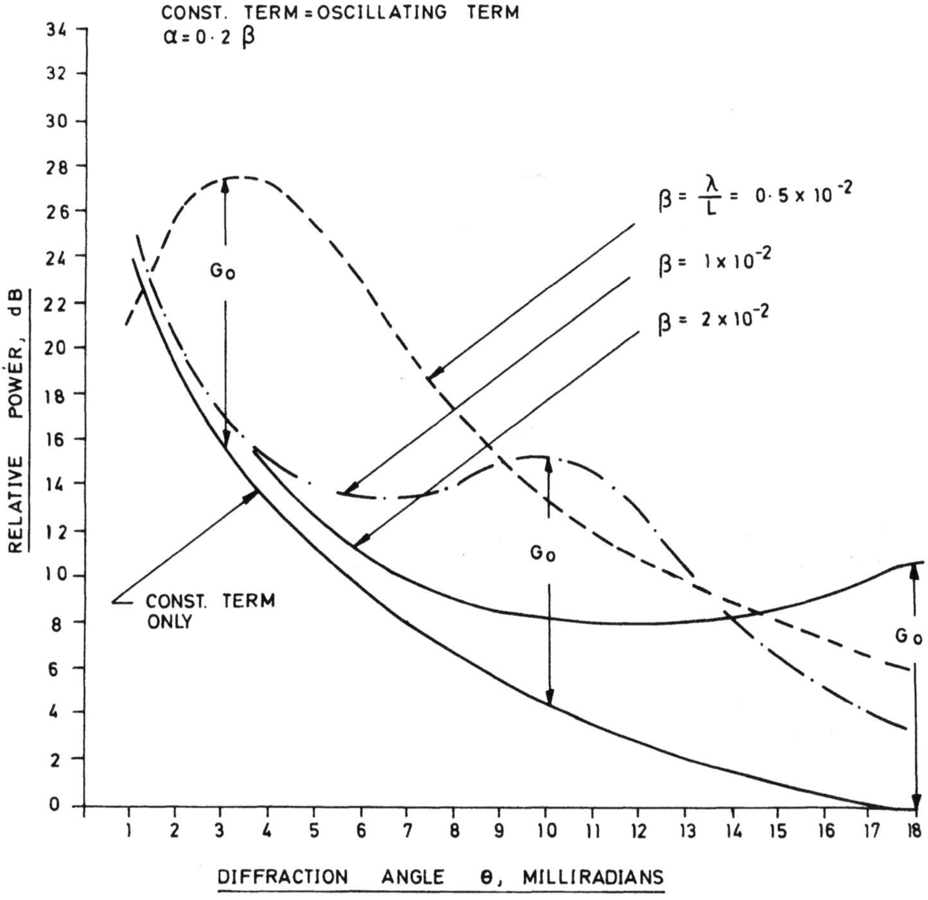

Figure 4 Angular diffraction spectrum for various periods of field strength oscillations

CONCLUSIONS

We have shown that a terrain obstacle within line of sight of a ground-based radio duct gives rise to a well-defined beaming of the diffracted power behind the obstacle. The direction in which the power is beamed is determined by factors such as the height above ground of the transmitting antenna. The path attenuation can be greatly reduced by adjusting the height and the radiation direction of the transmitting antenna.

The theoretical approach on which the numerical results are based is very simple. Knowing the path and obstacle geometry and knowing the properties of the duct and of the reflecting ground, the diffracted power spectrum can readily be calculated.

REFERENCES

Ratcliffe, J.A. - Some Aspects of Diffraction Theory and their Application to the Ionosphere, The Physical Society Reports on Progress in Physics 19, 188 (1956)

Gjessing, D.T. - Atmospheric Structure Deduced from Forward-Scatter Wave Propagation Experiments, Radio Science 4, 12, 1195-1210 (1969)

APPENDIX A

Huygen's Principle for Diffraction Problems

Huygen's principle is a well-known method for treating the propagation of waves. We shall now in a brief and semi-intuitive way consider its mathematical background, showing its close connection with Green's theorem.

The method is this: From each point of a given wavefront, at t=0, we assume that spherical wavelets start out. At the time t, each wavelet will have a radius ct and the envelope of these wavelets will form a new surface which, according to Huygen is simply the resulting wavefront at this later time t. Thus, if the original wave was plane, it is easy to see that the final one will be a plane distant by an amount ct, while if it is a sphere, the final wavefront will be a concentric sphere whose radius is greater by an amount ct. In either case this construction gives us the correct answer agreeing with the more usual methods of computation.

We may look at our process in a slightly different way, not used by Huygen but developed later when the interference of light was being worked out.

Suppose that, instead of taking the envelope of all the spherical wavelets, we consider that each of these wavelets has a certain amplitude of sinusoidal oscillations. We add these oscillation waveforms just as if our wave were being sent out by interfering sources of radio waves, and the resulting amplitude is taken to be that of the actual wave. This process can be shown to lead to essentially the same result, and it is this that can be justified theoretically. As a further generalization it is not necessary to take the original surface to be a wavefront; it can be any surface so long as we allow the scattered wavelets to have the suitable phase and amplitude.

Our final result, then, is this:

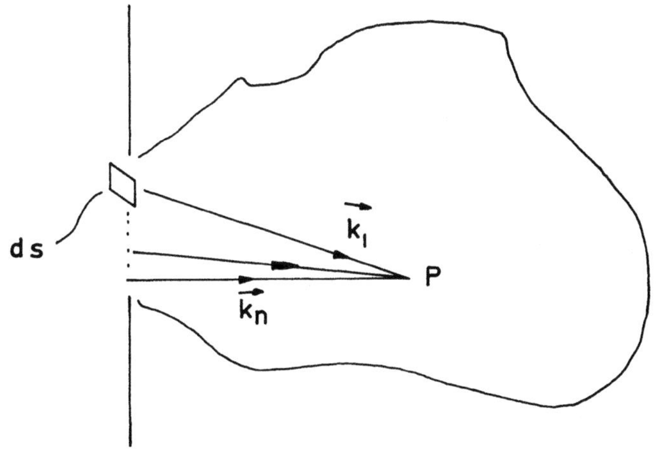

Figure A.1 Diffraction geometry

The field at P, due to the waves transmitted from the slot, is obtained by taking an arbitrary surface round P through the slot and integrating over this surface. The contribution to the field at P of a small element of area dS on this surface equals the amplitude of P of a spherical wave starting from dS at such a time so that it reaches P at a time t. That is, if the distance from dS to P is denoted by r, this wave is of the form

$$\frac{f(t - \frac{r}{c})}{r} \tag{A.1}$$

Now, the contribution from a given wavelet must surely be proportional to the field at dS, which we may call $f(\vec{r},t)$ and to dS. Hence for the field we have something like

$$f(\text{at P}) = \text{constant} \times \iint \frac{f(t - r/s)}{r}\, dS \tag{A.2}$$

where the surface integral is over the surface surrounding P.

This suggests the solution of Laplace's equation by Green's method, where we have the value of a function ϕ at an interior point of a region where $\nabla^2\phi$ is zero as a surface integral over the boundary.

As a matter of fact, an analogy to Green's theorem is the correct statement of Huygen's principle, and this replaces the formula which we have derived intuitively above, and which is not quite correct.

Now, performing a rigorous integration over the surface, setting the integral equal to zero except at the opening of the screen, we finally obtain an expression for the diffraction field of the form

$$F \sim \iint e^{-j \frac{2\pi}{\lambda}(r_1+r_2)} \, dS \qquad (A.3)$$

APPENDIX B

Summary of the Discussion

P.A. Mathews, University College, London

The validity of the assumption of knife-edge diffraction is not often justified in practice. Are the obstacles of interest knife-edge obstacles at the frequency of interest?

D.T. Gjessing

A purpose of my talk has been to demonstrate "obstacle gain" in relation to ducted propagation. I have shown that provided you can get a field strength profile above the obstacle which exhibits an oscillatory behaviour, you will experience an obstacle gain, i.e. field strength enhancement, relative to that experienced with no field strength oscillations above the obstacle (i.e. no duct).

For the purpose of simplifying the mathematics I referred the calculations to the simple knife-edge case. I could equally well have chosen a cylindrical or any other obstacle as the basis. In that case, however, you would have to include one more geometrical parameter, namely the diameter of the cylinder. This makes the calculations slightly more cumbersome. I am sure you are right, though, in questioning the degree to which knife edges are typical.

T. Kester, NATO Scientific Affairs Division

My question is about the truncation of the illumination function above the obstacle. Has this effect also been taken into account and what are the consequences? Can the zeros of the sinc-function ($\frac{\sin x}{x}$) be harmful for the geometry considered?

D.T. Gjessing

I have actually a truncated sinusoidal illumination function above the obstacle. I have, however, truncated the sinusoidal oscillations by an $e^{-\alpha z}$ function. In the damping factor α (or, if you wish, truncation factor α) I have included factors such as the illumination pattern of the transmitting antenna. If you truncate the damped sinusoidal function with a step function, you will have another convolution integral. The degree to which this truncation is important depends on the "width" of the step function relative to the width of the damped sinusoidal function.

R. Bolgiano Jr, Cornell University

It was not clear to me in just what way your results (as presented in your last diagrams) were affected by the presence of a duct. Was I misinterpreting your analysis?

D.T. Gjessing

In the beginning of my presentation I discussed the influence of the duct and of ground reflections on the characteristic properties of the illumination function above the obstacle. I stated, with reference to the presentations of Früchtenicht and Rotherham in particular, that the duct affects the damping factor α, the depth of the field strength oscillations E_2/E_1. The duct, apparently, has little or no effect on the period of oscillations β. Hence, in order to form an estimate of the influence of the duct, you should look at table 1 and figure 3.

H. Kruspe, Institute for Radiometeorology, Hamburg University

Can ocean waves form obstacles of the type considered in your presentation?

D.T. Gjessing

I think you could get obstacles of the type discussed from ocean waves. I doubt, however, that you could get the well-behaved field strength profile above the obstacle necessary to give a beamed diffraction pattern. The ground reflections surely would be most irregular, and I have difficulty in visualizing a horizontal duct, waveguide, under such sea conditions.

H.W. Früchtenicht, Institute for Radiometeorology, Hamburg University

What is the change in your results if the depth and the fine-structure of the interference pattern above the obstacle first decreases with height and then increases again?

D.T. Gjessing

If the depth of the "modulation" first decreases (e.g. as in the example according to the $e^{-\alpha z}$ law) and then increases again in exactly the same manner, you will get twice as much power diffracted. Otherwise the power versus diffraction angle remains as in the case discussed in the presentation. If, however, the period of the damped oscillations changes with height, this will have the effect of changing the direction of maximum diffracted power.

INTERFERENCE LOBES PATTERN CONNECTED TO A GROUND BASED RADAR DUCT

A. Nania and A. Pucciano

Italian Meteorological Service

Let us consider an antenna completely embedded in a radar duct (Fig.I). The field strength at point T is determined by the direct ray from the antenna and the ray reflected by the radar duct. This is of course just a scheme in order to analyse the problem. In effect we should speak of refraction rather than reflection mechanism.

Anyway according to Fig. I, for simplicity sake, the echo signal reradiated by point T arrives back to the radar via the same two paths and the received echo is composed of two signals travelling separate path. The magnitude of the resultant signals will depend upon the amplitude and relative phase difference between

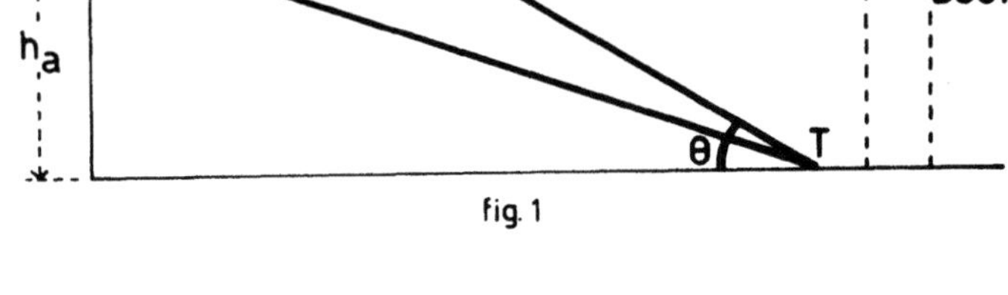

fig. 1

the direct and the duct reflected signal. At point T the modification of the field strength, caused by the presence of the duct, may be expressed by the ratio

$$\frac{\text{Field strength in presence of duct}}{\text{Field strength in free space}}$$

It is reasonable to assume that direct and reflected paths are approximately equal and that the amplitudes of the two signals are almost the same if there is no loss due to the reflection for the presence of the duct. If the two signals have different amplitudes this must be due to different influences of the duct on the direct and reflected rays. The difference in phase between the direct and reflected ray is due to the difference in the path length and the change in phase due to the reflection (phase shift: 180°). Following the geometry of figure I where A' is the mirroed image of A (antenna), the difference between the reflected and direct path is given by:

$$\ell = 2h_c \sin \theta \qquad (1)$$

where h_c is the distance of the antenna from the duct top and θ the angle by which the antenna image A' is seen from point T. Since $\sin \theta \simeq \tan \theta$ at small angles, we may write:

$$\ell = 2h_c \, (2h_c + h_a) \,/\, R \qquad (2)$$

where h_a is the antenna height from the ground and R is the range from the antenna to point T. The phase difference ψ corresponding to ℓ is then:

$$\psi = 2\pi\ell/\lambda \;=\; 2\pi/\lambda \bullet 2h_c \, (2h_c + h_a) \,/\, R \qquad (3)$$

or

$$\psi = 4\pi/\lambda R \;\bullet\; h_c \, (2h_c + h_a) \qquad (4)$$

where λ is the wave length. Finally we may say that two signals having a unity amplitude but a phase difference $\psi + \pi$ arrives at point T. The magnitude of the resulting signal is given by:

magnitude resulting signal =

$$= [1+1 + 2\cos(\psi+\pi)]^{\frac{1}{2}} = [2 \bullet 2\sin^2 \frac{\psi}{2}]^{\frac{1}{2}} = 2\sin \frac{\psi}{2} \qquad (5)$$

or, substituing the expression for ψ

magnitude resulting signal =

$$= 2\sin\left[\frac{1}{2}\frac{4\pi}{\lambda R} h_c(2h_c+h_a)\right] = 2\sin\left[\frac{2\pi}{\lambda R} h_c(2h_c+h_a)\right] \qquad (6)$$

Resulting signal is maximum when

$$2\pi h_c/\lambda \bullet (2h_c+h_a)/R = (2n+1)\frac{\pi}{2} \qquad n = 0,1,2\ldots \qquad (7)$$

or:

$$(2h_c+h_a)/R = (2n+1) \bullet \lambda/4h_c \qquad (8)$$

Resulting signal is minimum when:

$$\frac{2\pi}{\lambda R} h_c(2h_c+h_a) = n\pi \qquad n = 0,1,2,\ldots \qquad (9)$$

or

$$(2h_c+h_a)/R = n \bullet \lambda/2h_c \qquad (10)$$

As a result of the proposed mechanism, if enough energy hits point T, backscattering can send to the antenna echo signals stronger than the minimum detectable signal of the receiver and a ground (or sea) clutter appears on the scope. Figure 2 describes the structure of the interference lobes resulting from the proposed

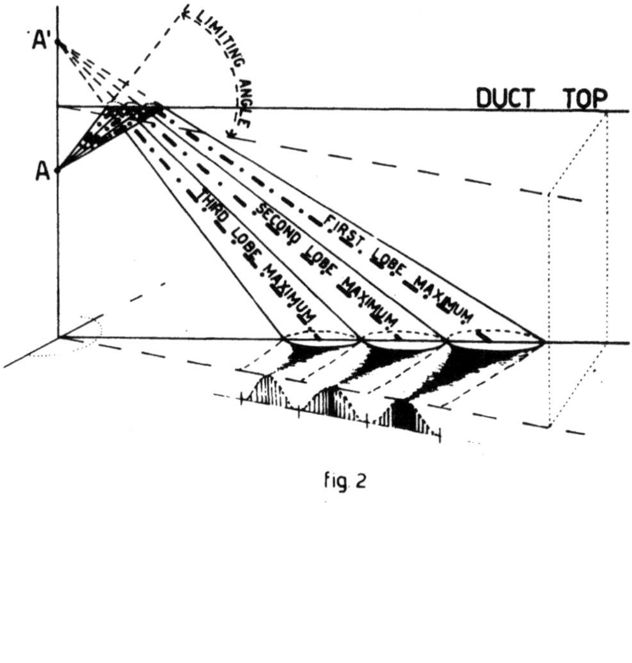

fig. 2

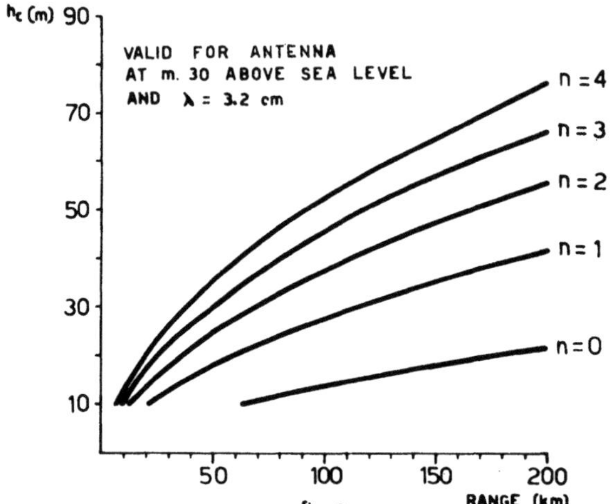

fig. 3

mechanism. Thus the presence of the reflecting duct causes the continuous elevation coverage to break up into a lobed structure as indicated in Fig. 2. It is easily understood that the greater the order of the interference lobe the smaller the amount of radar energy inside it. In other words the radar energy is mostly concentrated inside the first (n=0) lobe rather than in any other lobe.

Fig. 3 represents, depending on h_c, the range from the antenna site of points where maxima of radar energy hit the ground or sea surface. It must be pointed out that h_c is the height relative to the antenna at the points where the radar ray is parallel to the surface and it is then tightly connected to the limiting angle of the radar duct (largest zenithal angle of the ray still trapped by a duct). If the radar antenna is operating with such a tilt angle that its superior half-power ray (-3 dB relative to the main lobe axes) is emitted with an angle smaller than the limiting angle of the duct, then the most of the radar energy will undergo the interference mechanism and it is concentrated into the first lobe; viceversa, if the limiting angle is smaller, only a fraction of the radar energy will undergo interference mechanism.

It is well known that the radar energy per unit volume is proportional to the square of the field strength at the target; for reciprocity, the received power at the radar must be proportional to the fourth power of the field strength. In the free space the received power is:

$$P_r = P_t G^2 \lambda^2 \sigma / (4\pi)^3 R^4 \tag{11}$$

where P_r is the received power, P_t the transmitted power, G the antenna gain, σ the radar cross-section of the target. When interference mechanism takes place the previous equation must be modified including the resulting field strength given by (6). If (P_r) is the received power, modified for the presence of the radar duct

$$(P_r) = P_t G^2 \lambda^2 \sigma / (4\pi)^3 R^4 \bullet 16 \sin^4 [2\pi h_c / \lambda R \bullet (2h_c + h_a)] \qquad (12)$$

and then $(P_r)/P_r = 16$ as maximum value. Which means that the received signal, in the case of a radar duct, is several dB stronger than in case of free space. This should be proved experimentaly.

It must, be pointed out, again, that the proposed mechanism is just a rough simplification. As a matter of a fact refraction is the main phenomenon. The required convergence of radar rays (such as at point T) might be determinated, if refraction is considered, by different path lengths, in other words, the vertical gradients of atmosphere refractive index may have such a distribution to determine a focusing of a fraction of the radar beam at target points on the ground or sea surface. When such distribution is established the refraction process may be more easily assimilated to an equivalent reflection mechanism. From the area where the focusing takes place, radar energy is backscattered to the antenna and echoes are displayed on the PPI scope as extensive mantle echoes whose range and width should be in a reasonable agreement with the theoretic treatment of interference lobe pattern mechanism above discussed. Obviously, since the above proposed mechanism has been discussed only from a geometric point of view, no answer can be given about the presence and nature of the scatterers within the volume of air directly in contact with the sea surface illuminated by the radar beam, neither it is possible to infer conclusion about sea surface roughness and radar polarization. Temperature of sea water and marine currents are planned to be inestigated in our future research programs since it is believed that their influence is of paramount importance.

The measurements of range and width of mantle echoes (if present) could be highly influenced by weather echoes (clouds, rains) existing within the radar range. As said before, at times mantle echoes occur (Fig. 4) when no visible cloudiness is present as confirmed by reports from ground weather stations, pilot reports and satellite pictures. For such situations the interference lobe mechanism can furnish suitable esplanation for those mysterious echoes that appear on the radar scope since range and width measurements of mantle echoes during such situations seem to be in a reasonable agreement with expected results according the geometry of Fig. I. Many documentations of radar display have been collected during a large number of proper meteorological situations.

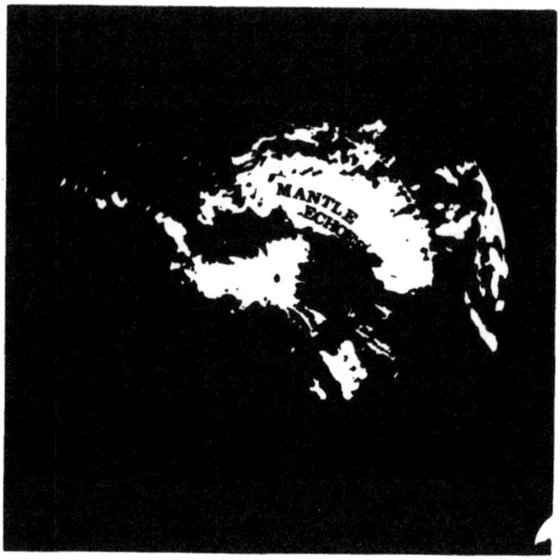

METEOR 50 BRINDISI 20·4·68 - 1917 TMG
fig.4

Radar coverages relative to a situation from August 9 to II, I972 have been plotted on the basis of vertical gradient of refractive index computed by radiosonde data, twice a day in Brindisi.

Figures 5, 6 show the focusing effect of the radar duct at distances from 20 to 75 miles from the antenna site. It is interesting to note that, during most of the period in consideration, a rather strong anti-duct was found at low levels (see, for example, Fig. 6 at levels between 60 and II0 m).

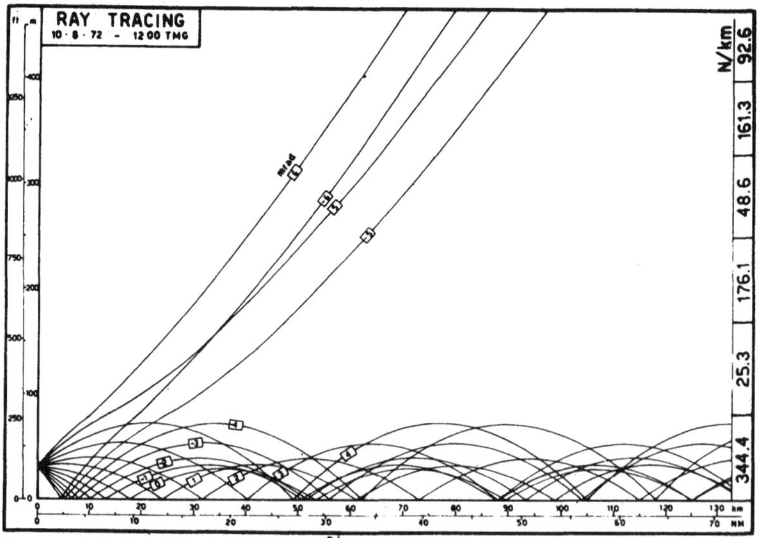

fig 5

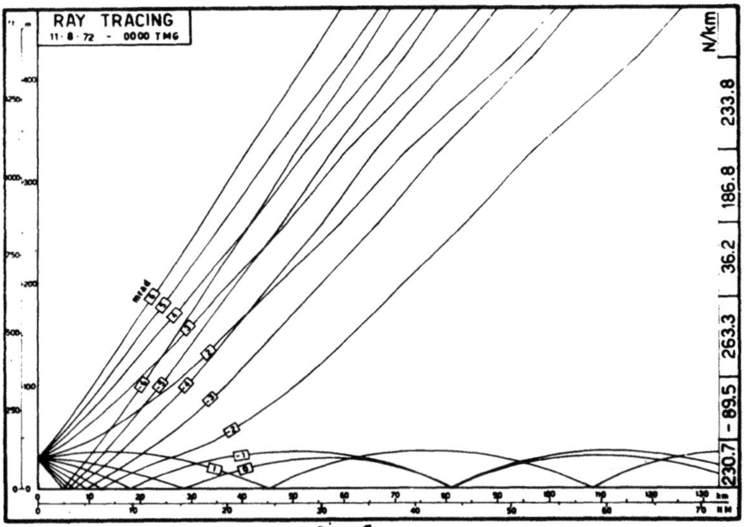

Fig 6

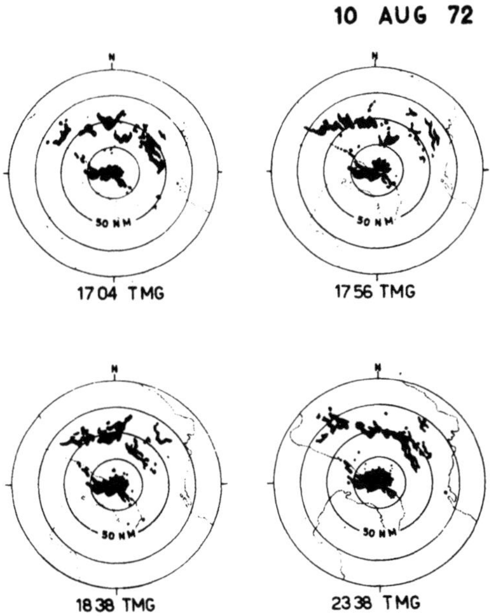

Fig.7 -(From the time lapse movie projected in Sorrento).

Figure 7 shows sketches of the mantle echoes displayed on the PPI. Time lapse movie will show in more details the suggestive evolutions of the phenomenon, with severe superrefraction and mantle echoes over the sea. Micro analysis of mantle echoes show often a wavy structure. Surface wind seems to be responsable of short period fluctuations and propagations of large and small scale organization of the mantle.

NOTE

Characteristics of radar used: SELENIA METEOR 50

Peak power	$>35.10^{3}$ W
Wave length	3.2 cm
Vertical beam width	0.65 deg.
I Pulse length	3 μsec
Pulse repetition frequency	200 p/sec
II Pulse length	0.5 μsec
Pulse repetition frequency	1200 p/sec

THE OCCURRENCE OF VERY HIGH FIELD STRENGTHS AT BEYOND THE HORIZON PROPAGATION OVER SEA IN THE FREQUENCY RANGE 60 - 5000 MHZ

Sture Wickerts and Lennart Nilsson

National Defence Research Institute
S-104 50 STOCKHOLM 80, Sweden.

SUMMARY

High field strength well, beyond the radio horizon at frequencies between 60 MHz and 5000 MHz can exist through the action of sporadic E-layers in the ionosphere in the low frequency end and through scatter, duct or reflection propagation in the troposphere in the entire frequency range studied. These mechanisms are sources of serious interference problems in connection with frequency sharing. The ionospheric propagation will not be treated in this paper.

The occurrence of enhanced fields has been studied over a sea path of 160 km at the frequencies 170 MHz, 460 MHz and 5000 MHz. For short periods also 60 MHz was used. Long-term measurements of field strength have been performed since 1968, which includes some period of meteorological measurement. For these measurements an airborne refractometer, captive balloon sondes and an acoustic sounder (SODAR) were used.

The received signals can be classified by the following three types: Standard atmosphere signal (tropospheric scatter signal), unstable high signal caused by reflection in elevated tropospheric layers and stable high signal caused by ducting. The fact that the signals at different frequencies are only occasionally correlated indicates that the propagation is governed by different mechanisms in different parts of the frequency spectrum.

A case study of propagation under various tropospheric conditions and the long-term statistics of the different signal types are given in this paper.

1. INTRODUCTION

Long-term measurements of field strength have been performed over a 160 km path entirely over sea in the Baltic area. For short periods the radio measurements have been combined with meteorological measurements in order to describe the structure of the refractive index field. The heigth of layers was continously determined by remote sensing techniques and their thickness and intensity by airborne and balloonborne instruments. From these data, field strengths have been calculated for scatter, reflection and duct propagation.

It will be shown that reflection propagation dominates at the low frequency end. Whilst at higher frequencies, very sharp layers are required for reflection propagation. At these frequencies, high signals are generally caused by duct propagation.

1.1 Instruments and plan for the experiments.

The radio measurements were performed with the following equipment.

5000 MHz: Transmitted power 500 kW (pulsed). Reflector antennas. Free space field strength over the path + 15 dBm.

460 MHz: Transmitted power 10 W (CW) 8 element Yagi antennas. Free space field strength over the path - 65 dBm.

170 MHz: Transmitted power 10 W(CW). 8 element Yagi antennas. Free space field strength over the path - 60 dBm.

The antenna heights for the transmitters and receivers were 100 m above sea level.

The field strengths were recorded on ink recorders and simultaneously sampled, digitized and evaluated on line in a minicomputer.

For the meteorological measurements, the following equipment were used:

Airborne microwave refractometer and thermistor, flown up to 2000 m.

Captive balloon borne wet and dry thermistors and aneroid, measuring heights below 800 m.

Acoustic sounder (SODAR), sound pulses (2000 Hz) vertically pointing horne antenna, measuring range 700 m.

Figure 1 shows a plan of the complete program and includes both radio and meteorological measurements.

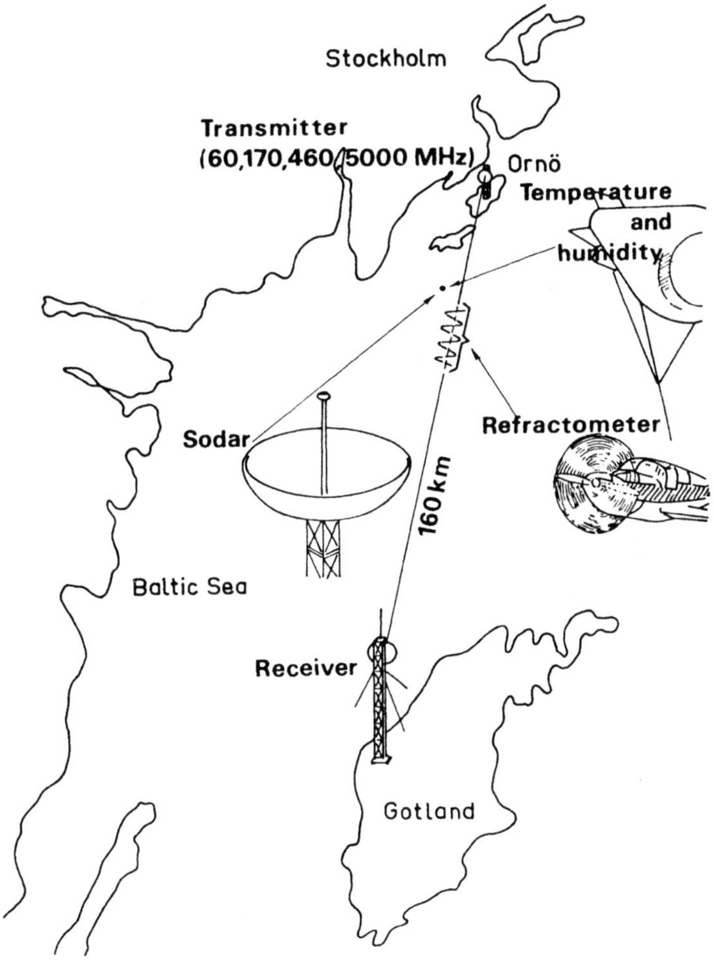

Figure 1. Plan of experiments.

1.2 Signal types

At the evaluation of the paper recordings the received signals were divided into three different types of signal. Signal type A is characterized by a low mean signal level with rapid superimposed fading. Compared with signal type A, signal type B has a higher mean level. In this case the fading is less rapid and has deeper minima than the signal type A. Signal type B is assumed to be caused mainly by reflection propagation. Signal type C, is a stable and very high signal which sometimes has deep fading minima of short duration. Examples of the three signal types at 5000 MHz and 170 MHz are shown in figures 2 and 3, respectively.

1.3 Calculation of received power

The received power P_R, at frequencies between 60 MHz and 5000 MHz far beyond the radio horizon is caused by scattering P_s, ducting P_d or reflection P_r. The calculation of P_s, in the case of narrow beam antennas, has been done according to Eklund and Wickerts (1968) from the following expression.

$$P_s = P_{FS} \cdot 0.76 \cdot 10^{28} \cdot C_n^2 \cdot \left(\frac{dM}{dz}\right)^{-14/3} \cdot x^{-11/3} \cdot \lambda^{-1/3} \cdot \alpha^3 \tag{1}$$

Where

x = distance in metres

P_{FS} = received power at free space propagation

C_n^2 = intensity parameter in the structure function

$\frac{dM}{dz}$ = refractive index gradient

λ = radio wavelength in metres

α = beam width of the antennas in radians

M is defined as

$M = (n-1) \cdot 10^6 + 0.157 \cdot z \qquad N = (n-1) \cdot 10^6$

where

n = refractive index

z = height

With broad beam antennas, like the Yagi antenna, it is only the lowest part of the common volume, which contributes to the scattered signal. In such cases the scattered signal can, according to Gjessing (1969), be calculated from equation 1 with an effective beam width defined as:

$$\alpha_e = \frac{1}{2} (x/a) (10^{1/k} - 1) \tag{2}$$

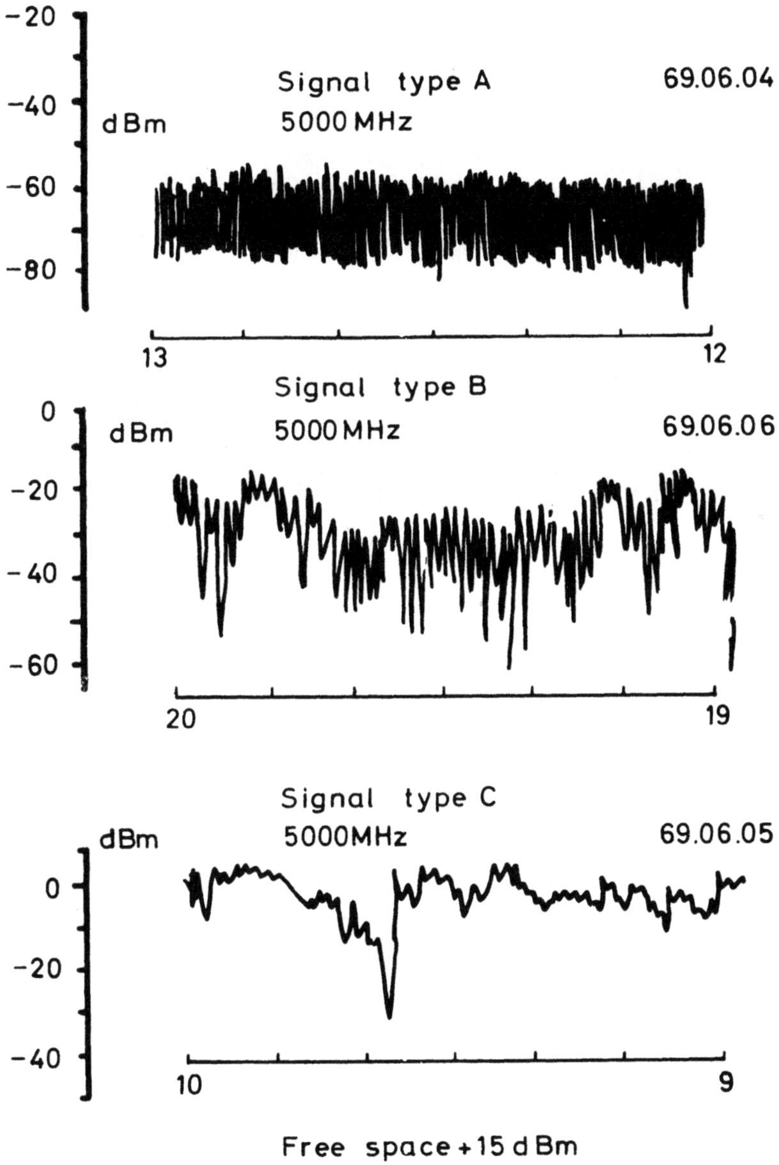

Figure 2. Signal types at 5000 MHz.

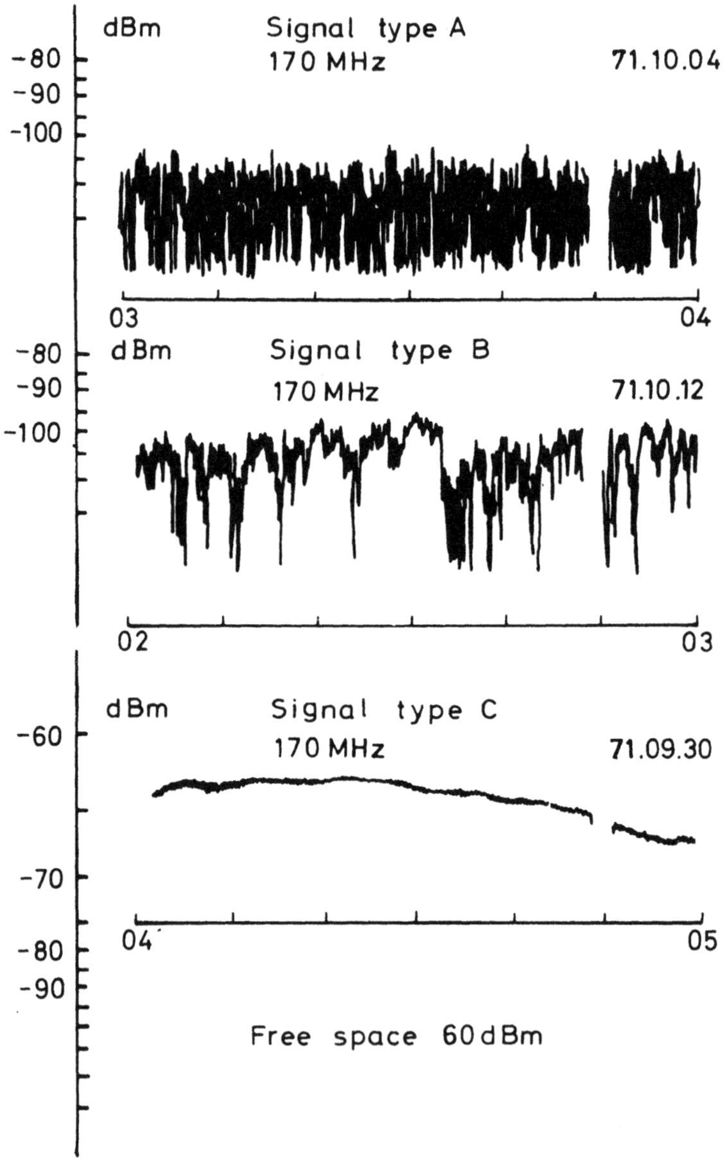

Figure 3. Signal types at 170 MHz.

where

a = radius of the earth

k = constant (= 11/3)

Duct propagation occurs when the antennas are situated in a duct layer or very close to it. The maximum radio wavelength λ_{max} which can be trapped in a duct layer has been estimated according to the following formula (Eklund, 1969)

$$\lambda_{max} \approx \frac{d\sqrt{\Delta M}}{530} \quad (3)$$

where

d = thickness of the duct layer in metres

ΔM = refractive index change through the duct layer (M-units)

λ_{max} = maximum trapped wave length in metres

In the case of reflection it is assumed that the active layer is horizontal and large in comparison with the beam dimensions. The expression used for calculating P_r is (Eklund et al op cit)

$$P_r = P_{FS} \frac{\Delta n^2}{4\Psi^4 \left[1+(\pi\frac{B\Psi}{\lambda})^2\right]^2} \quad (4)$$

where

Δn = total change of refractive index through the layer (n-units)

Ψ = the grazing angle in radians

B = thickness of the layer in metres

The ratio between received signal power and the received due to free space propagation defines the propagation factor F according to

$$F = 10 \log \frac{P_R}{P_{FS}}$$

The propagation factor for reflection propagation F_r is represented in figure 4 as a function of B and λ for $\Delta n = 10^{-5}$ and Ψ = 10 mrad. It can be seen from formula 4 that a change of Δn by a factor 2 will change the propagation factor by 6 dB.

Recorded data for the two days, Sept 6 - 7, 1972, which includes a complete set of both radio and meteorological data are available. The propagation factor for scattering (F_s) is calculated according to formulas 1 and 2 with the following values of the meteorological parameters.

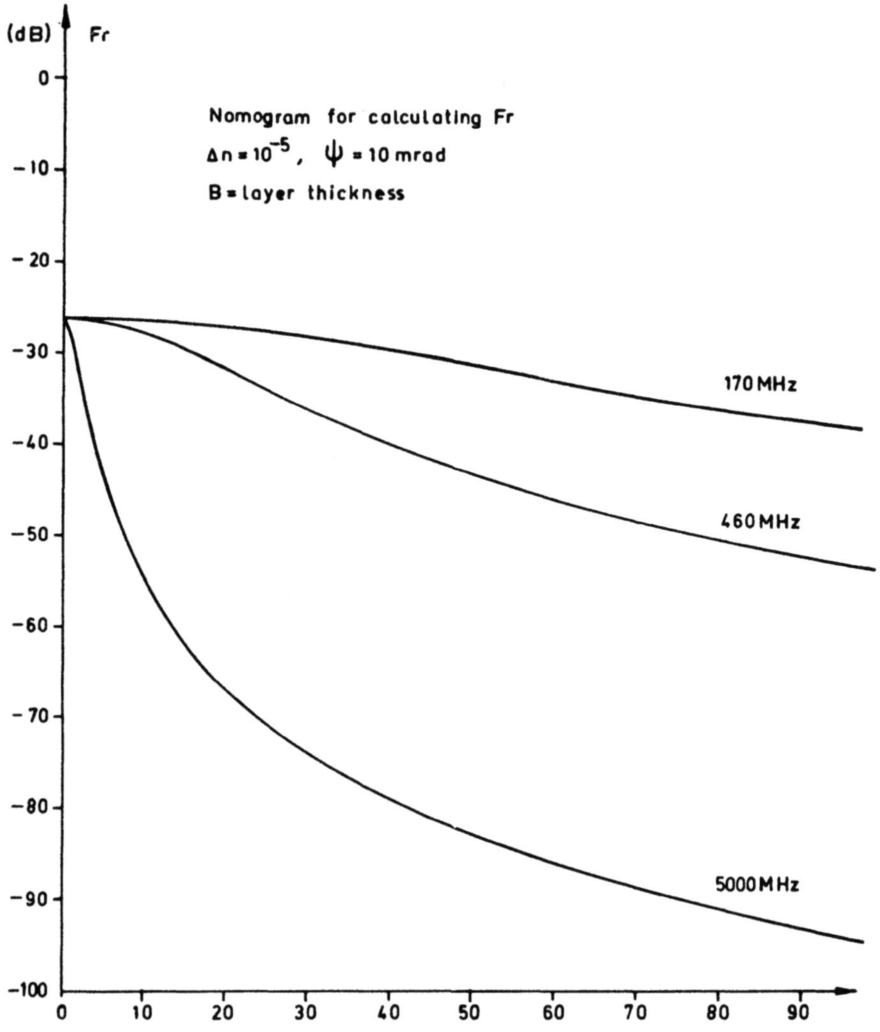

Figure 4. Nomogram for calculation of F_r.

$$C_n^2 = 10^{-16}$$

$$\frac{dM}{dz} = 0.118$$

The results are given in Table 1.

Table 1.

Frequency	α	F_s	P_s
170 MHz	110 mrad	-59 dB	-119 dBm
460	110	-57	-122
5000	40	-66	- 51

Between 2 pm and 11 pm on Sept 6 a duct layer was formed from sea level up to about 100 m. The generation and the dissipation of this duct layer is shown in the recordings from the SODAR measurements in figure 5. Direct measurement by airborne refractometer shows that the decrease of refractive index through the duct layer was 24 N-units as is seen in figure 6. Balloon borne temperature measurements made before, during and after the occurrence of the low duct layer are represented in figure 7. The heights of temperature inversions are strongly correlated to the heights of duct layers (Wickerts, 1970). From equation 3 it is found that the maximum radio wavelength which can be trapped in this duct is very close to 0.65 m (460 MHz).

A layer with a large refractive index gradient occured more or less all the time during the period at 400 - 600 m. It was found from refractometer measurement that the height of the duct was 500 m at 1030 pm on Sept 6, its thickness 50 m, and the total change of refractive index $1.5 \cdot 10^{-5}$ n-units, which is shown in figure 8. At 0245 pm Sept 7 the layer sharpened to a thickness of 10 m, and refractive index decrease of $2 \cdot 10^{-5}$ n-units at the height of 600 m, figure 9. A reflecting layer at 400-600 m gives a grazing angle of about 10 mrad. From figure 4 it can be seen that the weaker layer on Sept 6 gave reflected signals much higher than the scattered signals mainly at 170 MHz. The stronger layer on Sept 7 gave a higher reflected signal than the scattered at all three frequencies (5000, 460 and 170 MHz).

1.4 Comparison between observed and calculated received signals

Hourly median received signal powers at the three frequencies for the period Sept 6 - 7 are shown in figure 10. It can be seen that the 170 MHz signal is relatively higher than the other two between midnight Sept 6 and 2 pm Sept 6.

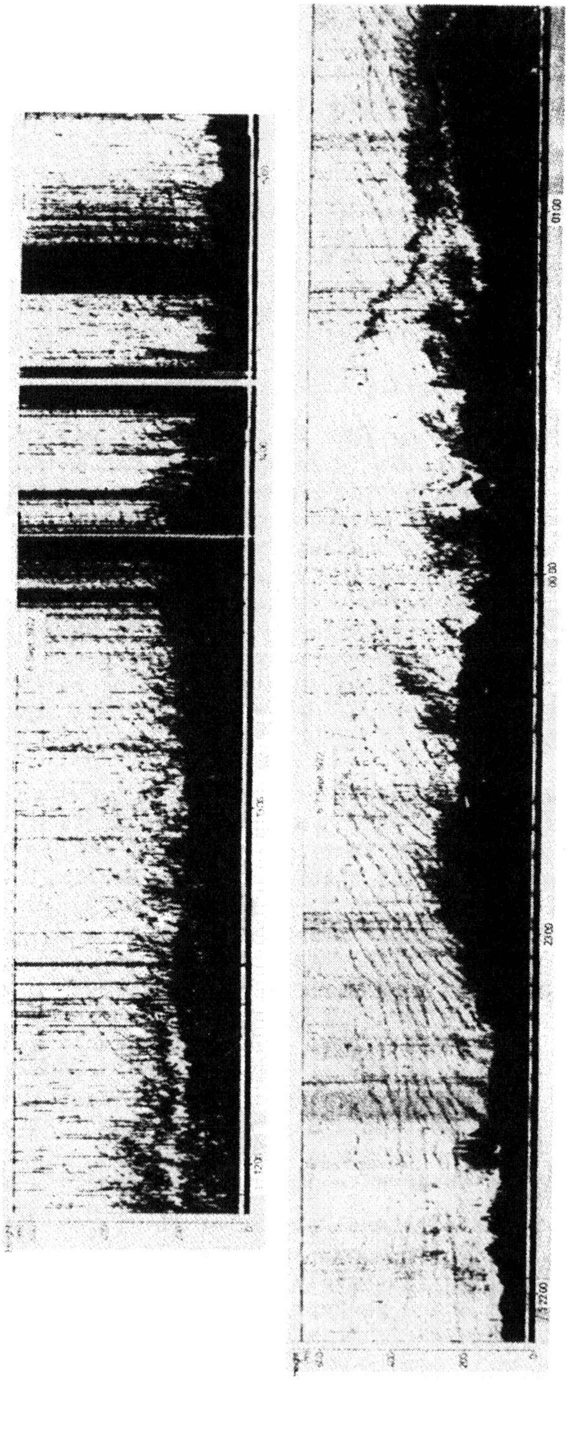

Figure 5. SODAR registration. The disturbances between 1340 and 1450 Sept 6 are caused by windgenerated noise in the wire to the captive balloon and by the refractometer airplane.

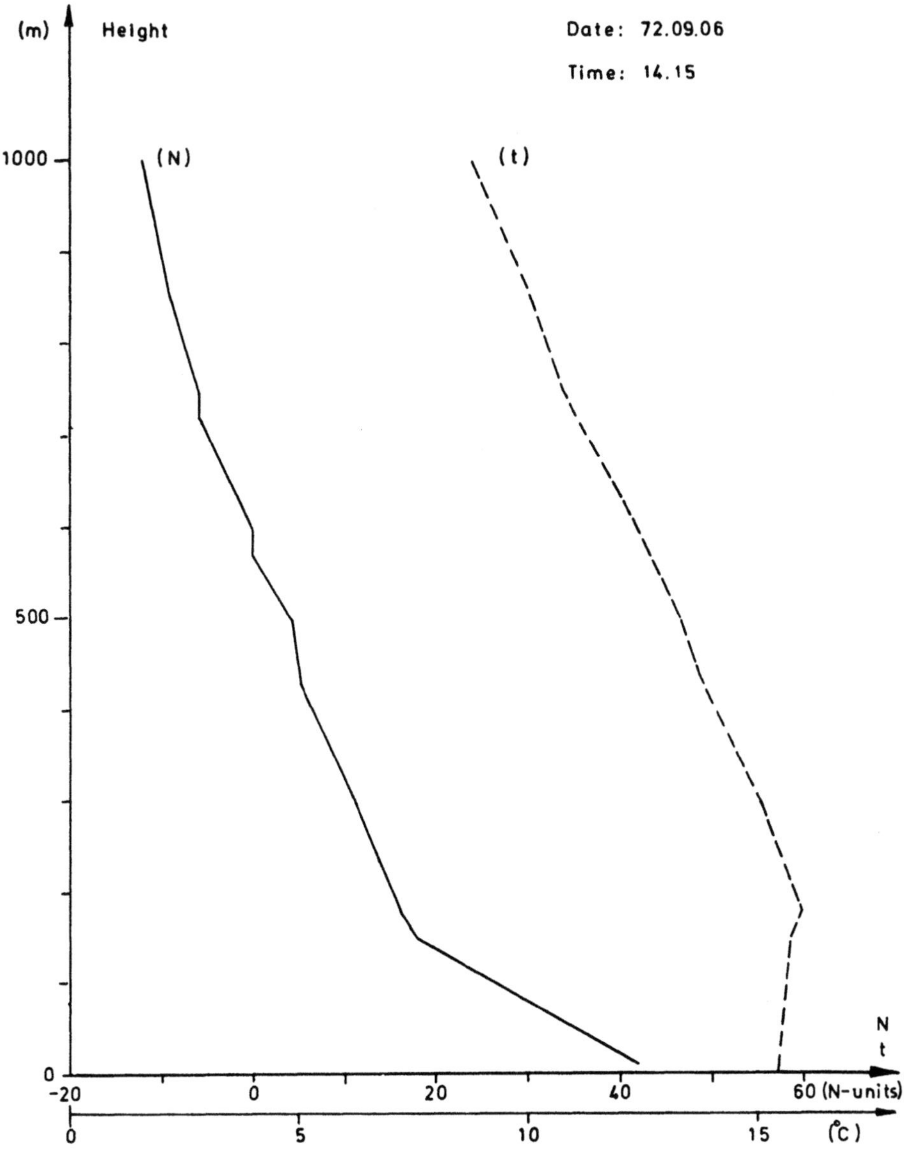

Figure 6. Refractive index profile.

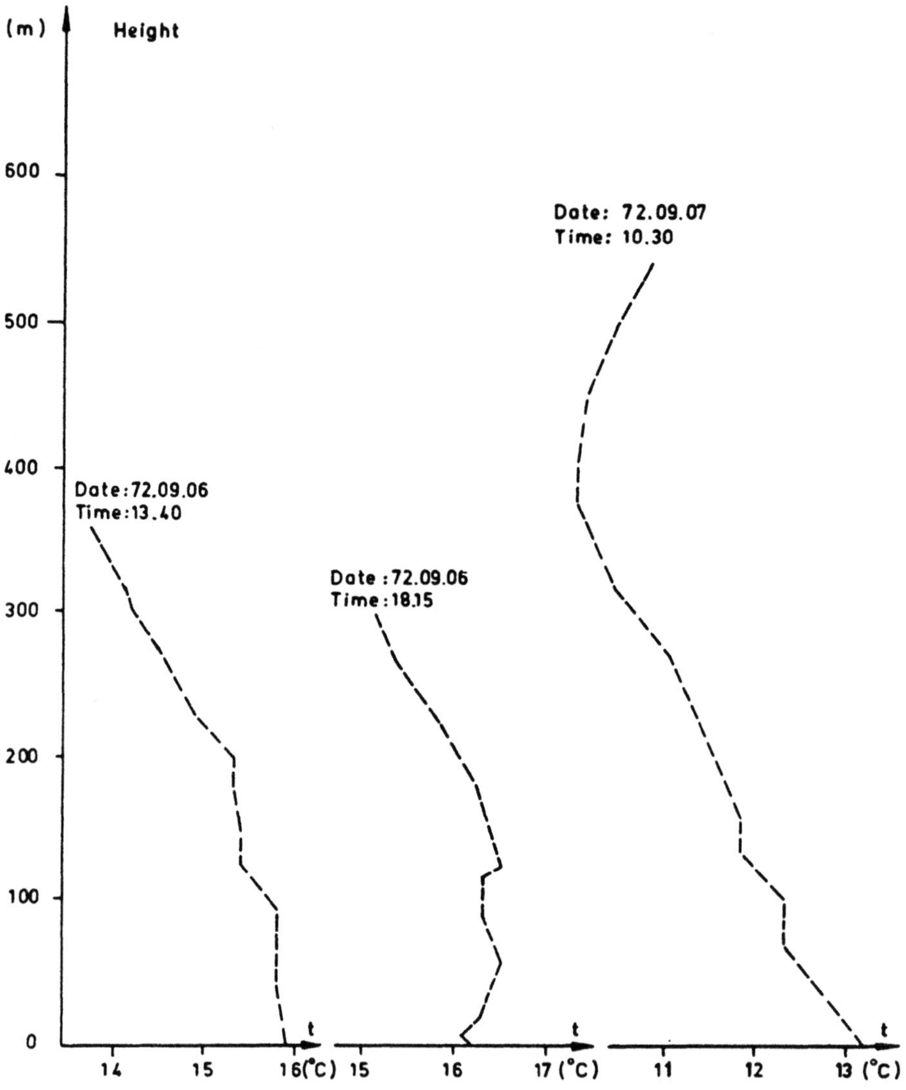

Figure 7. Temperature profiles.

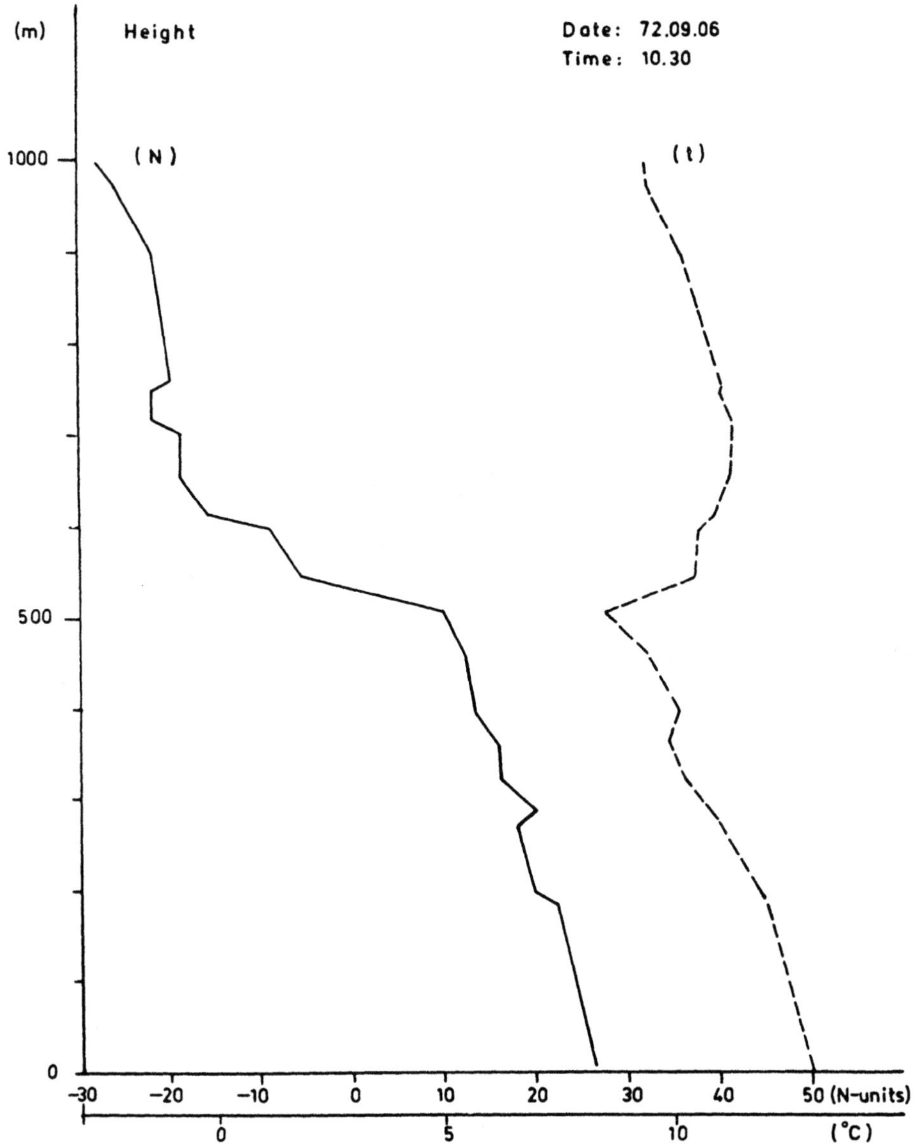

Figure 8. Refractive index profile.

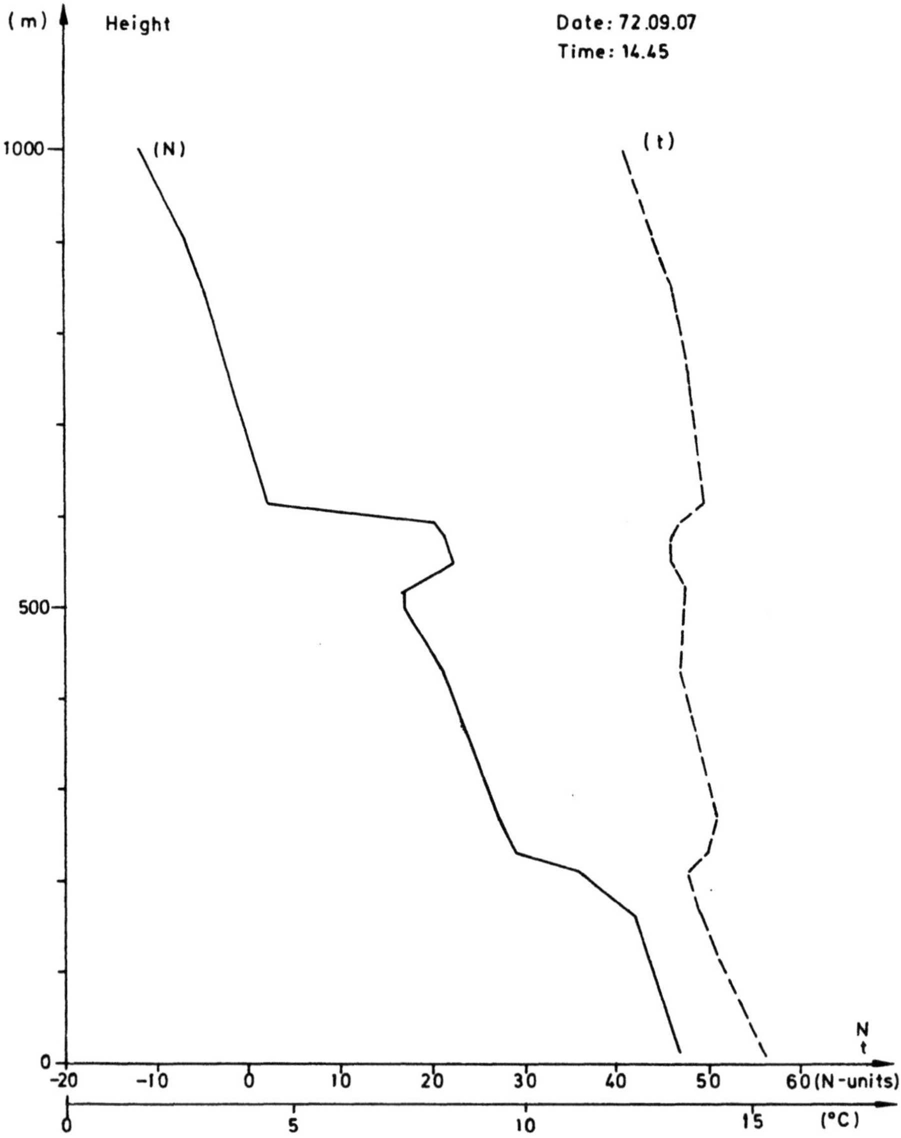

Figure 9. Refractive index profile.

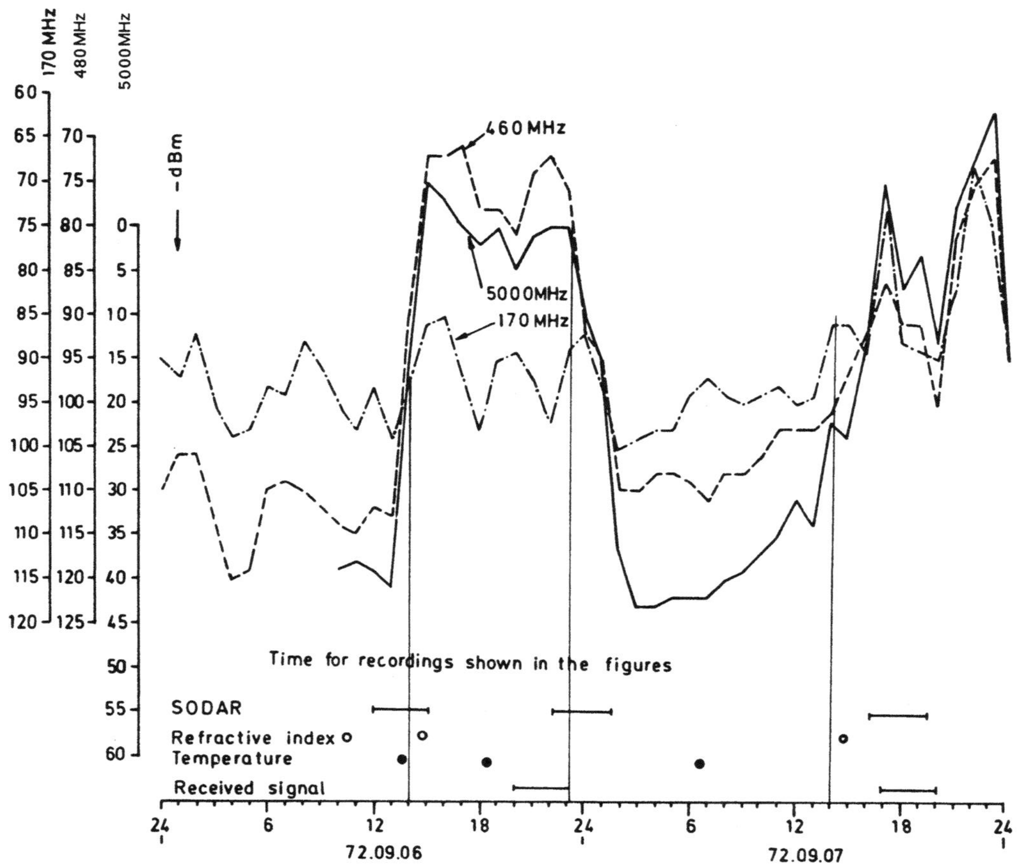

Figure 10. Hourly means of received signals at 5000, 460 and 170 MHz, Sept 6-7, 1972.

In table 2 the observed and calculated mean received signal powers for these times are given.

Table 2.

Frequency	Observed P_R	Calculated P P_s	P_r
170 MHz	- 87 - -99 dBm	-119 dBm	- 88 dBm
460	-106 - -120	-122	-104
5000	missing	- 51	- 64

Between 2 pm and 11 pm Sept 6 the low duct layer could trap the power at the two higher frequencies.

For the sharp layer observed at 4 pm Sept 7 at 500 m, the reflected signals at all the three frequencies were higher than the scattered. From figure 10 it is seen that this situation remained between 2 pm and midnight. The layer could be seen in the SODAR-recordings, when the height was less than 500 m, as shown in figure 11.

In table 3 the signals for this period are shown.

Table 3.

Frequency	Observed P_R	Calculated P_r
170 MHz	-68 - - 90 dBm	-80 dBm
460	-74 - -100	-87
5000	+13 - - 13	-34

The recorded signal levels at 460 MHz and 5000 MHz are shown from two periods in figures 12 and 13. In figure 12 a period is chosen when the low duct layer occured. Here both the signals are stable. In figure 13 reflected signals are shown. At both of these frequencies the signal levels are slowly changing. Periods of about 100 min and 6 min can be recognized in the recordings. This slow change may be caused by waves on the reflecting layer. Such waves give rise to slow variations in the grazing angle Ψ, and Δn.

The signal at 5000 MHz contains more fading than that at

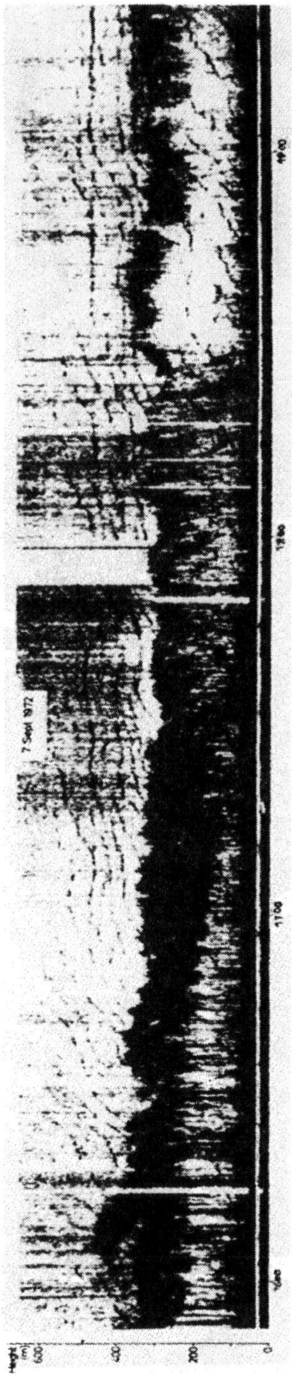

Figure 11. SODAR registration.

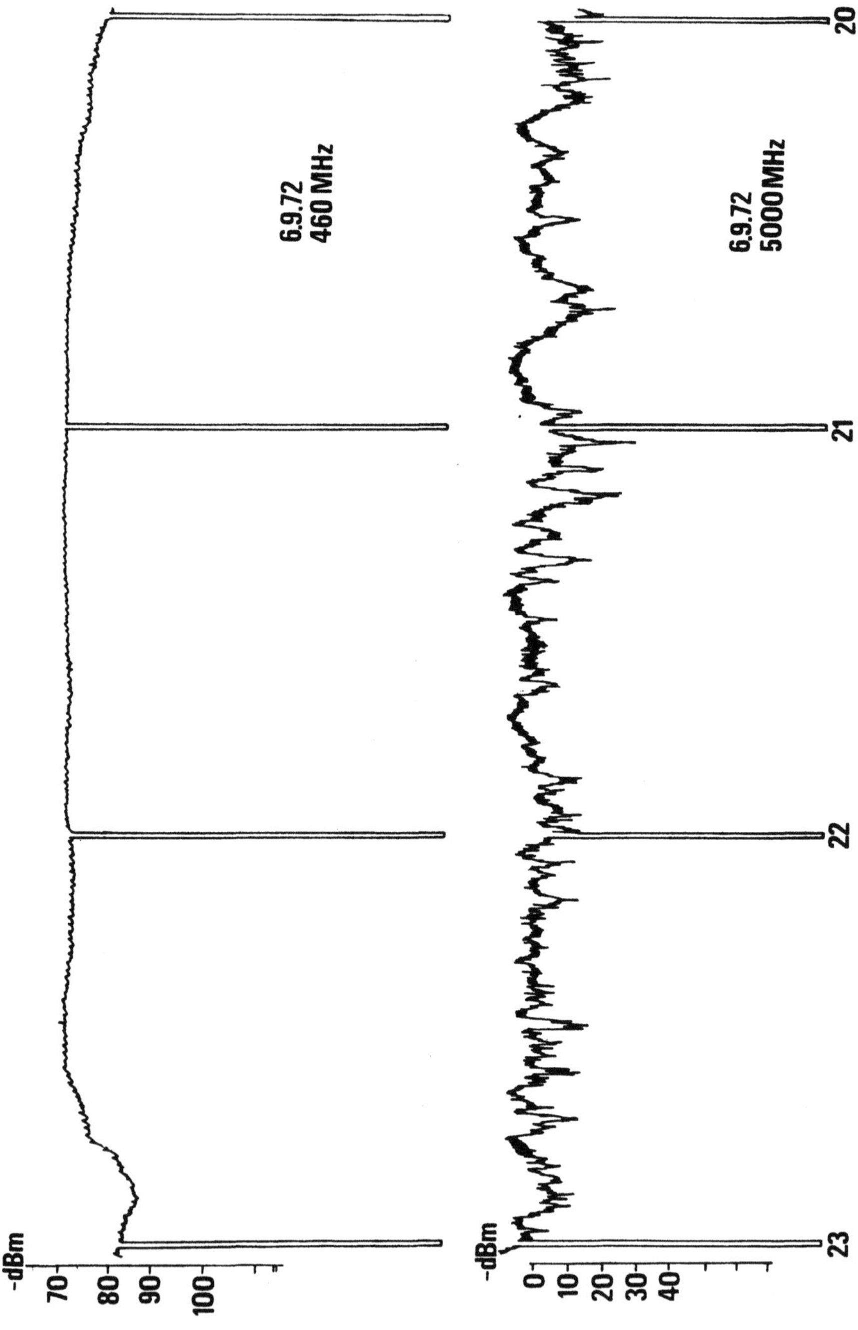

Figure 12. Recorded signals during duct propagation.

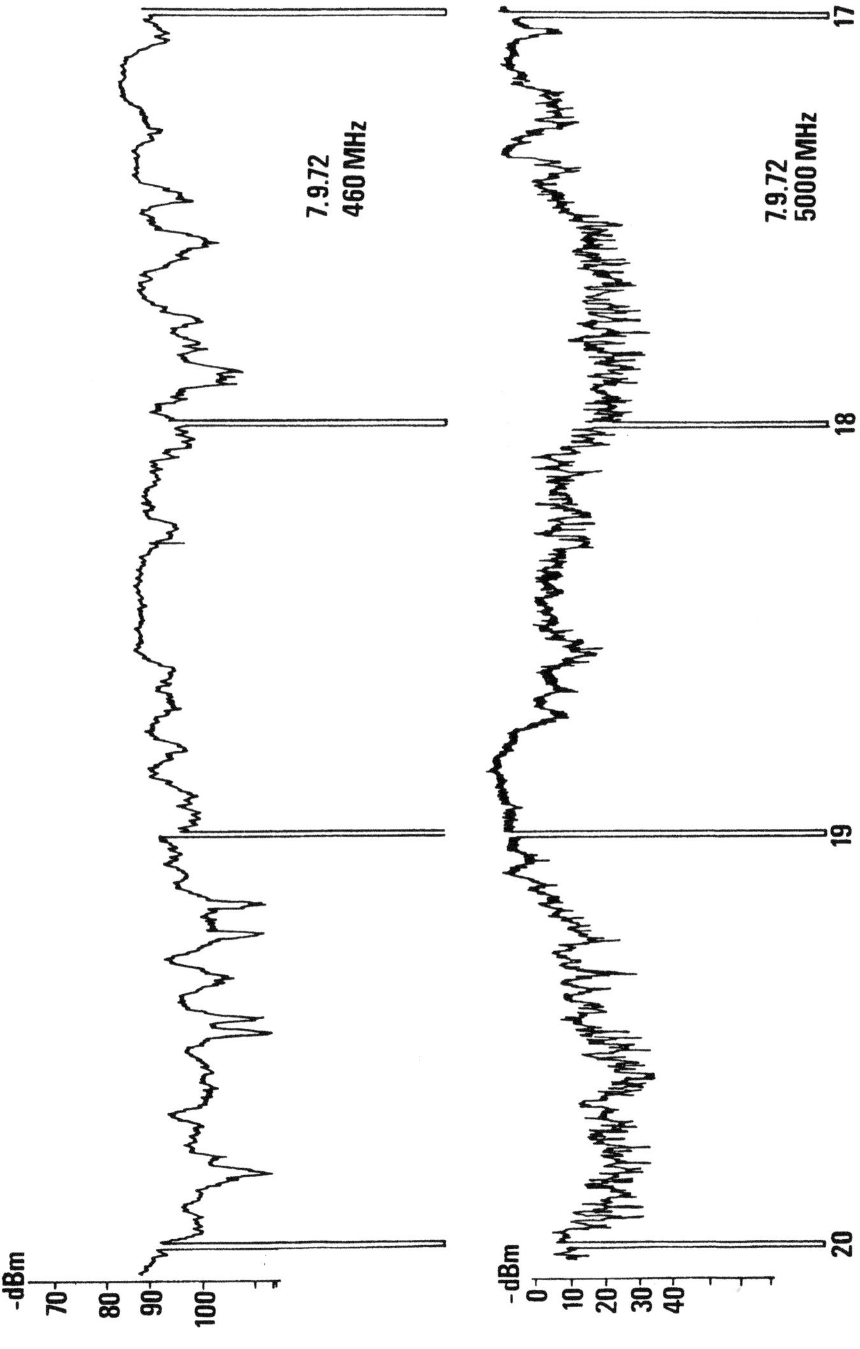

Figure 13. Recorded signals during reflection propagation.

460 MHz. The explanation for this is assumed to be that small-scale structures on the surface of the reflecting layer is only sensed by the shorter radio wavelengths, while the layer acts as a plane surface for longer wavelengths.

The agreement between observed signal powers and those calculated from the different propagation mechanisms is good. This is taken as a proof that it is possible to predict high signal levels far beyond the radio horizon from models of the tropospheric refractive index field.

1.5 Long-term statistics

In figures 14 and 15 are shown the annual occurrence of the three different signal types at 170 MHz and 5000 MHz. In figure 14 are plotted the weekly mean values for a period of three years 1970 - 1972. As can be seen, the scattered signal (signal type A) is more dominating at 5000 MHz than at 170 MHz during the winter period. Signal type B, which is mostly caused by reflection propagation, is more dominant at 170 MHz than at 5000 MHz. This is in accordance with the theory of reflection propagation. From figure 4 it is found that rather weak refractive index gradients through a layer is needed for obtaining reflection propagation at 170 MHz.

High signal levels either by duct or reflection propagation are very common in summer. At 170 MHz high signals occur during more than 90 % of the time and at 5000 MHz about 70 % of the time.

The occurrence of high and stable signal levels at 170 MHz, percentage of time per week is presented in figure 16.

2. CONCLUSIONS

It has been demonstrated that high signal levels far beyond the radio horizon are mainly caused by reflection at low frequencies (170 MHz). At high frequencies (5000 MHz) duct propagation is the dominating mechanism governing the high signals. Low signals are generated by scattering through turbulent fluctuations in the refractive index field.

Low level ducts occur mainly in summer. The occurrence of duct propagation at 5000 MHz is about 70 % of the time in summer and less than 10 % during wintertime. At 170 MHz relatively small refractive index gradients through a layer are required in order to cause reflection propagation. The optimum layer height to give strong reflections is just above the radio horizon at the midpoint of the path, which means 400 m on a 160 km path. From the meteorological point of view, such layers can be found in the Baltic very often all the year around.

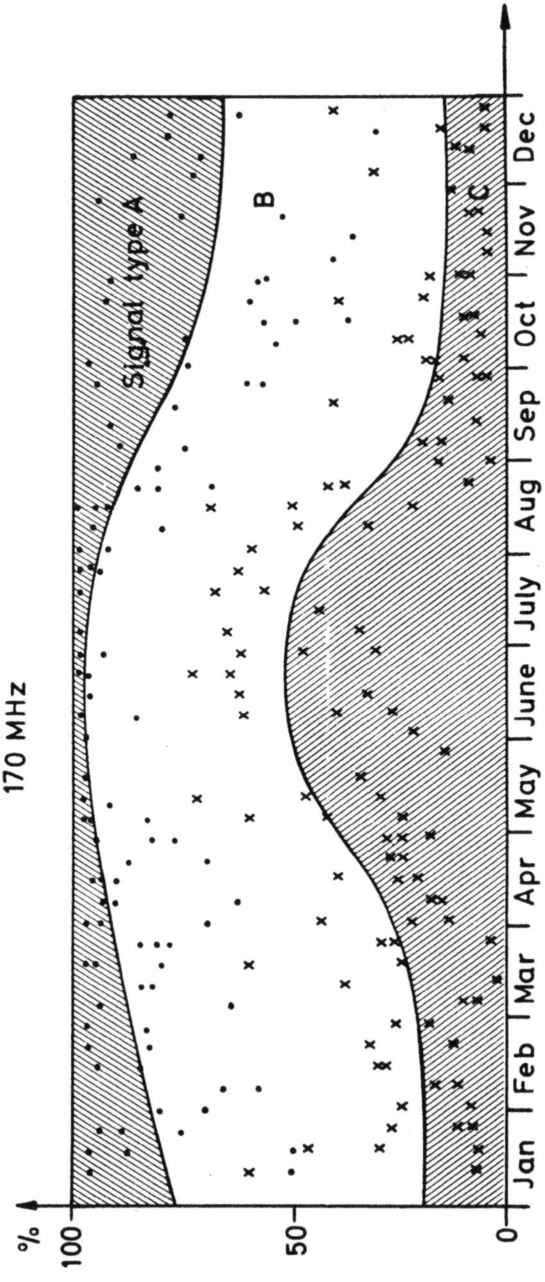

Figure 14. Annual occurrence of different signal types at 170 MHz.

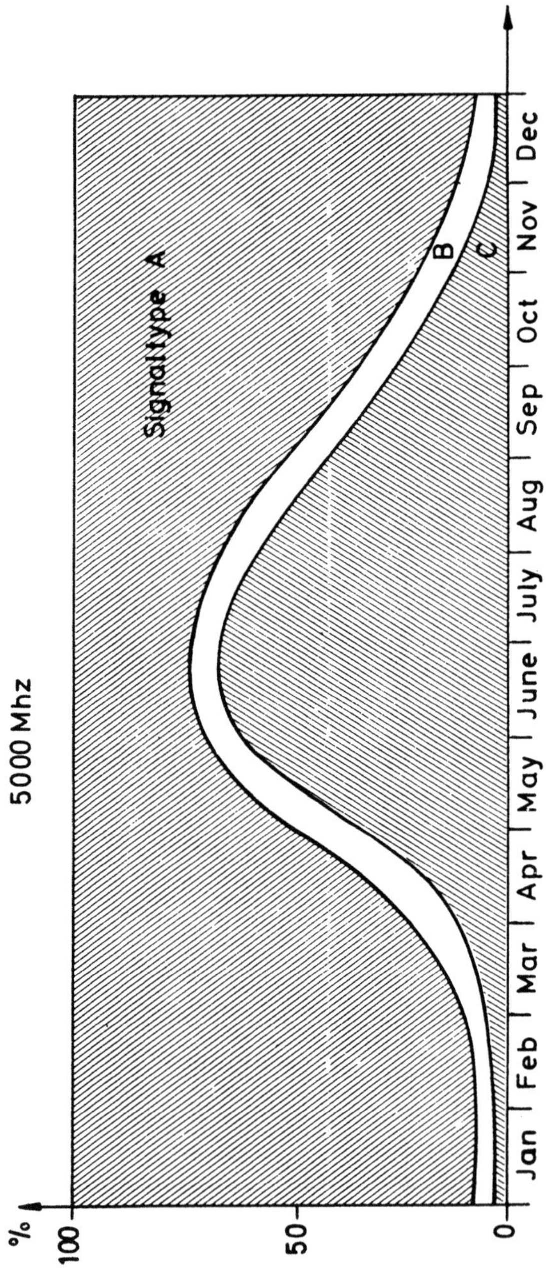

Figure 15. Annual occurrence of different signal types at 5000 MHz.

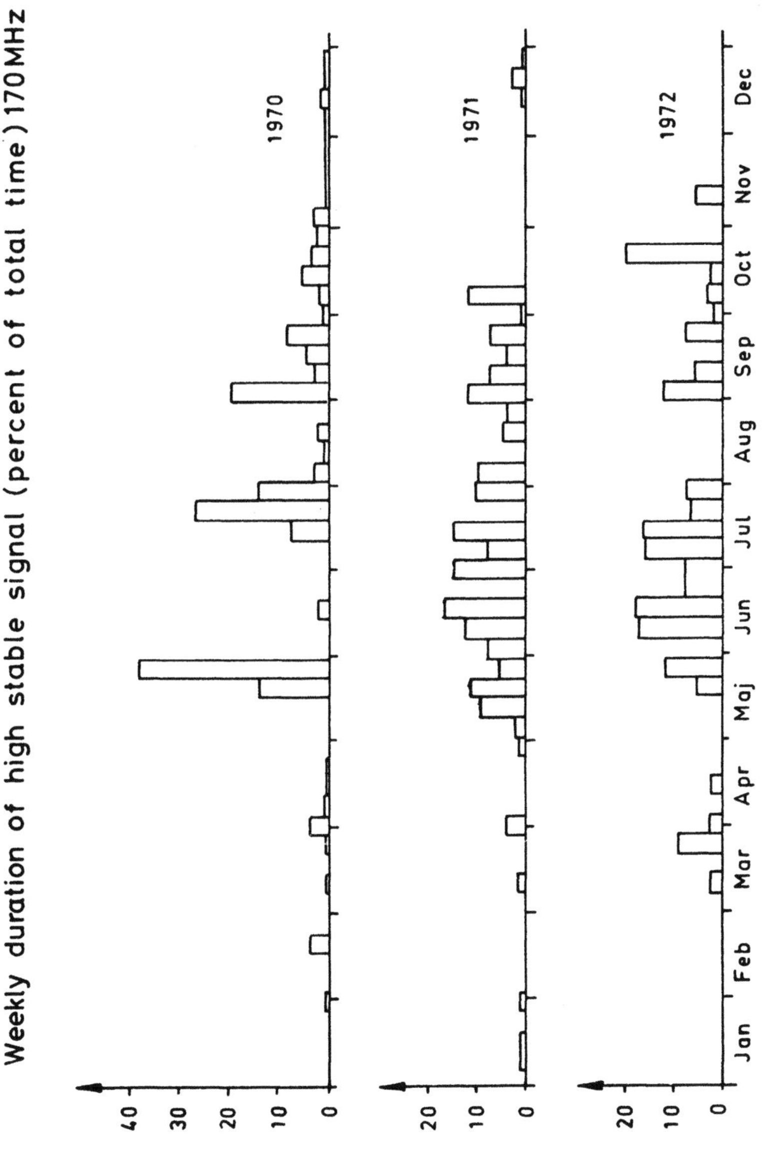

Figure 16. Occurrence of high signal levels.

Long-term statistics show that high signal levels occur more than 90 % of the time in summer and more than 50 % in winter at 170 MHz.

3. REFERENCES

Eklund, F and Wickerts, S: Wavelength dependence of microwave propagation farbeyond the radio horizon. Radio Science, vol 3 (1968) pp 1066-1079.

Eklund, F: Applied tropospheric propagation (in Swedish) Propagation of radio waves. Editor F Eklund, vol 25 B. Research Institute of National Defence (1969) p 10-84.

Gjessing, D: Tropospheric radio wave propagation (in Norwegian) Propagation of radio waves. Editor F Eklund, vol 25 A. Research Institute of National Defence (1969) p 9-67.

Wickerts, S: The refractive-index field in the lowest 2000 meters of the atmosphere. Research Institute of National Defence, FOA reports vol 4 no 3 (1970) p 1-16.

III GROUP

Atmospheric Influence on Radio Propagation above 10 GHz: Communication Aspects

COMMUNICATIONS REQUIREMENTS AT FREQUENCIES ABOVE 10 GHz AND THEIR CORRELATION WITH PROPAGATION ASPECTS

Bruno Peroni

Istituto di Comunicazioni Elettriche Università di Roma
Via Eudossiana 18, Roma, Italy

1 - INTRODUCTION

Radio-links at frequencies above 10 GHz can assume widely differing forms and characteristics according to the type of information to be transmitted and the type of service they are to perform. The time available for this report does not allow us to consider all the possible cases and compels us to choose a single class of radio-link.

We shall confine ourselves to long-haul, large-capacity transmissions of the point-to-point digital type, on frequencies of 18, 30 GHz and above.

We have made this choice because we maintain that, for many years, transmissions of this type will constitute the most important field of application of terrestrial radio-links at frequencies above 10 GHz and because the dependence of the link characteristics on the propagation phenomena appears more significant and interesting for these transmissions.

Long-haul, large-capacity point-to-point digital radio links will constitute an essential part of the integrated trunk network when digital techniques of switching and short-haul transmission become widespread in the individual urban areas.

This long-haul digital network will also comprise links employing means of guided propagation such as coaxial cables, waveguides and, in the more distant future, optical fibres.

SHF radio-links that employ a free means of propagation will be able to compete with the other means of transmission provided that technology and the design criteria succeed in overcoming the free propagation impairments to wich electromagnetic waves of frequencies over 10 GHz are subject.

2 - TRANSMISSION REQUIREMENTS OF THE VARIOUS TYPES OF MESSAGE

It is well known that, in digital transmission networks, with any means of transmission, the various messages-voice, sound programme, picturephone, television, data- are converted into sequences of symbols, each corresponding to a pulse waveform. The symbols and waveforms are the same for all the above-mentioned types of message, only the assemblies of symbols differing from one message to another

Any along message that is to be transmitted by a digital technique (pulse code modulation) is sampled at a frequency of not less than double the maximum frequency contained in its spectrum; then every sample is quantised into 2 m amplitude levels and converted into m bits per sample. Bits 1 and 0 correspond to two different forms of pulse. The pulses of several messages, having a duration less than the interval between one bit and the following bit in the same message, can be interleaved in time in order to obtain time division multiplexing. The number of bits flowing in any second along a section of the link in a given TDM assembly is called the "bit rate".

Coded pulse modulation may also be applied to groups of frequency-division multiplexed telephone signals (FDM-PCM).

For the transmission of data, which are delivered by the source already in digital form, there is no necessity for sampling and quantising. For their transmission it is merely necessary to provide for the block segmentation and their insertion into the time-division frames.

Table 1 shows, for the various types of message, the sampling frequency, the number m of bits per sample, and the channel groups (either already established or likely to be recommended). The Table also shows the maximum values assumed for the error rate and jitter, which, together with the quantisation distortion, determine the transmission quality of the digital messages.

2.1. Quantisation distortion (or noise)

This consists in the fact that in reception the messages does not assume the exact original value but the nearest approximate value chosen among 2 m levels. This distortion depends on the number m of bits per sample, on the amplitute statistics of the message

TABLE 1A

MESSAGE	Sampling Frequency HZ	Bits/Sample	Error Rate 80%	Error Rate 0,1%	Jitter ns
PCM TDM TELEPHONY	8000	8	10^{-7}	10^{-4}	900
FDM PCM TELEPHONY	110.10^3 570.10^3 9.10^6	9 ÷ 11	10^{-6}		0,2÷0,6
SOUND PROGRAMME	32.10^3	9 ÷ 10	10^{-9}	10^{-6}	50
PICTURPHONE	2.10^6	4	10^{-7}		
COLOR TELEVISION	$13,3.10^6$	9	10^{-8}	10^{-5}	0,15÷0,5
DATA			10^{-7}		

TABLE 1*B*

MESSAGE	Number of Channels	Bit Rate bits/sec	4 PSK R.F. Bandwidth MHz
PCM TDM TELEPHONY	1	64 10^3	
	30	2.048 10^6	~ 1.64
	120	8.448 10^6	~ 6.8
	480	~34 10^6	~ 27
	1920	~140 10^6	~ 112
	7680	~560 10^6	~ 450
FDM PCM TELEPHONY	12	1.3 10^6	1
	60	6.7 10^6	5.4
	960	99 10^6	80
SOUND PROGRAMME	1	0.32 10^6	0.25
PICTURPHONE	1	8 10^6	6.4
COLOR TELEVISION	1	120 10^6	96

and on the companding law. It is determined by the codec terminal equipment and does not depend at all on the transmission channel or on the propagation.

To have a low quantisation noise it is necessary to adopt a high number m of bits per sample, but this involves an increase in the bit rate and hence a reduction in capacity for a given frequency bandwidth. Beyond a certain limit, the cost of the increase in m is not compensated by any improvement in transmission. The values of m given in the table are those considered satisfactory for the different types of message.

2.2. Error rate

This consists in the fact that when some of the symbols reach the receiving end of the transmission channel they have deteriorated to such an extent that the decision device mistakes them for different symbols from the ones transmitted.

In binary transmissions, a symbol 1 may be mistaken for 0 and vice-versa.

In telephony, when an error occurs in one of the first two more significant binary digits of the 8-bit sequence into which a sample has been converted, a click is heard. Only if the clicks occur rarely - e.g. 1 every minute - do they constitute no disturbance. This means that 4 wrong bits can be accepted every 60 seconds: i.e. 0.065 wrong bits every second, on 64000 bits per second. Therefore an error rate of 10^{-6} on the whole link is admissible. On a hypothetical reference circuit, 2500 km long, the error rate may be fixed between 10^{-6} and 10^{-7} .

On the basis of subjective tests, the admissible error rate on particular reference circuits can also be established for the other types of message. The values shown in Table 1 refer to circuits of 2500 km in length and are the results of proposals put forward by various research and administrative organisations.

The actual error rate on a link depends on the transmission channel, consisting of the modems , the propagation medium and the repeaters.

2.3. Jitter

This consists in the fact that the individual symbols reach the receiver in time positions that are somewhat shifted with respect to the nominal positions. Beyond a certain value the time shifts give rise to distortion, crosstalk and error. PCM-TDM te-

TABLE II

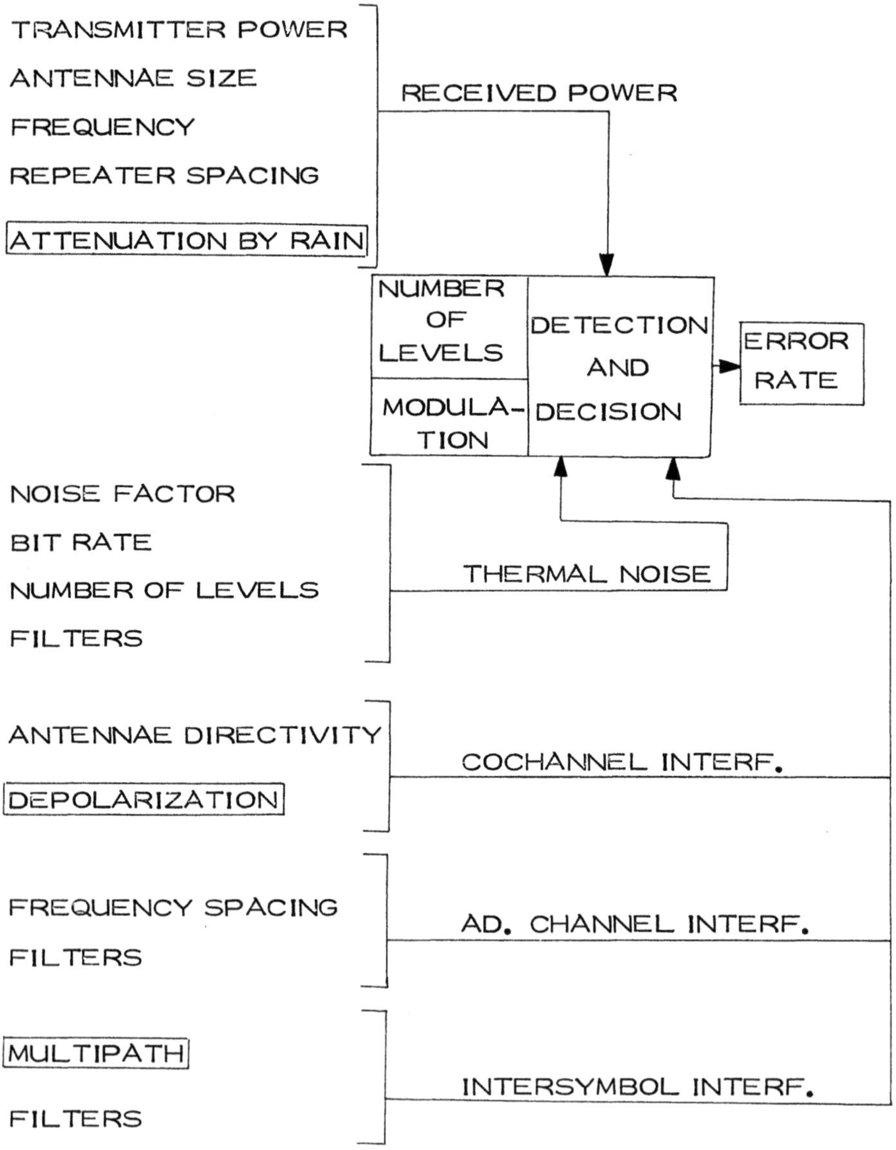
TRANSMITTER POWER
ANTENNAE SIZE
FREQUENCY
REPEATER SPACING
ATTENUATION BY RAIN
RECEIVED POWER
NUMBER OF LEVELS
MODULATION
DETECTION AND DECISION
ERROR RATE
NOISE FACTOR
BIT RATE
NUMBER OF LEVELS
FILTERS
THERMAL NOISE
ANTENNAE DIRECTIVITY
DEPOLARIZATION
COCHANNEL INTERF.
FREQUENCY SPACING
FILTERS
AD. CHANNEL INTERF.
MULTIPATH
FILTERS
INTERSYMBOL INTERF.

lephone messages are very tolerant of this type of degradation, but for FDM-PCM telephone messages, for sound programmes, for videophone and television messages and for data, the tolerances are more strict. Table I shows some of the values allowed.

It should be noted that jitter depends on regenerative type repeaters and hardly at all on the propagation medium.

Of the three characteristics that determine the digital transmission quality, only the error rate depends on the propagation medium; therefore we shall be concerned only with this, in what follows. In radio-links, in which propagation is subject to considerable variations in time, it is not sufficient to indicate the error rate desired during most (e.g. 80%) of the time. It is also necessary to indicate the maximum tolerable values of error rate (threshold error rate) and the percentage of time (outage time) in which the error rate increases beyond the maximum allowable value. For telephony, for example, the threshold error rate can be fixed at a value of 10^{-4}, corresponding to 100 clicks a minute.

The threshold error rate and the outage time percentage vary from one type of message to another. However, it is necessary to establish, for the transmission channel, single values for all the types of message, since any transmission channel may carry pulse stream belonging to all the types of message. This possibility constitutes one of the advantages of digital transmissions compared with analog transmissions.

For the value of the error rate on radio-links of 2500 km in length, the tendency is to assume 3.10^{-7} for 80% of the time and 10^{-4} (threshold value) for an outage time percentage of 0.1%.

This last value is considered unsatisfactory for certain types of message: for television and music transmissions and for data transmissions. For these types of message the tendency is to employ error correcting codes, where necessary.

3. FACTORS DETERMINING THE ERROR RATE

Table II shows the factors on which the error rate depends in a radio-link for digital transmissions.

The possibility is envisaged of using two radio carriers of the same frequency, or at least in the same frequency band, in order to double the transmission capacity per unit of frequency band. The two radio carriers consist of electromagnetic waves radiated with orthogonal polarizations. If, on account of the propagation, the polarization is altered, an interference is produced on the same frequency. The three factors associated with the propagation phenomena are en-

TABLE III

TRANSMITTER POWER	100 mW	RECEIVED POWER	
FREQUENCY	20 GHz		
ANTENNA EFF. AREA	0.5 m^2	$P_r = -22 - A$ dbm	
FILTERS FEEDER ATT.	3 db		
ATTENUATION BY RAIN	A db		
NOISE FACTOR	10 db	THERMAL NOISE POWER	
NOISE BANDWIDTH	250 MHz	$P_N = -80$ dbm	
MODULATION	C 4 PSK		

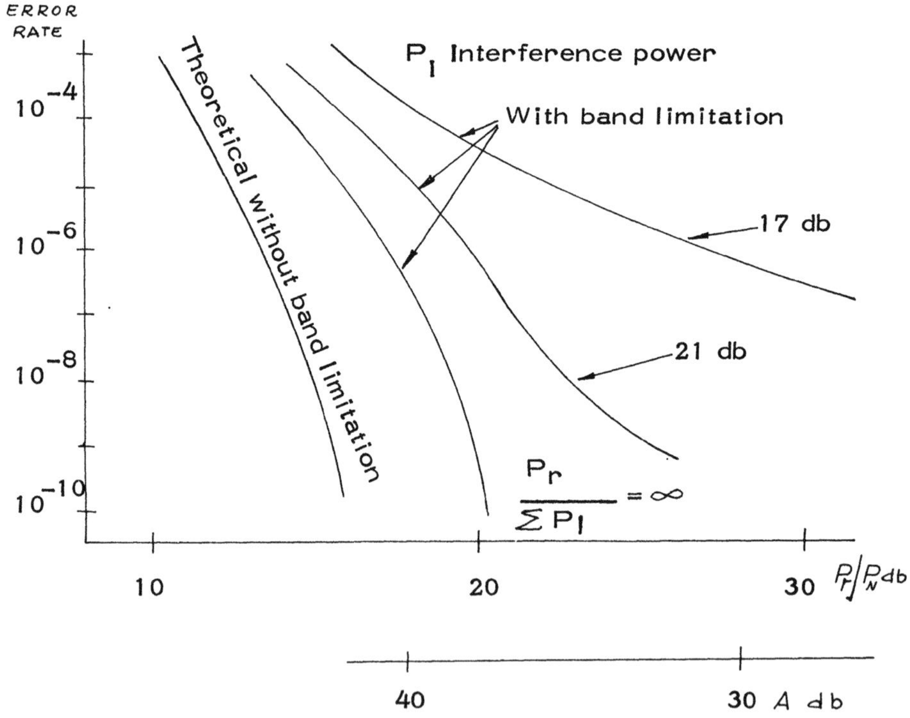

TIPICAL ERROR RATE CHARACHTERISTICS PATTERNS

closed in rectangles. They are: attenuation by rain, depolarisation; multipath propagation.

To give a brief illustration of how the three factors due to the propagation act upon the error rate, we refer to the case shown in Table III.

3.1. Absence of rain

The ratio between useful received power and thermal noise power is very high and has little effect upon the error rate. This is mainly determined by the interferences due to other radio channels of the same or of other paths, including satellite links. Most of these interferences are present simultaneously on all n hops of the 2500 km reference circuit.

If the error rate on this circuit in these conditions is to be 10^{-6} it is necessary that on every hop the error rate is of the order of $P_e = 10^{-6}/n$. For $n = 1000$, $P_e = 10^{-9}$.

This condition imposes limitations on the characteristics of the antennas and on the channel frequency spacing.

3.2. Heavy rain

The case considered here is that of the possibility that, on the generic hop, such heavy showers occur that the error rate rises to the threshold value. Such events are rate and of short duration, such that, as a first approximation, it may be assumed that in the whole circuit only one hop at a time is subjected to very heavy rain. The high attenuation due to rain on a hop has a deteriorating effect not only upon the ratio between received power and thermal noise but also upon the ratio between useful power and interfering signal power, when these signals follow a path other than the useful path attenuated by the rain.

It is found that, in the situation shown in Table III, the difference between the hop attenuation in threshold conditions for an error rate of 10^{-4} and the hop attenuation in the absence of rain is of the order of 40 dB. This difference is known as the "fade margin". It is desired that the sum of the times in which, in the individual hops, the attenuation due to rain exceeds the fade margin shall not exceed the outage time percentage laid down for this cause. This condition imposes a limitation on the hop length, according to the raininess of the area. The maximum limit of the hop length is found to be of the order of only a few km.

Heavy rain may also be responsible for another cause of degra-

TABLE IV
(from Ruthroff)

FREQUENCY GHz	MULTIPATH SHORT LENGTH Lo km
4	9,37
6	8,20
10	6,93
20	5,5
30	4,8
60	3,8

$$\tau = \frac{3}{8 f_{GHz}} \text{ SECONDS}$$

dation in the link. When the non-spherical raindrops fall oblique ly with respect to the planes of polarization of the e.m. waves, at an angle other than 0 or $\pi/2$, they produce a variation in the plane of polarization of the waves. If the waves used are of the same frequency and polarized orthogonally, this may cause co-channel interference

3.3. Multipath fading

In SHF line-of-sight radio-links, it is generally necessary to consider the possibility that multiple paths may occur between transmitter and receiver, which may give rise to fading and deformation of the pulse waveform. In the case of digital transmission this distortion causes interference between successive symbols and reduces the fading margin.

It is a known fact that the possibility of the occurrence of a multipath situation is associated with the length of the hop. For lengths of only a few km, such as are imposed by the necessity of not exceeding the fading margin, no multiple paths of any appreciable significance have been measured experimentally.

Ruthroff has proposed a theoretical model of the mechanism of formation of multiple paths, which explains quantitatively the statistical properties of fading for multiple paths measured in experiments carried out in New Jersey. According to this model it is possible, for every frequency, to establish a hop length L_o below which fadings for multiple paths are negligible.

Table IV shows the values L_o for frequencies between 4 and 60 GHz. For hop lengths less than L_o the echo delay may assume the maximum value of indicated in Table IV. For frequencies of 20 GHz. the maximum echo delay is of the order of 0.018 ns, which is small compared with the pulse duration of the order of 1 ns that occurs with the higher values of bit rate evisaged today.

4 - LIMITATIONS DUE TO ATTENUATION CAUSED BY RAIN

It is as well to examine more throughly the requisite regarding the exceeding of the error rate threshold of 10^{-4} for the percentage of time of 0.1% per year, equal to 8.75 hours or 32000 seconds.

First of all, it is to be observed that for radio-links at frequencies lower than 10 GHz, intended for analog transmissions (FDM-FM), the requisite regarding the worst transmission conditions is formulated with reference to a percentage of time of the worst month. In the case of radio-links at frequencies higher than

10 GHz, on the other hand, the tendency is to refer to the year, because most of the available meteorological statistics refer to the year. Since heavy rainfalls are generally concentrated in one or two months of the year, 0.1% of the time per year corresponds to higher percentages if referred to the worst month. If it were desired to ensure, for the radio-links in question, the percentage of 0.1% of the worst month, it would be necessary to fix a percentage near to 0.01% referred to the year. This possibility has actually been considered: though more satisfactory from the point of view of the quality of the service, it is excessively costly.

In fact, it may be observed that, if one assumes the complete independence of the events in which the error rate threshold is exceeded in the n hops included in the 2500 km reference circuit, if n is taken as 1000, and if the outage percentage of 0.1% is adopted over the whole circuit, on every hop the fraction of outage time is found to be 10^{-6}, equal to 32 seconds per year. For an outage percentage of 0.01% over the whole circuit in the same hypotheses, the fraction of time for each hop is found to be 10^{-7}, equal to about 3 seconds per year, which is considered excessively costly nowadays.

At any rate, from what has been stated, it is found to be necessary to consider the distribution functions of the attenuation, and hence of the intensity of precipitation, up to probabilities of the order of 10^{-6}, i.e. to fractions of a minute per year.

This involves a great difficulty in collecting statistical data on the rainfall intensity.

The statistics available refer to several years, but they give the total rainfall measured in the year, sometimes in the month, more rarely in the day. It is only in the last few years that measurements have been made that provide rainfall statistics for the short intervals of time with which we are concerned; these measurements, made with instant raingauges, are confined to very few geographical areas. Equally recent and sporadic is the direct measurement of attenuations, carried out with radio-electric measurements.

As regards the collection of the meteorological data to be used in the design of a radio-link, there are two tendencies.

The first consists in extrapolating, from the existing statistics for each geographical area, the distribution functions for very small values of probability. The total quantity of mm of rainfall in one year (or in the worst year of the last decade) is not representative of the distribution function. A better indication may be obtained by considering the ratio between annual rainfall and number of rainy days in the year, but even this indication is probably not sufficiently significant.

A further indication may be provided by the maximum daily rainfall in every month: an indication that is available for certain geographical areas.

A second tendency consists in using the attenuation on rainfall statistics recently measured down to small values of probability. Also in this case there are difficulties due to the great meteorological variability from one year to another and from one place to another, so that it becomes arbitrary to extrapolate, for different years and different places, the results of statistics collected in a particular geographical area and in a particular year.

When it is possible to know or presume the distribution functions of the rainfall intensity R in mm/h at all the points in the area that is to be crossed by a radio-link, the point attenuation may be estimated for every frequency on the basis of theoretical and experimental relations. For the frequencies of I8 and 30 GHz these relations are shown in Table V, which also gives the expressions of the fraction of outage time for a generic link of L km in length, on the assumption of independence of the events in which the error rate threshold is exceeded.

The whole length of the link L must be subdivided into n hops, each l_i in length (i=I,...n), such as to satisfy the conditions indicated in Table V.

The meteorological statistical measurements needed by the designer of the radio-links in discussion concern not only the distribution function of the rainfall at a given point but also the joint distribution functions taken in pairs of points at different distances, up to about IO km.

This is because very heavy rainfall appears to be intensely concentrated in cells of relatively small dimensions and in relatively short intervals of time. The idea is to benefit from this space-time non-uniformity of the rainfall to achieve a path diversity consisting in the establishing of the two radio-links in parallel and with sufficient distance between them (about IO km) in order that the peaks of rainfall intensity on the hops of one radio-link do not occur at the same time as those on the hops of the other radio-link.

Advantage is taken of this independence to space the repeaters of each radio-link more widely. For a certain total length and for a given outage time percentage, the number of repeaters is found to be approximately equal in the two cases of presence

TABLE V

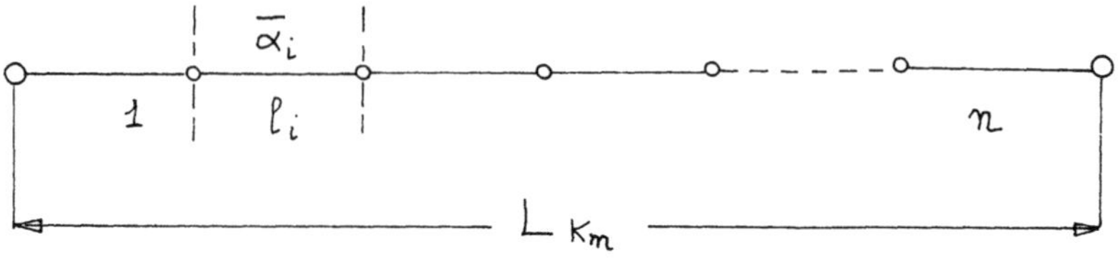

OUTAGE TIME 2500 km $= 10^{-3}$

" " L km $= 4 . 10^{-7}$ L km

$$\sum_{i=1}^{n} l_i = L$$

$$\sum_{i=1}^{n} P(\bar{\alpha}_i l_i \geq M) < 4 . 10^{-7} L$$ M FADE MARGIN

$\bar{\alpha}$ = AVERAGE ATTENUATION ON 1 km (db)

$\bar{R}$ = AVERAGE RAINFALL RATE ON 1 km (mm/h)

f = 18 GHz $\bar{\alpha} = 0.1\ \bar{R}$

f = 30 GHz $\bar{\alpha} = 0.2\ \bar{R}$

or absence of the path diversity, but in the case where there is path diversity there is the advantage of a useful reserve route, also in the event of any failure of the apparatus, and for maintenance.

5 - CONCLUSIONS

We have been considering the transmission requirements in terrestrial, point to point long haul, large capacity radio-links.

We showed that the only requirement depending on propagation phenomena, is the error rate.

We have seen that the propagation factors which determine the error rate are: multipath propagation, depolarisation by rain and attenuation by rain. This last factor is the most important for the design of SHF radio-links.

The design of a radio-link crossing a given area, requires the knowledge of the time-space distribution of rainfall rate in that area, and of the dependence of attenuation on rainfall rate. These subjects will be handled by Dr. Fedi and Dr. Mandarini. The depolarisation phenomena will be considered by Dr. Paraboni.

Thank you for your attention.

REFERENCES

- CCIT Com Sp. D N° 127-E Dec. 1971
- CCIT W. G 9-C N° 9/111-E 19- July 1972
- D.C. Hogg Statistics on attenuation of Microwaves by Intense Rain B.S.T.J. Nov. 1969
- T.L. Osborne Rain Outage Performance of Tandem and Path Diversity 18-GHz Short Hop Radio System B.S.T.J. Jan. 1971

QUESTION (Bolgiano)

Intermittency of errors (their occurrence in bursts) is a problem for communication theorists today. Do you believe that, on circuits involving radio propagation, this type of error will be particularly serious?

ANSWER (Peroni)

Yes. Burst errors are serious also on circuits involving radio propagation.

When they occur, the pulse stream of the disturbed radio channel is switched on a protection channel, whenever it is possible.

ASPECTS IN USING FREQUENCIES BANDS ABOVE 10 GHZ FOR SATELLITE COMMUNICATION SYSTEMS

F. Carassa

Istituto di Elettrotecnica ed Elettronica
Politecnico di Milano

1. INTRODUCTION

The rapidly increasing demand for communication through satellites requires a more efficient use of the bands presently available and of course the use of new frequency bands, e.g. of the bands above 10 GHz. The increasing demand referred to above is due not only to traffic increase on intercontinental routes, but also to the foreseen use of satellites for regional communication systems. The latter are characterized by a still higher request of telephone channels and by a regular requirement for television transmission, in the form of distribution or broadcasting.

Tables 1, 2 and 3 show the frequency bands assigned to satellite communication systems by the World Administrative Radio Conference for Space Communication (Geneva, 1971): some of these bands are exclusively assigned to the considered systems, whilst others have to be shared with different systems with consequent interference problems.

Frequencies above 10 GHz undergo atmospheric absorption and scattering specially due to rain but offer some advantages in comparison to lower frequencies for what concerns circuit dimensions and in particular antenna dimensions (in addition of course to the very wide band available). Highly directive antennas are in fact easily obtainable, and highly directive antennas are in turn one of the most important aspect of future satellite development. We will briefly discuss this point.

Highly directive antennas may be useful in television distribution allowing to concentrate power radiation only towards the re-

Table I

Allocation (GHz)	Bandwidth (MHz)	Notes
10,95÷11,2	250	↑↓ in RG 1; ↓ in RG 2 and 3
11,45÷11,7	250	↓
11,7 ÷12,2	500	Only for RG 2; ↓
12,5 ÷12,75	250	EXCL for RG 1,↑↓; ↑ RG 2; ↓ RG 3
14÷14,5	500	↑
17,7÷21,2	3500	↓; EXCL in the interval 19,7÷21,2 (1500 MHz)
27,5÷31	3500	↑; EXCL in the interval 29,5÷31 (1500 MHz)
40÷ 41	1000	↓ EXCL
50÷51	1000	↑ EXCL
92÷95	3000	↑ EXCL
102÷105	3000	↓ EXCL

Frequency allocations between 10 and 100 GHz to satellite communication systems with fixed earth stations.

(EXCL means exclusive allocation; if not so specified the allocation has to be shared with other services;↑ means earth-to-space;↓ means space-to-earth; RG means Region: the world is divided in 3 regions of which region 1 includes approximately Europe and Africa, region 2 America and region 3 Asia and Australia).

Table 2

Allocation (GHz)	Bandwidth (MHz)	Notes
11,7÷12,5	800	In RG 2 and 3 limited to 11,7÷12,2
22,5÷23	500	Only for RG 3
41÷43	2000	EXCL
84÷86	4000	EXCL

Frequency allocations between 10 and 100 GHz to broadcasting satellites.

Table 3

Allocation (GHz)	Bandwidth (MHz)	Notes
14,3÷14,4	100	Only radionavigation
43÷48	5000	EXCL
66÷71	5000	EXCL
95÷101	6000	EXCL

Frequency allocations between 10 and 100 GHz to satellite for aeronautical and maritime communication and radio-navigation.

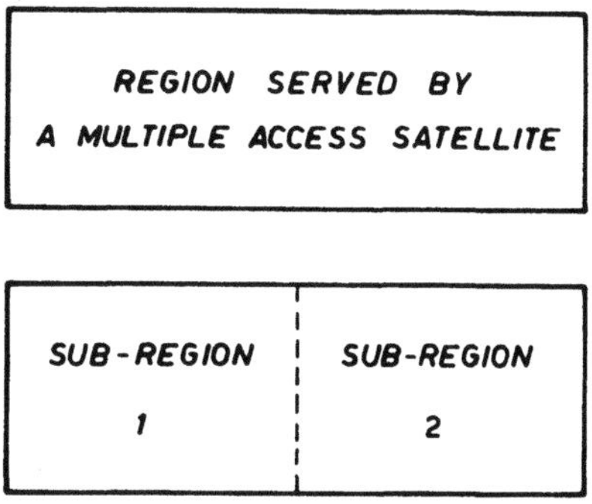

Fig. 1

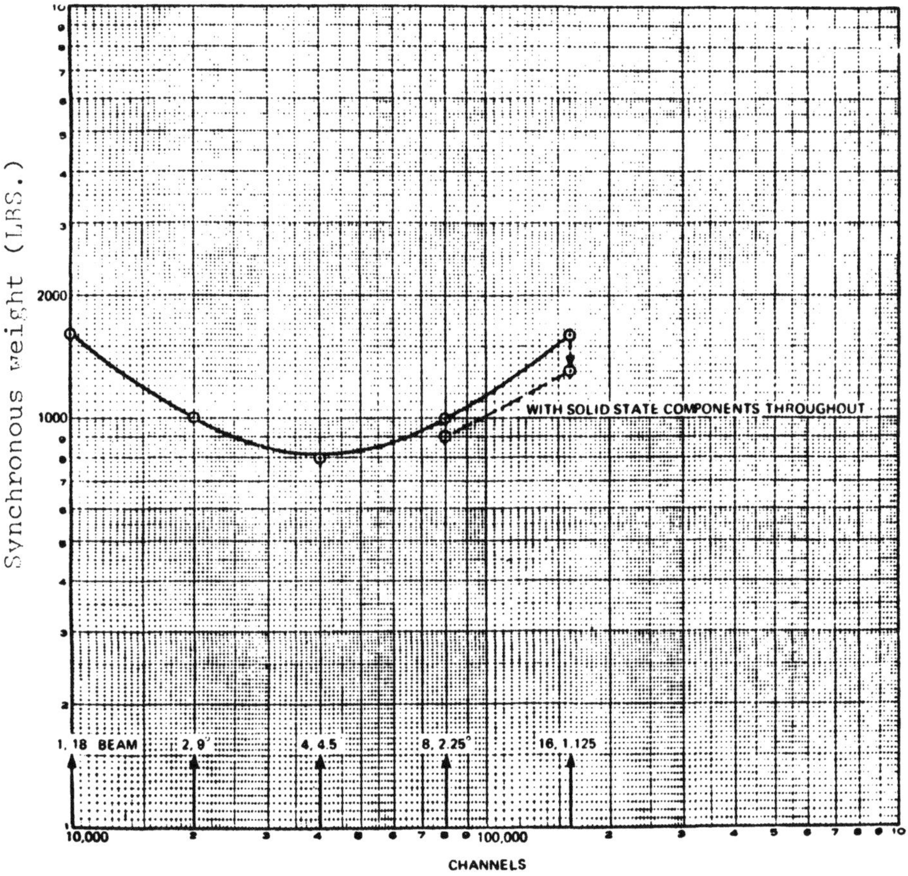

Fig. 2 Estimation of synchronous orbit weight versus satellite channel capacity

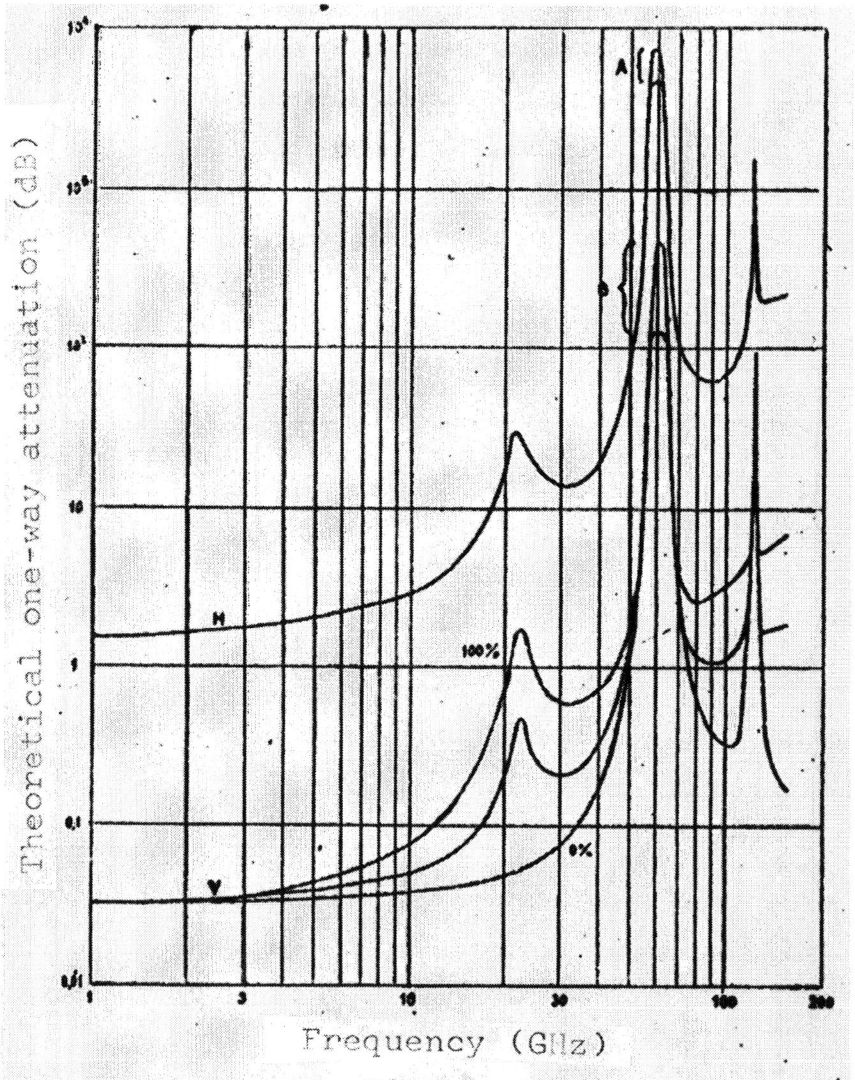
Theoretical one-way attenuation (dB)
Frequency (GHz)
A
B
H
V
100%
0%
10
1
0,1
1
3
10
30
100
200

Fig. 3

gion to be served, but obviously the easy availability of highly directive antennas is usable at the best in point-to-point communication. Consider a region at earth (fig. 1) containing the users (i.e., earth stations) of the system, which we suppose based on a multiple access satellite. If we subdivide the region in two sub-regions each one served by a different beam we can concentrate in each beam the previous transmission capacity, halving in each beam the total radiated power. This means that we double the total transmission capacity, with the same radiated power; moreover, if the beam are sufficiently decoupled, the same frequencies may be used in the two beams (this procedure is referred to as frequency reuse): therefore we double also the transmission capacity of the used frequency band. The procedure may be repeated, serving the region with n beams, obtaining n times the transmission capacity with the same total power and the same bandwidth; this result is achieved at the expence of some switching capability on board of the satellite and of an increase in the precision of satellite stabilization. In practice the advantage obtainable in the transmission power may be higher than the mentioned one, if the beams are not exactly adjacent. According to Metzger /1/ the previous concepts, if applied to present frequency bands, will drastically reduce the weight in orbit per channel as shown in fig. 2. (The number of beams and the aperture of the beams is also indicated).

Concerning frequency reuse the adoption of two cross-polarized waves in each beam may also double the capacity in a given bandwidth. Therefore, at frequencies above 10 GHz, depolarization due to rain, as considered in a different session of this meeting, is of importance.

More in general, as for any communication system, reliability of the system is a fundamental parameter: in the present paper we are concerned with reliability as affected by propagation in presence of rain. Rain can affect reliability through excess attenuation in the atmosphere, amplitude and/or phase distortion, scattering of other signals and depolarization.

2. ATMOSPHERIC ATTENUATION

In clear weather conditions, frequencies above 10 GHz undergo atmospheric attenuation due to absorption by gases, in particular oxygen and water vapour. We will not review in this paper the physical aspects of such absorptions and the contributions given to their study: exaustive reviews have been recently published, in particular by Millman /2/ and Crane /3/. Fig. 3, taken from /4/, gives the total atmospheric attenuation as a function of frequency for a vertical (V) and a horizontal (H) path. Among the V curves, the one labelled 0% refers to 0% relative humidity: it is the attenuation curve due to oxygen absorption. The solid curves are for a moderately humid atmosphere (7,5 g/kg at the surface), whilst the curve for 100% relative humidity is reported to give an idea of the maximum interval into which the attenuation due to water vapour absorption can vary. For elevation angles larger than 6 10 degrees the total atmospheric attenuation can be calculated multiplying the zenith attenuation for the cosecant of the evalation angle, thus introducing the increase in path length through the atmosphere (*). The peak attenuation in the vicinity of 60 GHz, due to oxygen absorption, is very high: as a consequence no frequencies in this band have been allocated for communication systems between space and earth: a band 5 GHz wide ($59 \div 64$ GHz) has instead been allocated for inter-satellite communication.

At the frequencies which have been allocated for communication systems between space and earth, the attenuation reported in fig. 2 is generally acceptable, and practically negligible at high elevation angles.

In addition to the previous permanent attenuation, attenuation due to clouds, fog and precipitations must be considered. Attenuation due to rain is the most important and is produced by absorption and scattering by raindrops. Reviews of the theoretical aspects of this problem can be found in the literature /2/, /3/, /5/, /6/, /7/. In the present paper we will simply resume some formulas, of interest for the following discussion.

a) Specific attenuation due to spherical hydrometeors:

$$A_s(\ell) = 4,34 \int_0^{\infty} N(a,\ell)Q(\varepsilon_k,a,\lambda)\, da \qquad (1)$$

where $A_s(\ell)$ is the specific attenuation, in dB/m, at a point defined by the line coordinate ℓ along the path; $N(a,\ell)da$ is, at the same point, the number of hydrometeors per cubic meter having diameter between a and a+da; Q is the extintion section of a hydrometeor having complex dielectric constant ε_k and diameter a. Q is a function of frequency also through ε_k which is a function of frequency; through ε_k Q is also a function of temperature. Q has been calculated and reported: the results of some calculation are summarized in /6/.

b) Total attenuation through the atmosphere:

$$A_t = \int_{path} A_s(\ell)\, d\ell . \qquad (2)$$

(*) This procedure is correct for a flat equivalent earth.

The quantity of interest in defining the reliability of a satellite link is the probability distribution of A_t, the latter being time-variable mainly due to the variations in time of the quantity $N(a,\ell)$. The fadings arising from rain attenuation can produce outages in the system: on this subject it is to be noted that such fadings have different characteristics in comparison to the familiar fadings experienced at frequencies below 10 GHz on terrestrial radio relay systems. The duration of fadings due to rain is in fact much longer so that it is appropriate, when the transmission requirements are no more satisfied during a fading, to call the corresponding interval an outage of the system.

The description of system behaviour, as far as propagation is concerned, mainly in terms of outages, is particularly appropriate when a digital modulation method is adopted: in fact, in front of a decreasing of the signal level in respect to noise level the digital transmission is characterized by a sharp transition from "regular" operation (very small error rate) to "break" of the system (excessive error rate).

The CCIR has not yet established a recommendation concerning the total amount of outage time to be accepted. However, in considering radio relay systems for the transmission of pulse-code modulation and other types of digital signal, report 378-1 specifies two error rates k and k' not to be exceeded for more than 0,01% and 0,1% of any month, respectively. According to this report the allowed outage time would be 0,01% of any month. A recent document of United Kingdom suggests possible areas of amendment of the above report: a first area concern the percent outage time which could be somewhat increased at an intermediate valued between 0,1 and 0,01%; a second area concerns the period of time to which refer to the above percentage: from the point of view of attenuation by rain a period of one year would be easier to consider, due to the fact that heavy rainstorms are rare events. A time interval between 0,1 and 0,01% of a year would amount to a few hours; this value is in agreement with the two hours per year of outage time considered by Tillotson in paper /8/: if we assign half of this outage time to propagation, we obtain 30 minutes for each path. This outage time can include only a few events per year and perhaps only one, as evidenced by Tillotson in his quoted paper. As we are considering rare events, the required performance would be better specified over several years; alternatively we may require that the probability of a given percent outage time (over one month, one year, etc.) be less than a certain value.

In any case acquisition of experimental statistical data on rain attenuation on a particular path needs several years of propagation tests.

So far, available experimental data are partly due to direct measurements of A_t and partly due to indirect measurements. Direct measurements have been carried out using the sun as incoherent microwave source or the NASA ATS-5 satellite as a coherent source at a constant elevation angle.

Indirect measurements may be carried out in various ways, measuring some integral quantity directly related to A_t, or measuring quantities appearing in the integrals (1) and (2) or quantities directly related to the above. $N(a,\ell)$ may be measured by photographic methods or using a distrometer /9/, but it is clearly impossible to measure $N(a,\ell)$ (and the temperature) at any point along the path. We must shift to some integral quantity of $N(a,\ell)$ such as the water content per unit volume (e.g., $M(\ell)= \pi/6 \int_{\infty}^{\infty} a^3 N(a,\ell)\, da$) or the rain rate (e.g., $R(\ell) = \pi/6 \int_0^{\infty} a^3\, v(a) N(a,\ell) da$ in which $v(a)$ is the velocity of a drop of diameter <u>a</u>) or the backscatter cross-section per unit volume (e.g., at frequencies below 10 GHz, $\beta_B(\ell)= \pi^5/\lambda^4 |K|^2 Z(\ell)= \pi^5/\lambda^4 |K|^2 \int_0^{\infty} a\; N(a,\ell) da$ in which $Z(\ell)=\int_0^{\infty} a^6\, N(a,\ell) da$ is the <u>reflectivity</u> of the rain and $|K|^2$ depends on ε_r and is nearly 1).

Clearly such quantities are strictly connected to A_s $_s(\ell)$ if an average temperature can be assumed and the distribution of raindrops diameters $N(a)$ does not vary to much for any given water quantity per unit volume, or rain rate of reflectivity. Average raindrop diameter distributions have been given by Laws and Parsons and by Marshall and Palmer. Table 4 reports the Laws and Parsons distribution /10/. Using these distribution it is possible to establish a direct correspondence between $A_s(\ell)$ and water content per unit volume, or rain rate or reflectivity, as reported in fig. 4 taken from /11/ with the addition of the reflectivity scale (*). Another presentation of the same results is given in fig. 5 taken from /3/. Note that the attenuation increases rapidly with frequency at the lowest frequencies and then has a broad maximum: at the radio frequencies of present interest it can be said that the attenuation tends to be constant increasing the frequency.

Instead of using average drop diameter distributions, calculations of the specific attenuation A_s can be carried out using actual measured drop diameter distributions in order to examine the dispersion of the obtained results. Fig. 6 taken from /3/ gives a scattergram of A_s at 16 GHz as a function of rain rate, obtained from nearly 5000 drop-size distributions obtained in North Carolina.

(*) Note that at the lowest frequencies having wavelength much larger than the drop diameter, the specific attenuation A_s is proportional to the water content per unit volume.

TABLE 4

DROP-SIZE DISTRIBUTIONS FOR VARIOUS PRECIPITATION RATES

Precipitation rate (mm/hour) / Drop size (cm) (mean in interval)	Percent of total volume								
	0.25	1.25	2.5	5	12.5	25	50	100	150
0.05	28.0	10.9	7.3	4.7	2.6	1.7	1.2	1.0	1.0
0.1	50.1	37.1	27.8	20.3	11.5	7.6	5.4	4.6	4.1
0.15	18.2	31.3	32.8	31.0	24.5	18.4	12.5	8.8	7.6
0.2	3.0	13.5	19.0	22.2	25.4	23.9	19.9	13.9	11.7
0.25	0.7	4.9	7.9	11.8	17.3	19.9	20.9	17.1	13.9
0.3		1.5	3.3	5.7	10.1	12.8	15.6	18.4	17.7
0.35		0.6	1.1	2.5	4.3	8.2	10.9	15.0	16.1
0.4		0.2	0.6	1.0	2.3	3.5	6.7	9.0	11.9
0.45			0.2	0.5	1.2	2.1	3.3	5.8	7.7
0.5				0.3	0.6	1.1	1.8	3.0	3.6
0.55					0.2	0.5	1.1	1.7	2.2
0.6						0.3	0.5	1.0	1.2
0.65							0.2	0.7	1.0
0.7									0.3

TABLE 5

ATTENUATION VERSUS RADAR REFLECTIVITY FACTOR MODEL

$$A = \alpha Z^{\beta}$$

Frequency (GHz)	α	β	rms error (percent)
7.5	0.0000573	0.792	35
9.4	0.0000929	0.819	30
16.0	0.000383	0.824	29
34.9	0.00315	0.772	41

TABLE 6

Separations Required for Space Diversity Systems Ottawa at 15.3 GHz

Condition	Elevation Angle 5	10	20	Average
(a) 10dB	6.5	4.3	6.1	5.6 km
(b) 15dB	8.6	6.9	9.4	8.3 km
(c) 10:1	10.7	10.1	9.4	10.3 km
(d) 40:1	13.1	12.9	12.7	12.9 km
(e) 40:1 10 dB margin	10.9	12.6	10.0	11.2 km

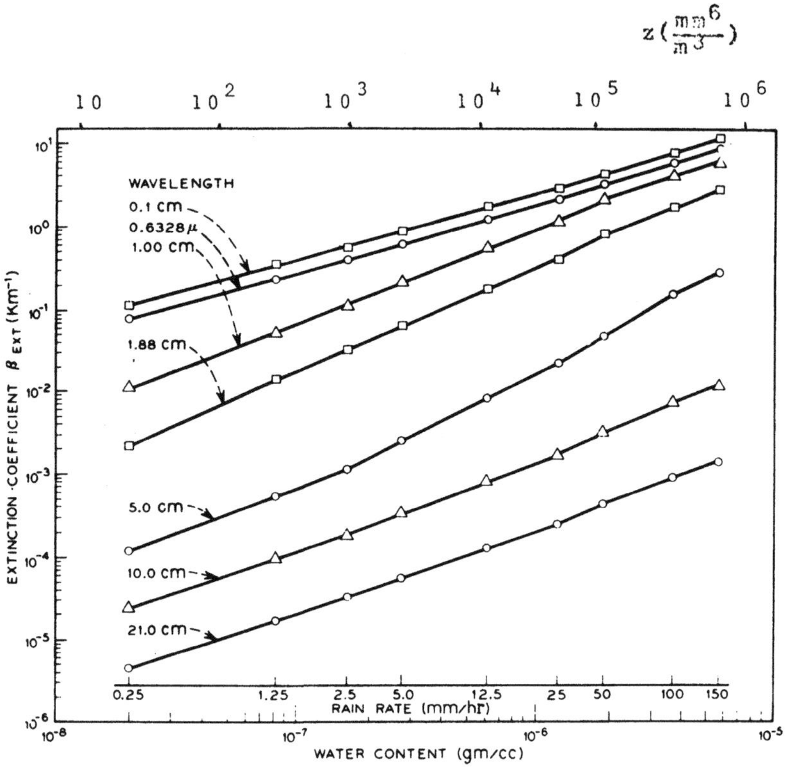

Fig. 4 -Rainfall water content and rain rate versus extinction coefficient β_{ext}.

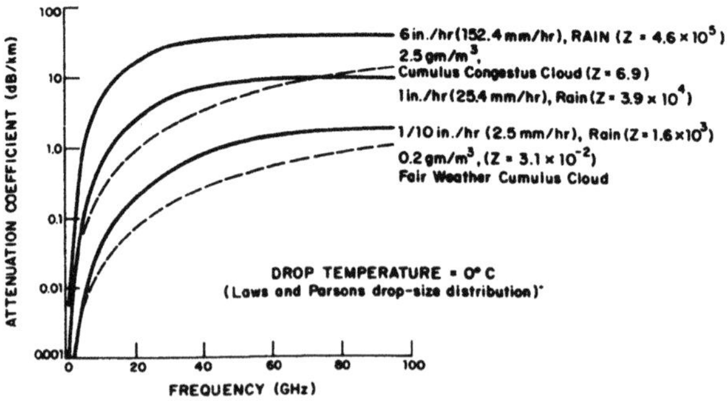

Fig. 5. Attenuation coefficient for liquid-water scatterers.

The constant (function of frequency) of the currently used expression

$$A_s = \alpha R^{\beta} \tag{3}$$

can be obtained from fig. 6 by drawing the logarithmic regression curve R of the scattergram.

A linear regression curve L (β = 1 in the previous expression) can also be derived. The root-mean-square error results to be 22% for the logarithmic fit and 28% for the linear fit: the relatively small difference in error corresponds to the fact that the exponent results to be near unity /3/. The constant α and β in (3) are function of frequency; their determination at other frequencies with the same method of fig. 6 shows that at lower frequencies the r.m.s. error increases, whilst at 35 GHz it reduces to 10% and 12% for the logarithmic and linear regression curve, respectively. Rain rate may be measured by rain gauges at ground: many rain gauges are needed to sample properly the field of surface rain rate during rainstorms, the latters having generally limited transversal dimensions (of the order of 1 km). For slant paths the rain rate along the path cannot be sampled, and only reference to ground measured values can be made /12/, /13/. In /12/ it is found that good correlation exists between zenith attenuation and rain rate measured at the receiving antenna on ground; in /13/ a correlation is tried between slant path attenuation and rain rate measured at ground under the path and properly weighted and averaged.

Measurement of reflectivity by radar is the most direct way for gathering data on rain along the path. Fig. 7 taken from /3/ reports a scattergram of specific attenuation versus reflectivity calculated for the same 5000 raindrop-size distribution using to derive fig. 6. In this case only a logarithmic regression curve

$$A_s = aZ^b \tag{4}$$

gives reasonable results. The r.m.s. error results to be about 30% to 40%. Table 5 shows the values of a and b as a function of frequency, obtained by the above method.

Values for a and b have been given by different authors for matching data in different climatic zones, or in different intensity ranges /14/ /15/. Fig. 8 taken from /15/ shows a reflectivity map obtained with a radar working at 2.9 GHz, e.g., at a sufficiently low frequency able to measure Z without incurring in appreciable attenuation (*).

(*) Investigations by radar operating at frequencies near or above 10 GHz have not given good results.

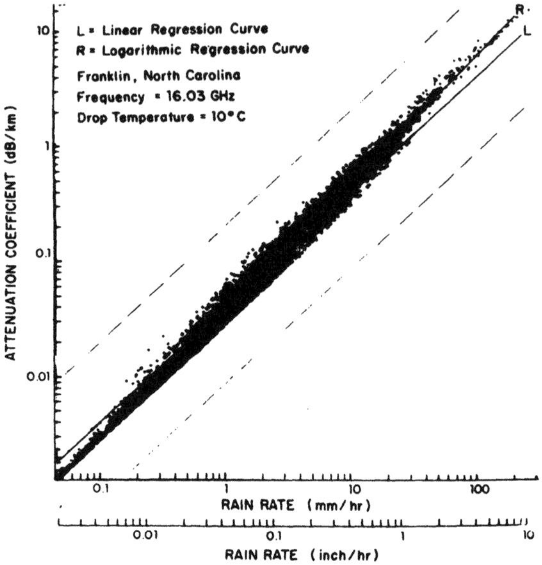

Fig. 6. Scattergram of attenuation coefficient versus rain rate, 16 GHz.

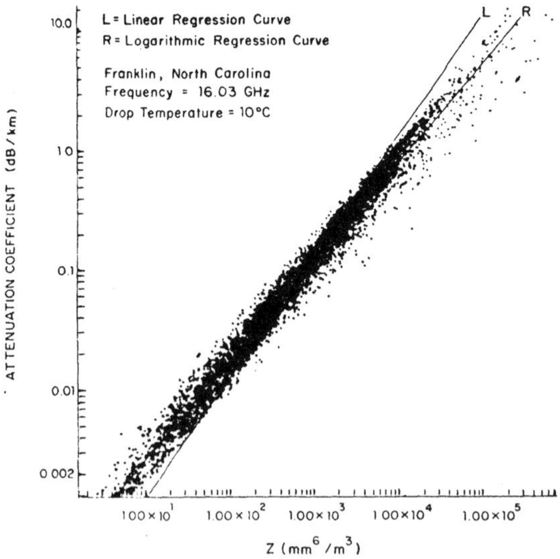

Fig. 7. Scattergram of attenuation coefficient versus Z, 16 GHz.

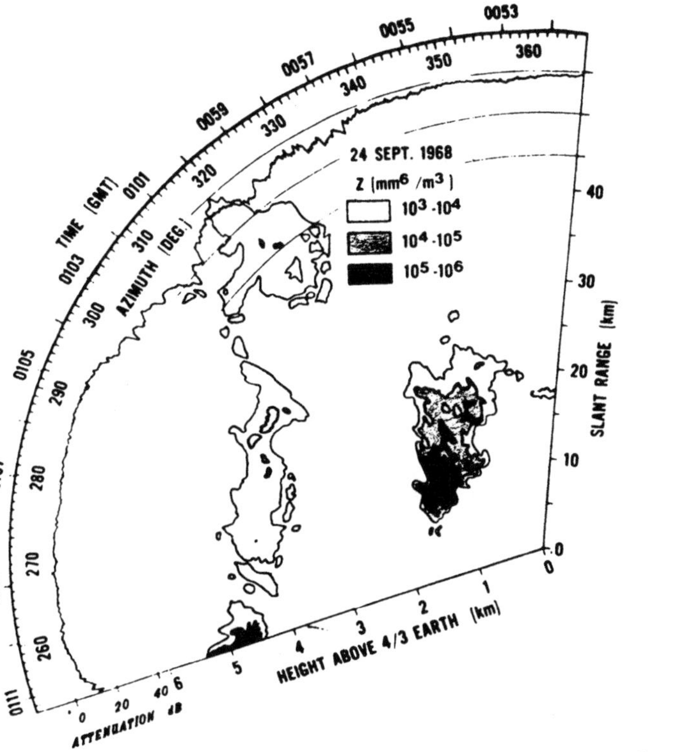

Fig. 8 Radar reflectivity measurements for 24 September 1968

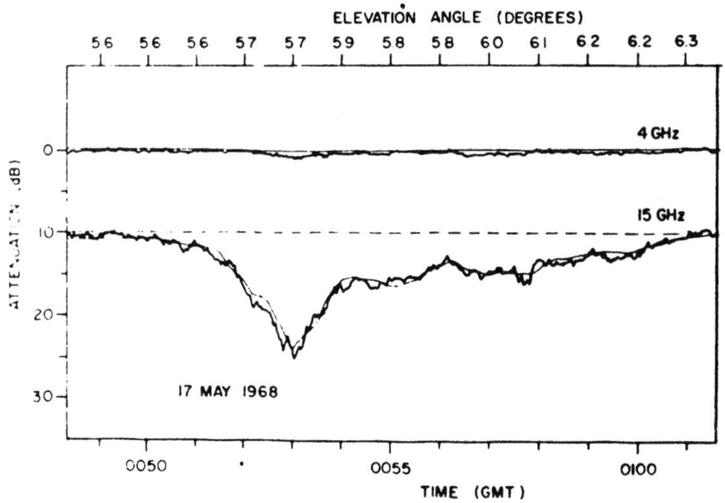

Fig. 9 Measured and calculated attenuation as a function of time

Fig. 9 from the same author (McCormick) shows a comparison between the measured attenuation on a slant path (using a transmitter on an airplane) and the attenuation calculated from radar measurements of Z along the same path. According to McCormick the different drop size distribution which may be encountered and affect the relationship between $A_s(\ell)$ and $Z(\ell)$ do not produce large errors in the evaluation of the attenuation because the radar averages Z over large volumes and moreover another averaging procedure is included in the integration of $A_s(\ell)$ along the path. Good correspondance between measured attenuation and attenuation calculated from radar backscatter measurement have been verified in /16/.

The estimation of attenuation from radar backscatter measurements requires absence of hail or snow which produce backscatter with much less attenuation than water. To avoid errors due to such causes it is necessary to disregard backscatter above the bright band (where snow exists), and to disregard Z values above 10^{-} mm^6/m^3 which very often indicate the presence of hail.

At this point we want to note that in discussing about more or less accurate correspondance between attenuation and radar backscatter measurements (or rain rate measurements) it is necessary to specify what correspondance we are referring to. An instantaneous correspondance like the one investigated in fig. 9 is of course the most desirable, but for many purposes a correspondance between the distribution curve of the calculated attenuation (from backscatter or rain rate measurements) and the distribution curve of the actual attenuation would be sufficient. The latter case may require a sample size on which to derive the statistical distribution larger than the already large sample size required in the former case, but radar backscatter measurements (and to a less extent rain rate measurement by means of a dense array of rain gauges) can give a much larger sample size than a direct attenuation measurement on a given path.

Another method for indirect measurement of attenuation consists in measuring a single integral quantity directly related to the total attenuation through the atmosphere: such a quantity is the temperature of the medium seen by the antenna of a radiometer. It is known that if an absorbing medium is interposed between the sky and a receiving antenna, the effective antenna temperature is

$$T_a = T'_a \frac{1}{L} + \int_0^{\infty} T(\ell) 2\alpha(\ell) e^{-2\int_0^{\ell} \alpha(\ell') d\ell'} d\ell \qquad (5)$$

where: $T(\ell)$ is the temperature of the absorber at distance from the antenna; T'_a is the effective antenna temperature in absence of the absorbing medium (i.e. the sky noise temperature);

$$L = e^{2\int_0^{\infty} \alpha(\ell)\, d\ell}$$

the total loss due to the absorber; $\alpha(\ell)$ is the attenuation coefficient of the absorber.

If the temperature of the absorbing medium has a constant value T_r formula (5) reduces to

$$T_a = T'_a \frac{1}{L} + T_r \left(1 - \frac{1}{L}\right) \tag{6}$$

If $T(\ell)$ is not constant, T_r may be interpreted as an apparent absorber temperature. The first term in (6) is of importance when the sun is used as a source of radiation in space as discussed previously; in this case the second term is balanced off during the measurements. On the contrary when the antenna is pointed away from the sun, only the second term is of importance, i.e.,

$$T_a = T_r \left(1 - \frac{1}{L}\right) . \tag{7}$$

Formula (7) allows to calculate the total loss L from measurements of the antenna temperature T_a, when T_r is known. Clearly a precise determination of L from T_a is possible only for moderate value of L. Formula (7) is used for radiometric measurements of the attenuation by rain: in this case, however, the attenuation is due not only to absorption but also to scattering so that expression (7) is rigorously valid only when scattering is small in comparison to absorption, i.e. in the lower range of the frequencies under consideration and/or for the lowest rain rates.

A very good correspondance between calculated attenuation from T_a and measured attenuation, up to attenuation of the order of 10 dB, has been reported by many sources /8/ /13/ /16/ /17/ /18/ /28/ (see for example fig. 10 taken from /8/) using appropriate values for T_r (between 270°K and 284°K). Therefore the radiometer represents a very simple and valuable mean to measure moderate atmospheric attenuations over slant paths without requiring any microwave source in space. According to Tillotson /8/ the measuring range of the radiometer can be extended above the mentioned range of 10 dB of attenuation, by carrying out the measurements at lower frequencies and then scaling the result to higher frequencies. This procedure requires of course a good correspondence between attenuation at various frequencies: from experimental data the ratio of the attenuation at two different frequencies appears to be often almost constant during a single storm but quite variable from one storm to another /13/ /19/ due to the different dimensions of storms: in fact for the same attenuation the storm having less transversal dimension has higher rain intensity and therefore different raindrop-size distribution. Therefore a good correspondence between attenuation at various frequencies can exist in the sense previously given referring to the distribution curves.

Statistical data on atmospheric attenuation presently available

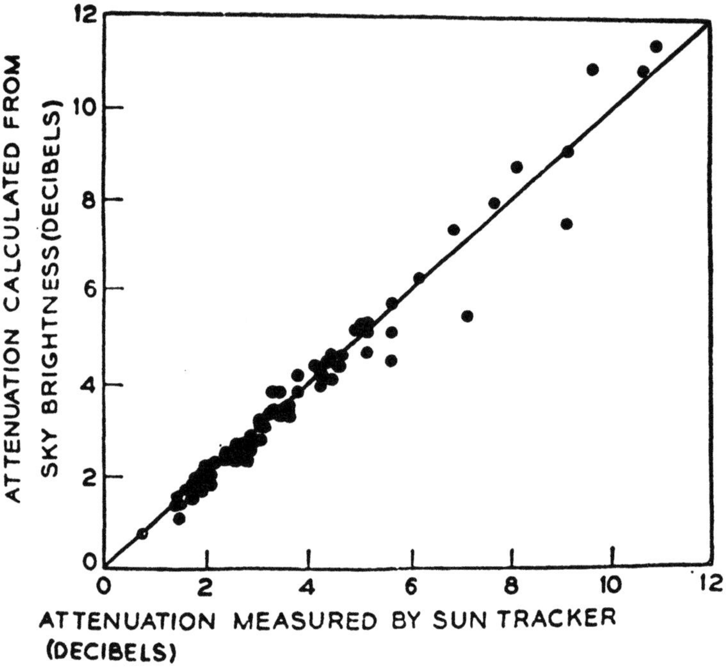

FIG. 10

for earth-space paths derive mainly from measurements of solar radiation attenuation and from measurements of the sky temperature /20/ /19/ /21/ /22/ /23/ /24/ /26/ /27/. Some data have been derived from radar backscatter measurements /25/ /28/. Data from ATS-5 satellite have been also given with reference to a smaller number of observations /13/ /29/.

Published data referring to several years, like the ones attained in Japan and in United Kingdom /20/ /22/ show a noticeable variation from year to year in the distribution curves of the attenuation, confirmigg the need of many years of experiemnts to acquire valid statistical information (see for example fig. 11 taken from /20/.

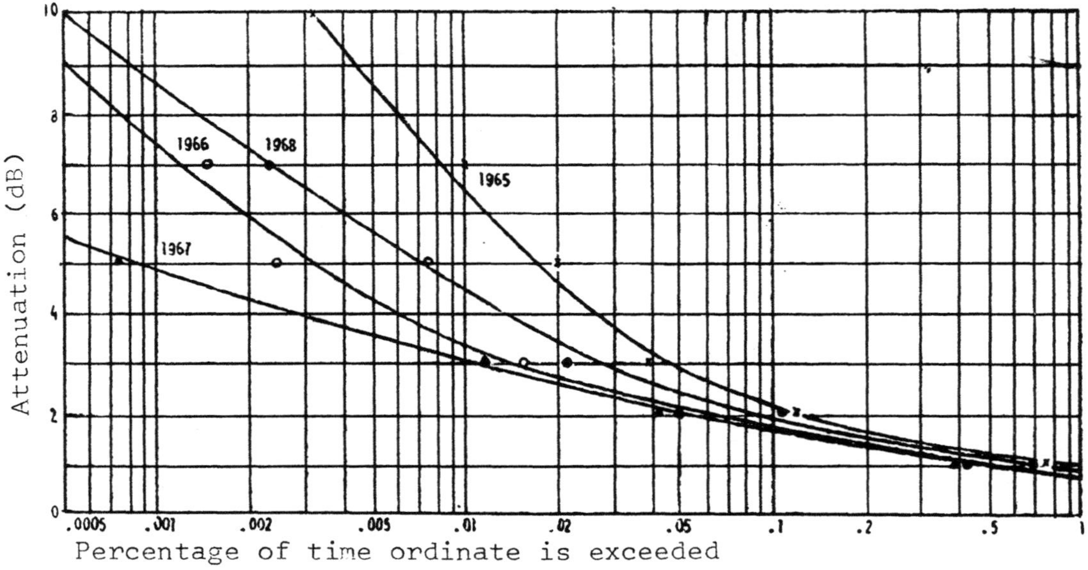

Fig. 11 Yearly distribution of attenuation due to precipitation: best-fit curves to measured values

TOYOKAWA 9.4.GHz

Comparison between results obtained in various parts of the world are given in paper /29/.(see fig. 12): it is evident that the distribution curves obtained in different locations are very different, so that a different fading margin has to be assumed in the design of satellite links in different parts of the world.

A general remark may be made: if for example in the design of a satellite communication system at frequencies above 10 GHz a margin of 10 dB for attenuation due to rain (*) is considered as acceptable in the economy of the system, this margin may give sufficient reliability in the lower part of the considered spectrum and/or in regions where rain is not very intense. It may suffice for example in the United Kingdom till 19 GHz, but not in the East Coast of the United States also at 16 GHz. When a very large margin would be required, some sort of adaptive operation may be considered /30/.

A higher power may be transmitted in the direction where a higher attenuation is experimented. For example a steerable narrow beam may allow to direct more power in the required direction. Among adaptive methods,a very attractive and flexible one is the space diversity method, which takes advantages from the limited spatial extent of heavy showers. Results of dual diversity experiments have been reported in /21/ and /31/, as shown in fig. 13 taken from /31/. The performance of a dual diversity system has been calculated by Strickland /28/ from radar reflectivity data: the results are reported in table 6. In this table rows a, b, c, d refers to a diversity system using two stations, each with a fading margin of 3 dB: a) and b) give the separation required between the two stations to obtain the same performance as a single station with a fading margin of 10 dB and 15 dB respectively; c) and d) give the separation required to reduce the outage time by a factor of 10 and 40 respectively; e) is as d) but for two stations each with a fading margin of 10 dB instead of 3 dB.

The advantage of diversity is very clear from the above; one has to consider also the possibility of diversity of higher order, to be used in a certain system according to the meteorological characteristics of the region in which an earth receiving system is to be installed.

(*) Remember that rain affects the performance of a link by increasing the attenuation and the antenna noise: as seen from (7) the antenna noise temperature reaches temperatures above 200°K for attenuations of a few dB: it follows that in general the use of a cooled low-noise amplifer is not economically justified.

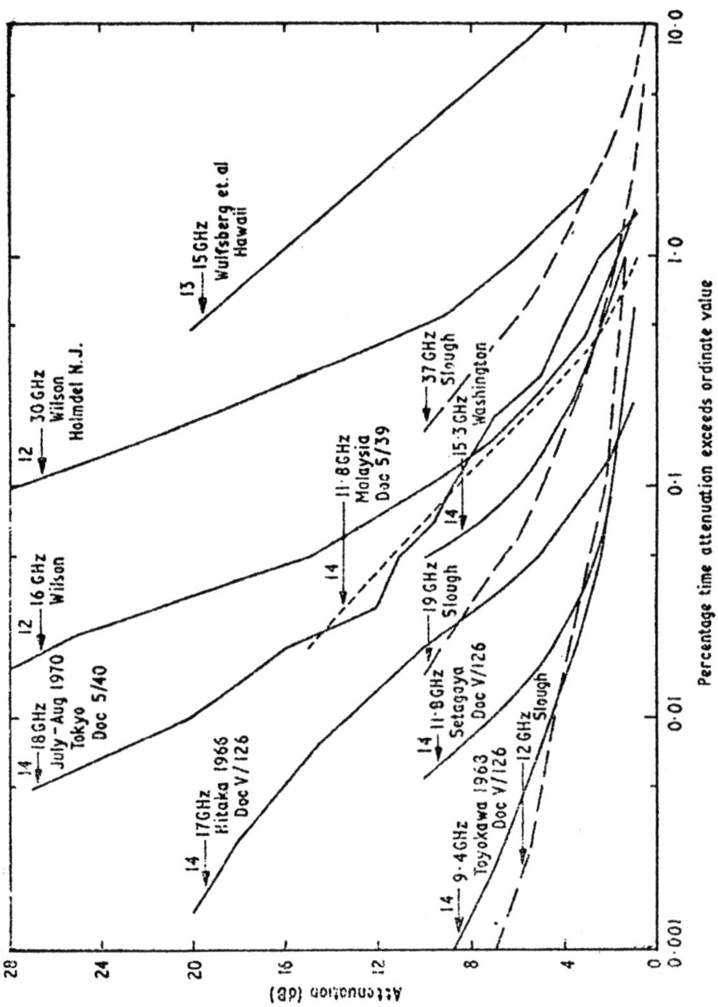

Figure 12 Comparison of the attenuation statistics for Slough with those obtained elsewhere.
Caption for ordinate scale : Attenuation (dB)
Caption for abscissa scale : Percentage time attenuation exceeds ordinate value.

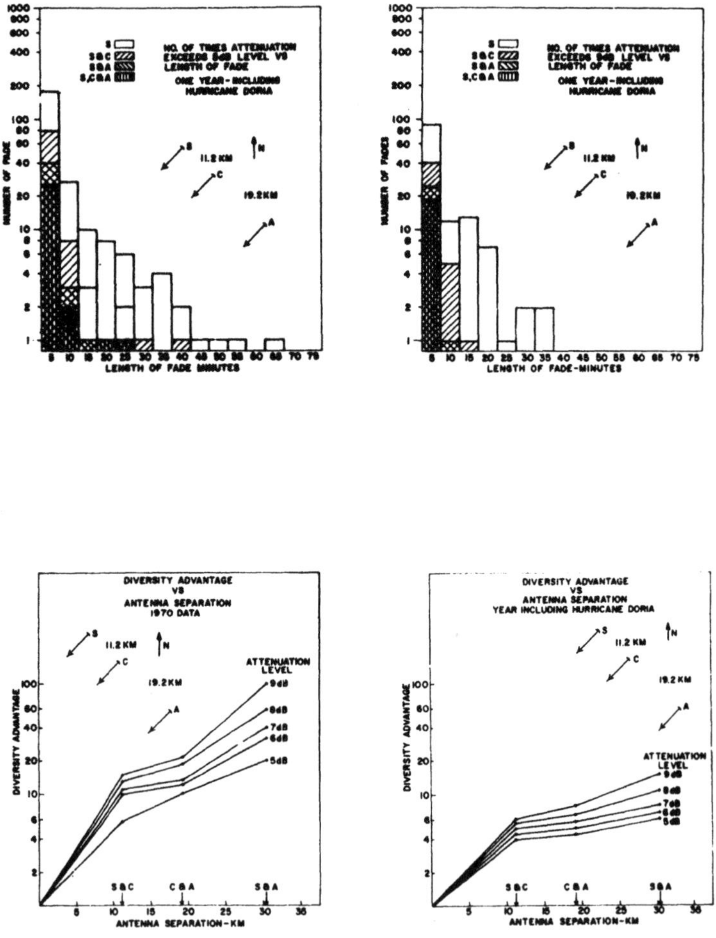
NO. OF TIMES ATTENUATION EXCEEDS 5dB LEVEL VS LENGTH OF FADE
ONE YEAR-INCLUDING HURRICANE DORIA
NUMBER OF FADES
LENGTH OF FADE-MINUTES
11.2 KM
19.2 KM
DIVERSITY ADVANTAGE VS ANTENNA SEPARATION 1970 DATA
DIVERSITY ADVANTAGE VS ANTENNA SEPARATION YEAR INCLUDING HURRICANE DORIA
ATTENUATION LEVEL
DIVERSITY ADVANTAGE
ANTENNA SEPARATION-KM

Fig. 13

3. OTHER PRECIPITATION EFFECTS

As mentioned before, other precipitation effects on propagation may be: - spectral distortion (still to be experienced); - depolarization (considered in another session of this meeting); - interference due to scatter by rain. The latter is a very important aspect to which much attention is to be dedicated for a proper design of actual systems /32/ /33/ /34/ /35/.

REFERENCES

/1/ S. Metzger: Communication satellite systems in the 70's Sumposium on long-term prospects for satellite communication, Istituto Internazionale delle Comunicazioni, Genova, June 1971.

/2/ G.H. Millman: Tropospheric effects on Space Communications - AGARD Conference Proceedings No. 70 on Tropospheric Radio Wave Propagation, 46; 1971.

/3/ T.K. Crane: Propagation phoenomena affecting satellite communication sustems operating in the centimeter and millimeter wavelength bands - Proceeding IEEE, 59, 2, p. 173, February 1971.

/4/ C.C.I.R. Rep. 234-2: Influence of tropospheric refraction and attenuation on space telecommunication systems - XII Plenary Assembly, New Delhi 1970, vol. II Part 1, p. 169.

/5/ R.G. Medhurst: Rainfall attenuation of centimeter waves: comparison of theory and measurement - IEEE Transactions on Antenna and Propagation, AP13, p. 550, July 1965.

/6/ F. Eklund: Radio propagation above 10 GHz: ground-to-ground communication - Presented at the XVII General Assembly of URSI, Comm. 2, Warsaw, September 1972.

/7/ B.C. Blevis: Radio propagation above 10 GHz: ground-to-space communications, including effects of precipitations - Presented at the XVII General Assembly of URSI, Comm. 2, Warsaw, September 1972.

/8/ L.C. Tillotson: Use of radio frequencies above 10 GHz for satellite communications - Symposium on long terms prospects for satellite communication, Istituto Internazionale delle Comunicazioni, Genova, June 1971.

/9/ J. Joss, A. Waldvogel: Ein Spectrograph für niederschlagstropfen mit automatischer answertung - Pure Applied Geophys. Vol. 68, p. 240, 1967.

/10/ J.O. Laws, D.A. Parsons: The relation of raindrop-size to intensity - Am. Geophys. Union Trans., Vol. 24, p. 452, 1943.

/11/ D.E. Setzer: Computed transmission through rain at microwave and visible frequencies - Bell Syst. Techn. Journ., Vol. 49, p. 1873, October 1970.

/12/ K.N. Wulfsberg, E.E. Altshuler: Rain attenuation at 15 and 35 GHz - IEEE Trans. on Antenna and Propagation, AP-20, 2, p. 181, March 1971.

/13/ L.J. Ippolito: Effects of precipitations on 15,3 and 31,65 GHz earth-space transmission with the ATS-V satellite - Proc. IEEE, 59, 2, p. 189, February 1971.

/14/ R.R. Rogers: Application of weather radar data to propagation questions - AGARD Conference Proceedings No. 70 on Tropospheric Radio Wave Propagation, 46, 1971.

/15/ K.S. McCormick: Simultaneous Measurements of Precipitation Attenuation and Radar Reflectivity at Centimeter Wavelengths - AGARD Conference Proceedings No. 70 on Tropospheric Radio Wave Propagation, 46, 1971.

/16/ A.W. Straiton, B.M. Fannin: Comparison of 15 GHz propagation data from the ATS-5 satellite with ground based radio and meteorological data - AGARD Conference Proceedings No. 70 on Tropospheric Radio Wave Propagation - 46, 1971.

/17/ E.E. Altshuler, K.N. Wulfsberg, V.J. Falcone, Jr.: Rain attenuation at millimeter wavelengths - AGARD Conference Proceedings No. 70 on Tropospheric Radio Wave Propagation, 46, 1971.

/18/ C.C.I.R. Doc. 4/72-E Doc. 5/72-E, February 1972, Canada: Comparison of direct and indirect measurements of rain attenuation at 15,3 GHz.

/19/ R.W. Wilson: Sun tracker measurements of attenuation by rain at 16 and 30 GHz - Bell system Technical Journal, 48, 5, p. 1383, May-June 1969.

/20/ C.C.I.R. Doc. IV/316-E V/126-E June 1969, Japan: Statistics of attenuation of radio waves at frequencies above about 10 GHz at higher angles of elevation in Japan.

/21/ R.W. Wilson: A three radiometer path-diversity experiment - Bell System Techn. Journ., 49, 6, p. 1239, July-August 1970.

/22/ P.G. Davies: Radiometer studies of atmospheric attenuation of solar emission at 19 GHz - Proc. IEEE, 118, 6, p. 737, June 1971.

/23/ P.G. Davies: Comparison of attenuation statistics at 19 and 37 GHz for sun-earth paths - Electronics Letters, 7, 2, p. 51, January 1971.

/24/ C.C.I.R. Doc. 5/39-E, December 1971, Japan: Statistics of attenuation due to precipitation of radio waves in the 12 GHz band at higher angles of elevation in a tropical zone.

/25/ R.R. Rogers, G. Drufuca, I.I. Zawadazki: Attenuation statistics from weather radar observations.

/26/ P.G. Davies: Comparison of attenuation statistics at 19 and 37 GHz for sun-earth-paths -Electronics Letters, 7, 51, 1970.

/27/ P.G. Davies, J.A. Lane: Statistics of tropospheric attenuation at 19 GHz from observation of solar noise. - Electronics Letters, 6, 522, 1970.

/28/ J.I. Strickland: Comparison of direct and indirect measurements of precipitation attenuation at 15,3 GHz - AGARD Conference on Telecommunication. Aspects on Frequencies between 10 and 100 GHz, Gausdal (Norway) 1972.

/29/ P.G. Davies: Slant path attenuation at frequencies above 10 GHz, ibid.

/30/ W.P. Robin: Techniques for dealing with the effects of bad weather in satellite communication systems.

/31/ R.W. Wilson, W.L. Mammel: Results from a three radiometer path diversity experiment. IEEE Conference on Propagation of Radio Waves at Frequencies above 10 GHz - Londra, 1973.

/32/ A. Buidge, J. Levatich: Measurement of precipitations scatter effects on propagation at 6, 12 and 18 GHz. AIAA 3rd Comm. Satellite Systems Conference, Los Angeles, April 1970.

/33/ A. Buidge, J. Levatich: Propagation experiments above 10 GHz for application to communications satellite systems. AIAA 3rd Comm. Satellite Systems Conference - Los Angeles, April 1970.

/34/ C.C.I.R. Report 339-1 (Rev. 72), "Influence of Scattering from Precipitation on the siting of Earth and Terrestrial Stations", and Report AC.5, "Simplified Propagation Model for the Determination of Coordination Distance in the Frequency Range 1-40 GHz", Conclusions of the Interim Meeting of Study Group 5, Propagation in non-ionized media (International Telecommunications Union, Geneva, 1972).

/35/ R.K. Crane: Analysis of data from the Avon-to-Westford Experiment. Tech. Rep. 498 Lincoln Lab., January 1973.

IV GROUP

Atmospheric Influence on Radio Propagation above 10 GHz: E.M. Propagation Phenomena and Results

ATTENUATION BY RAINFALL IN THE SUBMILLIMETRE WAVE REGION

D.T. Llewellyn Jones

Radio and Space Research Station
Ditton Park
Slough
Buckinghamshire, England

INTRODUCTION

In this paper discussion will mainly be concerned with the frequency range between 100 and 1000 GHz. The physical basis for the Mie theory (ref. 1) of scattering by spherical particles and its relation to rainfall attenuation is briefly reviewed. The results of theoretical calculations of attenuation will be presented and compared with available experimental data. It will be seen that although attenuation does not continue to increase with frequency in the submillimetre region the dropsize distribution plays a crucial role in determining the attenuation expected.

THE MIE THEORY

The following discussion is principally qualitative, quantitive treatments are given in detail, for example, Stratton (ref. 2) and Kerr (ref. 3). Consider a dielectric sphere in an electromagnetic field where the wavelength is long compared to the dimension of the sphere. The sphere will acquire an induced dipole moment which performs forced oscillations driven by the electric field, and as an oscillating dipole it radiates power which it has extracted from the electromagnetic field. In other words it scatters some of the incident radiation. The case where the wavelength is long compared to the dimensions of the sphere is known as that of Rayleigh or dipole scattering. The analysis is well-known and leads to the result that the scattering cross-section of the sphere increases as the fourth power of the frequency.

In many cases the sphere will have a complex dielectric constant, it will not only refract but also absorb radiation. The absorption will have the effect of damping the forced oscillations and some of the energy from the electromagnetic field will be used to heat the sphere and the amount of re-radiated, or scattered, power reduced. In the practical case of attenuation by rainfall, where the dielectric 'spheres' are raindrops, some of the energy is scattered and some absorbed, but it is important to realize that absorption in this case is a modification of the scattering process and not an independent one.

As the wavelength of the incident radiation decreases and becomes comparable to the size of the sphere the problem rapidly becomes more complex. The dipole scattering reaches maximum efficiency when the circumference of the sphere is exactly one wavelength and this condition is called dipole resonance. Similarly as the circumference approaches two wavelengths a quadrupole resonance occurs, followed in turn by hexapole, octupole, and so on until the cross section becomes a highly complex linear superposition of these multipole radiators which finally converges, in the short wavelength limit, to the limiting or 'geometrical' value.

This process is represented in fig. 1 which shows the normalized scattering cross-section of a sphere as a function of the proper variable, circumference/wavelength. The transition, as this ratio increases, from the Rayleigh scattering region to the geometrical limit is clearly seen. Also shown are the induced electric field patterns within the sphere which correspond to each of the first four resonances, as presented by Mie.

The occasional observation of the "blue sun" (the opposite of the red sunset) is a spectacular manifestation of the negative slope (cross-section decreasing as frequency increases) which follows the first resonance. This was observed and analysed spectroscopically in September 1950 by R. Wilson (ref. 4) at the Royal Observatory, Edinburgh, who identified the phenomenon as Mie scattering due to fine, and presumably highly uniform, particles (which were subsequently discovered to have come from Canadian forest fires).

ATTENUATION BY RAINFALL

It has already been mentioned that the effect of dielectric absorption within the scattering sphere is to damp the forced multipole oscillations. This has the effect of "smoothing out" the undulation in the cross-section curve of fig. 1. In the case of rain there is a distribution of drop sizes which also contributes to this.

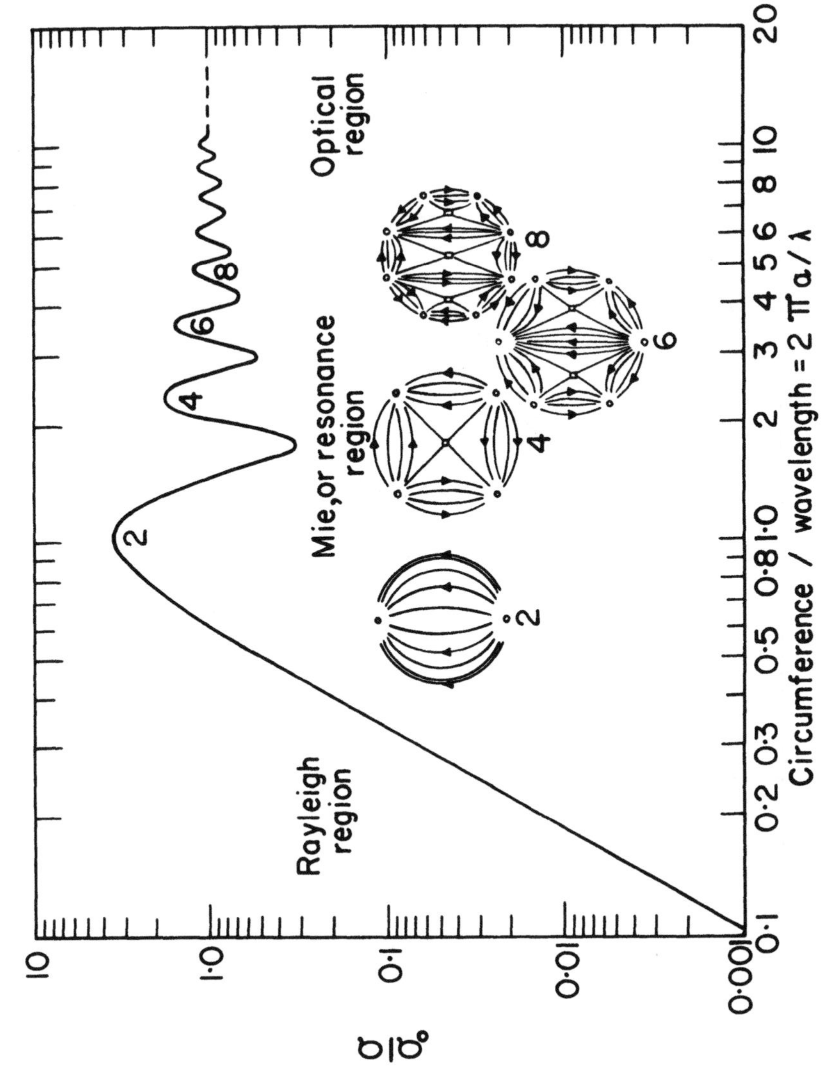

Fig. 1. The normalized scattering cross-section of a dielectric sphere, and its first four resonant electric field patterns

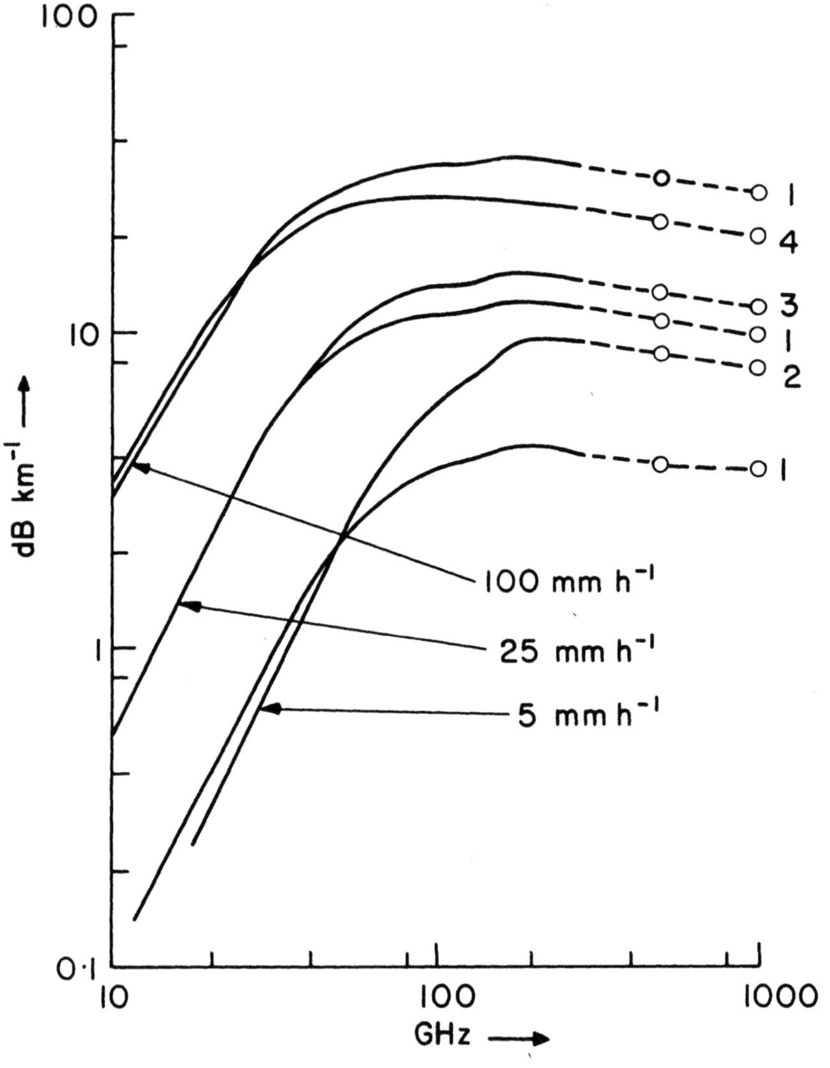

Fig. 2. Calculated values of attenuation due to rainfall in the millimetre and submillimetre regions.

Key to dropsize distributions: Laws and Parsons, 1; Joss et al, "drizzle" 2, "widespread" 3, "thunderstorm" 4.

Dropsize distributions have been studied by Laws and Parsons (ref. 5) and later by Marshall and Palmer (ref. 6) and by Joss et al. (ref. 7). Calculations of expected attenuation can be made on the basis of these empirical and commonly accepted dropsize distributions and application of the Mie theory. Such calculations were first performed by Ryde and Ryde in 1946 and refined by Medhurst (ref. 8) among others. In fig. 2 such calculations have been repeated and extended to a frequency of 1000 GHz. The resemblance between these curves and those of fig. 1 is obvious. The dropsize distributions used are indicated on the figure. They are the Laws and Parsons, used for each of the three rainfall rates considered, and three Joss et al. distributions, based on the Marshall and Palmer formulation, which apply to "thunderstorm", "widespread" and "drizzle" type rain. Although only one of the Joss distributions is shown for each rainfall rate in fig. 2 it is generally true that above 70 GHz the Laws and Parsons distribution yields attenuation levels between those due to "widespread" and "thunderstorm" type rain and that "drizzle" type rain leads to the highest values of predicted attenuation.

The important feature of this set of curves is the strong dependance of attenuation upon dropsize distribution at frequencies above 70 GHz. In fact at low rainfall rates, up to 5 mm h^{-1}, where Joss et al.'s "drizzle" type rain is likely to occur it is clearly reasonable to expect a variability of a factor of 2 in attenuation due to a given rainfall rate. Therefore it is clear that it is hard to perform definitive experiments of rainfall attenuation at frequencies above 70 GHz unless the dropsize distribution is properly determined. This situation is contrasted by the region near 35 GHz, where the influence of dropsize distribution is almost insignificant, and observations have been reported by Norbury and White (ref. 9) which show satisfactory agreement with the values shown on fig. 2.

OBSERVATIONS

There are two reported cases of measured rainfall attenuation in the submillimetre wave region. The first are due to Babkin et al. (ref. 10) at a wavelength of .96 mm and of rainfall rates up to 12 mm h^{-1}. In table 1 their results are summarized and compared with the calculated values from which fig. 2 is derived.

Rainfall Rate mm h^{-1}	Measured Attenuation (dB km^{-1})	Calculated Attenuation			
		1	2	3	4
5	4.8	4.1	9.2	5.2	3.2
12	7.8	7.3	16.4	9.2	5.8

Table 1. Observations of Babkin et al. compared to calculated values from various dropsize distributions (1, Laws and Parsons; 2, Joss et al. "drizzle"; 3, Joss et al. "Widespread"; 4, Joss et al. "Thunderstorm")

As these observations fell comfortably within the limits of the calculated values they in no way contradict the Mie theory.

The other observations, by Llewellyn Jones and Zavody, (ref. 11) were of simultaneous values of attenuation, over the same propagation path, at frequencies of 110 GHz and 890 GHz and at rainfall rates up to approximately 36 mm h^{-1}. The results indicated values of attenuation at 890 GHz that were about 25% higher than those at 110 GHz. This is not consistent with the values shown in fig. 2 unless it is suggested that all the rain was "drizzle" type, or, more plausibly, that there are rather higher concentrations of airborne small drops than is usually indicated from the "catch" of a ground-sited dropsize analyzer.

SUMMARY AND CONCLUSIONS

There are very few observations of rainfall attenuation in the submillimetre wave region, but they do not give any grounds to question the validity of the Mie theory or its application to this problem. On the contrary the need is stressed to make careful observations not only of rainfall rate and attenuation, but also of airborne dropsize distribution. This last quantity obviously presents practical difficulties, and there is a case for also making simultaneous observations at several frequencies in order to observe the shape of the curves shown in fig. 2, thereby eliminating to some extent the variable dropsize distribution.

The author wishes to thank Dr. J. R. Norbury of R.S.R.S. for making available his calculations of attenuation as a function of rainfall rate and also Dr. J. A. Lane (of R.S.R.S.) for his helpful advice.

REFERENCES

(1) Mie, G.; Ann. Physik, 25, 377, 1908.

(2) Stratton, J.; "Electromagnetic Theory", McGraw-Hill, 1941.

(3) Kerr, D.E.; Propagation of Short Radio Waves, Dover Publications, 1965.

(4) Wilson, R.; Mon.Not.R.astr.Soc.,111,477,1951.

(5) Laws, J.O., and Parsons, D.A.; Trans.Amer.Geophys.Union, 1943, 24, 432, 1943.

(6) Marshall, J.S., and Palmer, W.Mck.; J.Meteorol.,5,165,1948.

(7) Joss, J., Thams, J.C., and Waldvogel, A.; Proc.Int.Conf.on Cloud Physics, Toronto, 1968, pp.369-373.

(8) Medhurst, R.G.; I.E.E.E.Trans.,AP-13,550,1965.

(9) Norbury, J.R., and White, W.J.K.; Electron.Lett.,8,91,1972.

(10) Babkin, Yu.S., et al. Radio Engng. and Electron.Phys.,15, 2164,1970.

(11) Llewellyn-Jones, D.T., and Zavody, A.M.; Electron.Lett.,7, 321,1971.

RECENT STUDIES OF VISIBILITY AND ATMOSPHERIC ATTENUATION AT OPTICAL WAVELENGTHS

D.R. Hay

University of Western Ontario
London, Canada

ABSTRACT

The transmission of optical waves through the atmosphere is governed by atmospheric refractivity and extinction coefficient. Attenuation is mainly through scattering by aerosol particulate, air molecules and precipitated water. There is growing evidence of the effects of high humidity on particle scattering, and several laboratory studies of particle scattering are underway. Optical visibility through the atmosphere is being examined in lidar experiments. Optical scintillation associated with thermal turbulence is important to ocular hazards in laser signal transmission, and as a means of remote wind indication.

The requirement to see through the earth's lower atmosphere has taken on new significance in recent years. This has been prompted partly by the need for new frequency allocation in communications, and partly by the special advantages of optical wavelengths for some types of atmospheric measurement. This paper will review briefly some recent contributions to optical radiometeorology, where attention is confined to wavelengths in the region 0.4 - 0.7 μm.

(a) OPTICAL TRANSMISSION

Consider a monochromatic, plane optical wave passing through a homogeneous volume of the atmosphere. If the complex refractive index of the medium is

$$m = n - ik \tag{1}$$

where n is the phase refractive index and k is the absorption index, then the electric field of the wave varies with time (t) and penetration of the medium (x) as

$$\exp\{i\omega(t - \frac{mx}{c})\} = \exp(\frac{-2\pi kx}{\lambda}) \exp\{i\omega(t - \frac{nx}{c})\} \quad (2)$$

Hence, the wave intensity decreases with penetration according to the Bouguer-Beer-Lambert law:

$$I = I_o \exp(-\alpha x) \quad (3)$$

where α is the extinction coefficient of the medium, and αx is defined as the turbidity.

We note that k in Eq(1) is related simply to extinction coefficient in Eq(3):

$$\alpha = \frac{4\pi k}{\lambda} \quad (4)$$

This simple model describes extinction of the wave by absorption only. When the model is extended to include re-direction of part of the incident wave through scatter by particles and molecules, then Eq(3) still applies and the extinction coefficient includes both absorption and scattering.

At optical wavelengths, the extinction coefficient of the lower atmosphere depends upon four characteristics of the atmospheric components. These are their shape, size (relative to the optical wavelength), concentration and complex refractive index. Experiments demonstrate that α may vary widely over the range of optical wavelengths and through the normal course of weather changes.

(b) SOURCES OF OPTICAL EXTINCTION

Gases in a clean atmosphere are responsible for only weak extinction through absorption. Specifically, molecular resonance absorption by oxygen, ozone and water vapour are the major contributors (Kingslake 1965, p. 131; Harvey 1970, sect. 25.3; Pressley 1971, p. 50). For example, the light from a broad-band source is attenuated by 5% over a 1 km path at the ground, for an air temperature 15°C and relative humidity 10%. The attenuation increases to only 20% as the relative humidity approaches saturation. At this stage, however, details in the atmospheric absorption spectrum are not readily derived from laboratory measurements on the individual molecular resonances (Derr 1972, ch. 9). An indication of the complexity of detail in atmospheric absorption is seen in the absorption spectrum for the wavelength interval 0.6933 - 0.6946 μm as derived from radiometer measurements.

A number of theoretical and experimental studies are underway currently to investigate optical extinction by particulate matter and precipitated water in the air. These significant sources of attenuation have been confirmed by long-term radiometer records (Irvine et al. 1970). The particulate may be in the form of dust, plant spores, silicates, sulphates, carbon etc.; precipitated water appears as smog, clouds and fog, mist, drizzle and rain (Bond and Straub 1972). Since the particle diameters are within the range 0.001 - 100 μm, Rayleigh and Mie scattering theory are used to calculate optical extinction coefficients (for example, Rensch et al. 1970, Bergstrom 1972). Laboratory measurements of scattering by various types of particulate are being made to test the theory (Covert et al. 1972, Quiney et al. 1972). Table 1 gives some representative extinction coefficients, in the second column.

TABLE I

Representative Optical Extinction Coefficients

Particulate	$\alpha(m^{-1})$	V(approx.)
pure air	0.000016	240 km
pollutant particles ($5x10^{10}m^{-3}$)	0.0005	8 km
haze	0.0005	8 km
fog	0.01	400 m
cloud	0.16	25 m
rain (20mm hr^{-1})	0.0015 (red)	2.6 km

The extinction of a HeNe laser beam through rain has been studied by Setzer (1970). Mie scattering theory with the geometrical optics approximation has been applied, using forward scatter enhancement of the direct transmitted signal. A limited comparison shows reasonably good agreement between the predicted and measured rain attenuation.

Covert et al. (1972) have provided further evidence of the strong interaction between particulate extinction coefficient and environmental humidity. It is believed that the particulate diameter grows through the adsorption of water vapour. Laboratory measurements show that the interaction is relatively weak until relative humidity exceeds about 80%; thereafter, the extinction coefficient increases very rapidly.

The inverse problem of deriving information on the physical characteristics of the particulate from their optical scattering also is receiving attention. Cohen (1972) and Westwater et al. (1973) indicate theoretically that the particle size distribution may be obtained from the measured scattering intensity over a range of wavelengths.

(c) VISIBILITY

The advent of lasers has renewed interest in the familiar concept of optical visibility. For many years, trained meteorologists have estimated horizontal visibility by observing selected dark targets against the horizon sky through a uniformly illuminated atmosphere. Koschmieder analyzed this procedure by taking into account atmospheric scattering and extinction along the observing path, the apparent brightness of the dark object and the adjacent sky background, and the brightness threshold of the eye, to show that (see, for example, Berry et al. 1945, p. 242)

$$\text{visual range (V)} = \frac{3.91}{\alpha} \tag{5}$$

The third column of Table I gives the appropriate visibilities for the indicated particulate loading of the air. Recently, Horvath (1971) has examined the accuracy of Eq(5); he has concluded that the derived α is precise within 10% if the targets are sufficiently large and they are viewed in their own shadow, and if α is interpreted as the mean along the optical path. Further, Bertolotti et al. (1969) have demonstrated the feasibility of measuring V over a wide range of distances with the aid of a pulsed ruby lidar (laser radar) and special targets.

The laser radar* offers potential in the measurement of α-profiles through the atmosphere. In the author's companion paper, it is noted that the lidar will indicate the variation in α with distance (R) along an atmospheric path with certain assumptions, according to an equation of the form:

$$\alpha(R) = \alpha(R_o)\ f_1(P_r, R) \tag{6}$$

where R_o is a nearby reference range, P_r is received power at the lidar, and f_1 is a specified function. A local measurement of visibility or α is required to establish $\alpha(R_o)$, through Eq(5) or

---

*For a more detailed discussion see the author's companion paper: "Some recent lidar experiments on particle scattering in the lower troposphere", NATO Study Institute, Sorrento, June 1973.

(3). Then profiles of $\alpha(R)$ are computed from the lidar signatures. Collis et al. (1971) and Viezee et al. (1973) describe an application of this principle to study the dispersal of fogs at an airport by an array of propane heaters on the ground. The author also is employing this principle to examine visibility over an urban area, and its relationship to local weather (Hay et al. 1973). Sample profiles of $\alpha(R)$ as observed over the city of London in the latter experiment show local increases in atmospheric extinction coefficient above roadways, a river, parking lots etc.

(d) OPTICAL SCINTILLATION

A third aspect of optical transmission of current interest is the effect of atmospheric turbulence upon the progressing wave. Here, we are concerned with the real part of the atmospheric refractive index, n, in Eq(1). At optical wavelengths, the refractivity is essentially governed by air pressure (p) and temperature (T) according to

$$(n-1)10^6 = N = 79 \frac{p(\text{mb})}{T(^{\circ}\text{K})} \tag{7}$$

Fluctuations in refractivity are directly related to temperature fluctuations in the air turbulence (see, for example, Lawrence et al. 1970; Derr 1972, ch. 25).

One effect of atmospheric turbulence upon optical propagation is the production of scintillation in the observed intensity. Here, local "hot spots" appear momentarily within the beam, with 10-100 times the average beam intensity. These scintillations are of importance to ocular safety, when the light source is a powerful laser. Dabberdt (1972, 1973) has observed fluctuations in laser beam intensity over several atmospheric paths, in conjunction with auxiliary measurements of wind and temperature fluctuations under various atmospheric stabilities. This has provided statistics on optical scintillation and a guide for evaluating atmospheric effects on laser eye safety.

Measurement of the average transverse wind along an optical path may be obtained by observing optical scintillation. The natural turbulence provides a wind indicator (Gray et al. 1970). When two detectors at fixed spacing observe the beam of a remote laser, the time-lagged covariance function of the recorded signals is computed, and its slope at the time-lag origin represents the turbulence drift velocity across the optical path (Lawrence et al. 1972).

(e) REFERENCES

Bergstrom, R.W. Jr., 1972. Atmos. Environ. 6, 247-258.
Berry, F.A., E. Bollay, N.R. Beers, 1945. Handbook of meteorology. McGraw-Hill, New York.
Bond, R.G., C.P. Straub (editors), 1972. Handbook of Environmental Control. Vol I: Air pollution, CRC Press, Chemical Rubber Co., Cleveland.
Cohen, A., 1972. Appl. Optics, 11, 1657-1658.
Collis, R.T.H., W. Viezee, E. Uthe, J. Oblanas, R.A. Roberts, 1971. J. Geophys. Res., 76, 5104-5109.
Covert, D.S., R.J. Charlson, N.C. Ahlquist, 1972. J. Appl. Meteorol., 11, 968-976.
Dabberdt, W.F., 1972. "Guide for the evaluation of atmospheric effects on laser eye safety," Tech. Report on Project 1341, Stanford Res. Inst., Menlo Park, California.
Dabberdt, W.F., W.B. Johnson, 1973. "Analysis of multiwavelength observations of optical scintillation," accepted for publication in Appl. Optics.
Derr, V.E. (editor), 1972. Remote sensing of the troposphere. Supt. of Documents, U.S. Govt. Printing Office, Washington.
Gray, D.A., A.T. Waterman Jr., 1970. J. Geophys. Res., 75, 1077-1083.
Harvey, A.F., 1970. Coherent light. Wiley-Interscience, London.
Hay, D.R., V.E. Sells, J.G. Tillotson, J.H. Aitkenhead, 1973. Report TP-10, Centre for Radio Science, Univ. of Western Ontario, London, Canada.
Horvath, H., 1971. Atmos. Environ., 5, 177-184.
Irvine, W.M., F.W. Peterson, 1970. J. Atmos. Sci., 27, 62-69.
Kingslake, R. (editor), 1965. Applied optics and optical engineering. Academic Press, New York, vol. 1.
Lawrence, R.S., G.R. Ochs, S.F. Clifford, 1972. Appl. Optics, 11, 239-243.
Lawrence, R.S., J.W. Strohbehn, 1970. Proc. IEEE, 58, 1523-1545.
Pressley, R.J. (editor), 1971. Handbook of lasers. Chemical Rubber Co., Cleveland.
Quiney, R.G., A.I. Carswell, 1972. Appl. Optics, 11, 1611-1618.
Rensch, D.B., R.K. Long, 1970. Appl. Optics, 9, 1563-1573.
Setzer, D.E., 1970. Bell Syst. Tech. J. 49, 1873-1892.
Viezee, W., J. Oblanas, R.T.H. Collis, 1973. Report AFCRL-TR-73-0052, Stanford Research Inst., Menlo Park, California.
Westwater, E.R., A. Cohen, 1973. "Application of Backus Gilbert inversion technique to determination of aerosol size distribution from optical scattering measurements," submitted to Appl. Optics.

SOME RECENT LIDAR EXPERIMENTS ON PARTICLE SCATTERING IN THE LOWER TROPOSPHERE

D.R. Hay

University of Western Ontario
London, Canada

ABSTRACT

During the past 10 years, the optical lidar (laser radar) has demonstrated competence as a remote probe of atmospheric particles. Volume resolution of a few cubic meters is available. Daytime observations are limited in sensitivity by the sky irradiance. Current applications involve pattern recognition in studies of atmospheric convection, stratification and diffusion, and examination of particle size-distributions and number concentrations. Interpretation of lidar soundings requires various assumptions and degrees of approximation; the interpretation is improved with auxiliary soundings and extended sophistication of the lidar.

The laser radar (lidar) as a probe for studying the atmosphere is celebrating its tenth anniversary. It combines the principle of the microwave radar, that has been well known to radar meteorologists for a much longer time, together with some of the special interactions between atmospheric particles and laser waves. This paper will review briefly the principle of the lidar and the characteristics that distinguish it from the microwave radar. Then some current applications of the lidar in studies of the lower atmosphere will be noted. More detailed discussions may be found in several excellent reviews elsewhere (for example Derr et al 1970, Kildal et al 1971, Derr 1972 ch. 23).

(a) THE LIDAR EQUATION

Four components are basic to the lidar. These are the

transmitter, the scattering region of the atmosphere, the receiver, and the interpreter of the lidar signature. The latter usually dictates the choice of lidar parameters, while the lidar equation relates the other three components. The instantaneous received power (P_R) is governed by the transmitted power (P_T), the range to the scattering volume (R), the total extinction coefficients of the atmosphere along the transmitted and received paths (α_1, α_2 respectively) and the volume back-scattering coefficient (β_θ), through the lidar equation:

$$P_R = P_T \exp[-\int_0^R \alpha_1 (r)dr]\ \beta_\theta\ L\ \gamma_R \exp[-\int_0^R \alpha_2 (r)dr] \quad (1)$$

where L is the effective length of the scattering volume, and γ_R is the effective solid angle subtended by the receiver at the scattering volume. We note that P_T and γ_R include efficiency factors for the transmitting and receiving optics, and that γ_R is proportional to R^{-2}. For monostatic lidar applications, the transmitter and receiver are either adjacent or collinear, and hence α_1 and α_2 are assumed identical in these cases.

(b) TYPICAL LIDAR PARAMETERS

Closer examination of the parameters in Eq(1) shows the distinctive features of a lidar. Comments on typical values of these parameters will be given in the following sub-sections.

(i) Transmitted power (P_T).

Single pulses from a ruby or neodymium laser will contain energy (E) of 1-20 joules, distributed over the pulse duration (τ) of about 10^{-8} second. Transmitting optics efficiency is of the order 80%. P_T is related to E and the number of photons per pulse (N_T) by Eq(2):

$$P_T = \frac{E}{\tau} = \frac{N_T\ hc}{\tau\lambda} \quad (2)$$

where h is Planck's constant, c is the velocity of light and λ is the laser signal wavelength. Typically $P_T \simeq 10^8$ watts and $N_T \simeq 3 \times 10^{18}$ photons. These pulses are produced at a rate of about one per second, for pulse-by-pulse observations on the atmosphere. Higher repetition rates with lower peak power also are available.

(ii) Scattering volume of the atmosphere.

This is defined by the common volume between the transmitter beam and the receiver field-of-view, and by the duration of the transmitted pulse. The region of complete overlap extends radially between ranges R_o and R_{max}, with a typical divergence of 0.5 mrad. For Mie and Rayleigh scattering at atmospheric particles, the radial depth of the scattering volume (L) is determined by the transmitted pulse length (τ) or by the matching receiver bandwidth (β) such that

$$L = \frac{c\tau}{2} \simeq 0.220 \frac{c}{\beta} \tag{3}$$

Hence, at a range of 5kms the instantaneous scattering volume is only about 2m in diameter and 2m in radial depth.

(iii) Field-of-View of the Receiver.

Two competing factors usually enter into the choice of receiver field-of-view. One is the requirement that the acceptance angle be sufficiently large to include the transmitter beam over the full range interval R_o to R_{max}. The other is the need for reducing the field-of-view until the sky background noise entering the receiver is at an acceptable level. In practice, the acceptance angle is at least as large as the transmitter beam divergence, and for the non-collinear arrangement of transmitter-receiver generally is around 2 mrad.

The sky background provides the limiting noise signal for daytime observations with an optical lidar. Under clear daytime conditions, the spectral radiance of the sky (η) is approximately 10^{-2} watts m^{-2} A^{-1} $ster^{-1}$. (Pratt 1969). If the receiver aperture has typical diameter 8 in. (0.032 m^2), optical efficiency is 25%, field-of-view is 3.1 x 10^{-6} ster. (2 mrad.), and optical bandwidth is 10A, the sky background noise at the receiver is of the order

$$P_R \min \simeq 4\text{x}10^{-8} \text{ watt} \tag{4}$$

Nighttime observations in the absence of moonlight usually are noise limited by the dark current of the photo detector. In this case, P_{Rmin} is reduced by 2 or 3 orders of magnitude. Signal integration may be used to improve the received signal-to-noise ratio when P_T is low and atmospheric scattering is very weak.

(iv) Extinction coefficient (α) and volume back-scattering coefficient (β_θ).

The above discussion suggests that the lidar equation may be rewritten to emphasize the role of the atmospheric parameters. Eq(1) becomes:

$$P_R(R)\ R^2 = A \exp[-2 \int_o^R \alpha(r)dr]\ \beta_\theta\ (R) \qquad (5)$$

where A is the lidar calibration that takes into account the design details in Sections (i) - (iii). In principle, Eq(5) is solved for $\alpha(R)$ and β_θ (R) by appropriate measurements of $P_R(R)$ (the "lidar signature").

Exact solution of Eq(5) is impractical except in special cases because of the complexity of α and β_θ. As noted in the writer's companion paper* optical extinction by molecular absorption (O_2, O_3, H_2O) is less than 20% per kilometer, when laser transmissions at resonance wavelengths are avoided. Rayleigh scatter by air molecules and Mie scatter by suspended particulate (especially in the lowest 4km of the atmosphere) are mainly responsible for $\alpha(R)$ and $\beta_\theta(R)$.

A unique relationship exists between α and β_θ for optical scattering by air molecules. Rayleigh theory gives

$$\beta_{\theta M} = \frac{3}{8\pi}\,\alpha_M \qquad (6)$$

Then Eq(5) is rewritten:

$$P_R(R)\ R^2 = A \exp\{-2 \int_o^R [\frac{8\pi}{3}\ \beta_{\theta M}(r) + \alpha_A(r)]dr\}$$

$$\times\ [\beta_{\theta M}(R) + \beta_{\theta A}(R)] \qquad (7)$$

where subscript M implies molecular (Rayleigh) scattering, and subscript A implies scattering by aerosol particulate (Mie scattering, at optical wavelengths). For most applications, $\beta_{\theta M}(R)$ is obtained with sufficient precision from standard

---

*"Recent studies of visibility and atmospheric attenuation at optical wavelengths", NATO Advanced Study Institute, Sorrento, June 1973.

atmosphere tables or from local radiosonde profiles. The major difficulty in solving Eq(7) arises because

$$\alpha_A = f(\text{particle size, wavelength, particle shape, particle-size distribution, particle number density, particle refractive index, particle orientation}), \quad (8)$$

and similarly for $\beta_{\theta A}$.

(c) APPROACHES TO THE SOLUTION OF EQ(7)

All lidar experiments require simplifying assumptions regarding atmospheric particles for the solution of Eq(7). Two approaches will be noted briefly here to illustrate the techniques.

(i) Pattern recognition

This method is used to examine patterns of particulate distribution as an aid to studying thermal convection and stratification in the atmosphere. Accuracy in determining number concentration of particulate is of secondary importance. For this purpose, an empirical approximation relating $\beta_{\theta A}$ and α_A is invoked:

$$\beta_{\theta A} = k_1 \, \alpha_A^{k_2} \quad (9)$$

where k_1 and k_2 are constants and the particulate is assumed to be polydisperse. In addition, the lidar calibration (A) requires an auxiliary measurement of $\alpha(R_o)$ at reference range R_o, or a separate measurement on a calibrated target. Then Eq(7) reduces to

$$P_R(R)\, R^2 = \text{constant} \cdot f[\alpha_A(R)] \quad (10)$$

and a profile of $\alpha_A(R)$ is derived from the lidar signature. By defining a reference particle whose scattering cross-section (σ_r) is representative of those found commonly in the lower atmosphere, the approximate number concentration of particles (ρ) may be derived from the $\alpha_A(R)$ profile through the relation

$$\rho(R) = \frac{\alpha_A(R)}{\sigma_r} \quad (11)$$

Examples of this approach are found in Collis et al (1972), Viezee et al (1973a and b), Hay et al (1973).

(ii) Quantitative measurement

More precise information on the particulate loading of the atmosphere will require added sophistication in the lidar measurement. As Eq(7) and (8) suggest, this involves several simultaneous and independent lidar observations on the same volume of the atmosphere. Reagan et al (1972) and Herman et al (1971) describe this type of experiment with monostatic and bistatic lidars.

The monostatic lidar is directed vertically, and profiles of $\alpha_A(R)$ are derived essentially according to Eq(9), (10). The accuracy of this derivation is improved, however, through iterative solution that forces the $\alpha_A(R)$ profile to conform to the total extinction as observed independently by a solar radiometer.

The bistatic lidar operates in parallel with the monostatic lidar. The transmitter and receiver of the former are separated by several km, with the common scatter volume in the atmosphere above the monostatic lidar. The position of this common volume is changed five times during an atmospheric sounding, and for each position the intensity and polarization of the scattered light are recorded (i.e. the Stokes parameters are ascertained). Hence, twenty independent measurements of Eq(7) are obtained (four Stokes parameters at 5 scattering angles). By assuming spherical scattering particles with refractive index 1.54 and a trial size-distribution function, iterative procedure is used to derive the actual particle size-distribution. One of the major requirements on the lidar system in this application is an accurate measurement of the transmitted power as part of the lidar calibration.

McNeil (1973) has reported recently on a variation of this technique. He employed a monostatic lidar to measure the Stokes parameters of signal scattering along the transmitted beam. Provision is made for simultaneous scattering at two lidar wavelengths (0.6943 and 0.3472 μm). Preliminary analysis indicates several categories of atmospheric particulate, that are associated with different urban environments.

(d) REFERENCES

Collis, R.T.H., E.E. Uthe, 1972. Opto-Electronics, 4, 87-99.
Derr, V.E. (editor), 1972. Remote sensing of the troposphere. Supt. of Documents, U.S. Government Printing Office, Washington, D.C. 20402.
Derr, V.E., C.G. Little, 1970. Appl. Optics, 9, 1976-1992.

Hay, D.R., V.E. Sells, J.G. Tillotson, J.H. Aitkenhead, 1973. Report TP-10, Centre for Radio Science, University of Western Ontario, London, Canada.

Herman, B.M., S.R. Browning, J.A. Reagan, 1971. Atmos. Sci., 28, 763-771.

Kildal, H., R.L. Byer, 1971. Proc. IEEE, 59, 1644-1663.

McNeil, W.R., 1973. Lidar atmospheric polarization diagnostics. Ph.D. Thesis, York University, Toronto, Canada.

Pratt, W.K., 1969. Laser communication systems. Wiley and Sons, New York.

Reagan, J.A., B.M. Herman, 1972. AIAA Journ., 10, 1401-1407.

Viezee, W., R.T.H. Collis, J.D. Lawrence Jr., 1973a. J. Appl. Meteorol. 12, 140-148.

Viezee, W., R.D. Hake, P.B. Russell, 1973b. Lidar measurements of stratospheric constituents. Quarterly Report on Project 2217, Stanford Res. Inst., Menlo Park, California.

RADIOMETRY, THE STRUCTURE OF ATMOSPHERIC SPECTRUM LINES, AND RELATED EXPERIMENTS

David L. Croom

S.R.C. Radio and Space Research Station
Slough, Bucks. England

Several atmospheric gases have molecular rotation emission and absorption lines in the wavelength range 100-0.1 mm (frequencies of 3-3000 GHz). The principal molecules concerned are H_2O and O_2, though others (for example O_3, OH, CO, N_2O) also have resonances in this range and at least one of these minor constituents (O_3 at 3 mm) has been detected in the radio band. In the relatively high pressure lower atmosphere the resonance lines are considerably smeared out by intermolecular collisions (pressure-broadening). Above about 75 km the atmospheric pressure is so reduced that the random thermal motion of the molecules (Doppler broadening) becomes the dominant factor in determing the line-width. Recent interest in these atmospheric spectrum lines has been in their potential for remote probing of the atmosphere, particularly in relation to temperature profiles and water-vapour content. Below 1 mm wavelength (sub-millimetre region) there are large numbers of these lines, and the further development and use of Fourier spectroscopic techniques has resulted in considerable improvements in our knowledge of this part of the atmospheric spectrum.

The last few years have also seen an increasing use of solar radiometry as a tool for obtaining statistical attenuation data, particularly in relation to rain effects. Efforts in this direction have been encouraged by the need to open up the mm part of the e.m. spectrum for communications.

REFERENCES

Basharinov, A. E., Gurvich, A. S., Yegorox, S. T., Kurskaya, A. A., Matvyev, D. T. and Shuto, A. M. 1971 The results of microwave sounding of the earth's surface according to experimental data from the satellite Cosmos 243. Space Research XI Akademie-Verlag, Berlin.

Croom, D. L. 1969 Atmospheric thermal radiation due to the 22.235 GHz (13.5 mm) and 183.311 GHz (1.64 mm) H_2O rotational lines. Proc. NATO Advanced Study Inst. of Effects of atmospheric water on electromagnetic wave propagation "University of Western Ontario" (Ed. D. R. Hay), p. 53.

Croom, D. L. 1971 The 2.53 mm molecular rotation line of atmospheric O_2. Planet. and Space Sci., 19, p. 777.

Haroules, G. G. and Brown, W. E. 1969 A 60 GHz multifrequency radio-metric sensor for detecting clear air turbulence in the troposphere. IEEE Trans. AES-5, p. 712.

Harries, J. E. and Ade, P. A. R. 1972 The high resolution millimetre wavelength spectrum of the atmosphere . Infra-Red Physics, 12, p. 81.

Harries, J. E., Swann, N. R. W., Beckman, J. E., Ade, P. R. 1972 High resolution observations of the submillimetre stratospheric emission spectrum. Nature 236, p. 159.

Kislyakov, A. G. 1972 On the atmospheric transparency spectrum in the millimetre waveband. Infra-Red Physics 12, p. 61.

Nolt, I. G., Radostitz, J. V. and Donnelly, R. J. 1972 Upper atmosphere emission spectrum at millimetre wavelengths. Nature 236, p. 444.

Rosenkranz, P. W., Staelin, D. H., Barath, F. T., Blinn, J. C. and Johnston, E. J. 1971 Indirect sensing of atmospheric temperature and water-vapour using microwaves. Proc. 7th Int. Symp. on Remote Sensing of the Environment, Ann. Arbor, Mich.

Staelin, D. H. 1969 Passive remote sensing at microwave wavelengths. Proc. IEEE, 57, p. 427.

Toong, H. D. and Staelin, D. H. 1970 Passive microwave spectrum measurements of atmospheric water-vapour clouds. J. Atmos. Sci., 27, p. 781.

Viktorova, A. A. and Zhevakin, S. A. 1967 Absorption of microwaves in air by water-vapour dimers. Soviet Phys. Doklady 11, p. 1065.

Voronov, V. N., Kislzaykov, A. E., Kukina, E. P. and Naumov, A. I. 1972 On the CO and N_2O content of the earth's atmosphere by observation of their rotational spectral lines. Izv. Vuz. Fizika Atmosferi.i Okeana. 8, p. 29.

DEPOLARIZATION PHOENOMENA ON ELECTROMAGNETIC WAVES THROUGH RAIN

A. Paraboni

Centro Telecomunicazioni Spaziali del CNR
Istituto di Elettrotecnica ed Elettronica
Politecnico di Milano

1. FOREWORD

The depolarization, e.g. the polarization change of an E.M. wave is a very frequent phoenomenon ancountered whenever the propagation is influenced by the presence of real media.

Many examples can be made: the reflection of E.M. waves on slant or corrugated surfaces, the scattering by obstacles or small particles dispersions, the crossing through anisotropic media (like magnetized plasmas or ferrites) or, similarly through dispersions of non spherical obstacles.

This subject is concerned in several applications, both in physics and engineering;sometimes depolarization is regarded as a positive factor as it carries many useful informations about the depolarizing medium; in other cases it is a negative factor due to the degradation inferred on the transmissive system.

Our attention will be devoted to the effects of rain or, more generally, of atmospheric precipitations on the polarization of traveling E.M. waves.

Such effects are very relevant in the centimeter or millimeter wave transmission, expecially because they are able to deteriorate the performances of radio links operating with frequency re-use.

As a consequence, a strong coupling between cross polarized traveling waves can happen: this phoenomenon, called dyaphony , is always negative because it gives rise to a noise enhancement in the radio channels.

Generally speaking the depolarization experienced by E.M. waves through rain is harmful in any application where an initial polarization degree has to be estimated or preserved.

Many examples can be made: the radar detection of obstacles on the basis of the asymmetry of their cross section and, similarly, the radar detection of hail based on the back-scatter depolarization |Barge, B.L. 1971| are applications in which the anisotropic propagation can give raise to a strong degradation of the system performances.

The depolarization effects suffered passing through intense rain can be very considerable: values up to 16 dB have been measured; for what concerns the telecommunication aspects such values widely exceed those allowed by the CCIR for large capacity communication links (20÷25dB).

Also in the radar detection such values are prohibitive being largely comparable with the values to be measured as originary polarization.

2. THE POLARIZATION FACTOR

There are many representations of the polarization degree of a wave: according to the specific use some are more suitable than others.

A general T.E.M. wave is fully described by the complex notation

$$\vec{E} = E^x \vec{U}_x + E^y \vec{U}_y \tag{1}$$

where $\vec{U}_x$ and $\vec{U}_y$ are orthogonal unit vectors along the x,y transverse axes and E^x, E^y are the complex components:

$$E^x = |E^x| \; e^{j\phi x} \qquad E^y = |E^y| \; e^{j\phi y} \tag{2}$$

describing the harmonic time-variation:

$$e^x = \sqrt{2}\; |E^x| \; \cos(\omega t + \phi_x) \qquad e^y = \sqrt{2}\; |E^y| \cos(\omega t + \phi_y) \tag{3}$$

Of course E^x and E^y depend on the longitudinal coordinate z alone. The polarization status is fully identified by the transverse coordinates ratio

$$p = \frac{E^y}{E^x}$$

which will be referred as 'polarization complex factor'.
Any information about absolute levels and absolute phases is irrelevant in polarization problems.

The field strength vector, in its motion describes an ellipse whose pertinent parameters are the axial ratio, the tilt angle of the axes with respect to the reference system and the sense of percurrence (right or left).
These informations are fully contained in the polarization factor.

Whichever is the type of representation, two real parameters are needed: relationships among the various pairs of descriptive parameters can be found in many basic textbooks about the subject |Beckman P. 1968|.

3. ORTHOGONAL POLARIZATIONS

Two overlapping waves are said orthogonal if the total carried power equals the sum of the two contributions carried separately by the two waves.

It is easily found that orthogonality exists if the two polarization factors p_1 and p_2 fulfil the relationship:

$$p_1 \, p_2^{*} = -1 \tag{5}$$

The frequency re-use technique consists in using two orthogonal waves carrying two independent radio channels at the same frequency.

4. THE TRANSFER MEDIUM

If the medium were an ideal one it would preserve the orthogonality so that at the receiving front end the two signals could be shared by means of suitable polarizers.

Unfortunately, the real medium infers some amount of coupling between the modes and dyaphony takes place.

The behaviour of the medium can be fully described by a linear system which gives the two output signals E_{out} as linear combination of the two input ones E_{in}.

$$\begin{aligned} E^{I}_{out} &= t_{11} \, E^{I}_{in} + t_{12} \, E^{II}_{in} \\ E^{II}_{out} &= t_{21} \, E^{I}_{in} + t_{22} \, E^{II}_{in} \end{aligned} \tag{6}$$

For sake of compactness the (6) can be rewritten in matrix

form

$$\begin{Vmatrix} E^{I}_{out} \\ E^{II}_{out} \end{Vmatrix} = \begin{Vmatrix} t_{11} & t_{12} \\ t_{21} & t_{22} \end{Vmatrix} \cdot \begin{Vmatrix} E^{I}_{in} \\ E^{II}_{in} \end{Vmatrix} \qquad (7)$$

The transfer matrix (7) is replaceable with any other one related to each pair of polarization factors.

Assuming as basis two linearly polarized waves along the x and y axes the (7) can be rewritten

$$\begin{Vmatrix} E^{x}_{out} \\ E^{y}_{out} \end{Vmatrix} = \begin{Vmatrix} t_{xx} & t_{xy} \\ t_{yx} & t_{yy} \end{Vmatrix} \begin{Vmatrix} E^{x}_{in} \\ E^{y}_{in} \end{Vmatrix} \qquad (8)$$

Transfer formulas to get any required transfer matrix are easily obtained resolving the two general orthogonal waves

$$\vec{E}^{I} = E^{I}(\vec{U}_x + p\vec{U}_y) \quad , \quad \vec{E}^{II} = E^{II}(\vec{U}_x - \frac{1}{p^*}\vec{U}_y) \qquad (9)$$

along the x and y axes.

The E^x and E^y cartesian components of the overall wave $\vec{E} = \vec{E}^{I} + \vec{E}^{II}$ can be obtained using the (9).
The resulting transformation is

$$\begin{Vmatrix} E^{x} \\ E^{y} \end{Vmatrix} = \begin{Vmatrix} 1 & 1 \\ p & -\frac{1}{p^*} \end{Vmatrix} \begin{Vmatrix} E^{I} \\ E^{II} \end{Vmatrix} \qquad (10)$$

Inverting the (10) we get

$$\begin{Vmatrix} E^{I} \\ E^{II} \end{Vmatrix} = \begin{Vmatrix} \frac{1}{1+|p|^2} & \frac{p^*}{1+|p|^2} \\ \frac{|p|^2}{1+|p|^2} & \frac{-p^*}{1+|p|^2} \end{Vmatrix} \begin{Vmatrix} E^{x} \\ E^{y} \end{Vmatrix} \qquad (11)$$

The transfer matrices (7) (8) (10) and (11) will be indicated with $\|T_{12}\|$, $\|T_{xy}\|$, $\|p\|^{-1}$ and $\|p\|$.

The general transfer matrix T_{12} is given by the product $\|p\| \cdot \|T_{xy}\| \cdot \|p\|^{-1}$, according to the relationships:

$$\begin{Vmatrix} E^{I}_{out} \\ E^{II}_{out} \end{Vmatrix} = \overbrace{\|p\| \cdot \|T_{xy}\| \cdot \|p\|^{-1}}^{\|T_{12}\|} \cdot \begin{Vmatrix} E^{I}_{in} \\ E^{II}_{in} \end{Vmatrix} \qquad (12)$$

Let us define the cross-talk ratios δ_{12} and δ_{21} as the diaphony signal components with respect to the main signals:

$$\delta_{12} = \left.\frac{E^{I}_{out}}{E^{II}_{out}}\right|_{E^{I}_{in}=0} \qquad \delta_{21} = \left.\frac{E^{II}_{out}}{E^{I}_{out}}\right|_{E^{II}_{in}=0} \tag{13}$$

According to the (7):

$$\delta_{12} = \frac{t_{12}}{t_{22}} \qquad \delta_{21} = \frac{t_{21}}{t_{11}} \tag{14}$$

The cross-talk ratios qualify the medium and must not be confused with the polarization factor even if in some cases they can identify itself with it.

The complete identification of the medium, with respect to its depolarization capability in any polarization, requires three complex parameters instead of four because the terms of the overall transfer matrix can be known unless an arbitrary complex constant.

5. DEPOLARIZATION BY RAIN

In this section we will determine the cross-talk ratios in the case of strong rain.

Because the transfer characteristics of the medium are strictly dependent on the forward scattering properties of the hydrometeors, a previous analysis on this subject is needed.

We will assume that each particle has rotational symmetry: because most hydrometeors are nearly ellipsoids or cones (including rain droplets and hailstones), this condition appears to be satisfactory fulfilled.

With this assumption we can define two orthogonal planes, called principal planes, along which a linearly polarized wave causes an equally polarized forward scattered field.

These planes are identified by the symmetry axis and the propagation direction as it appears in fig. 1.

The non-depolarizing properties of these two planes are easily understood by supposing to solve the scattering problem using only that field strength (electric or magnetic) which is parallel to the symmetry plane called second principal plane as in fig. 1.

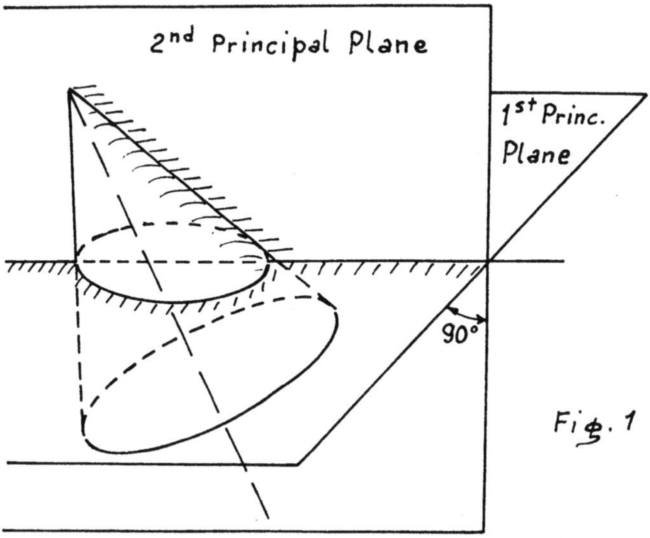

Fig. 1

As the scattered field can be supposed to be caused by equivalent surface currents (electric or magnetic depending on the assumed strenght) it must preserve the original polarization due to the symmetry of the re-radiating sources.

The principal planes are assumed as reference in defining the single scatter properties.

Incident and scattered fields fulfill the relationship

$$E_{sc}(R) = E_{in}(0)\,\frac{e^{-j\beta_o R}}{j\beta_o R}\;S \qquad (15)$$

where E_{sc}, E_{in} and R are evaluated according to the fig. 2, β_o

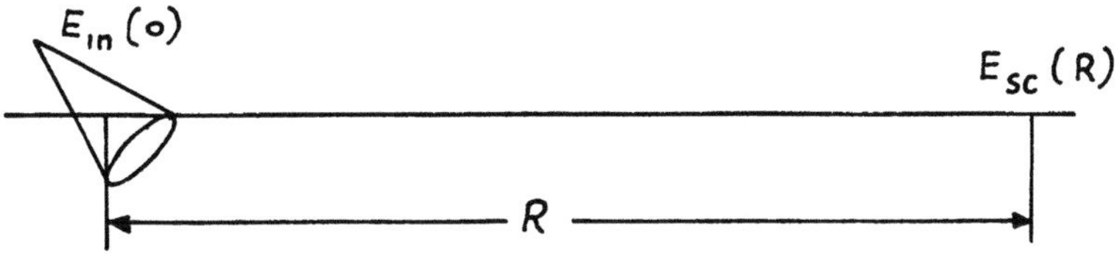

Fig. 2

is the free space wave number and S is a complex dimensionless constant, called scattering amplitude function.
S is independent of R and of the assumed origin.

We will call S' and S" the two amplitude functions along the principal planes.

The forward scatter for a general incident polarization can be expressed resolving the wave along the principal planes. In particular, using two waves linearly polarized along the reference axes, the computation requires only the use of rotational matrices.

Calling θ the tilt angle of the first principal plane with respect to the X axis, (see fig. 3), we get:

$$\begin{vmatrix} E^x_{sc} \\ E^y_{sc} \end{vmatrix} = \underbrace{\|-\theta\| \cdot \frac{e^{-j\beta_o R}}{j\beta_o R} \begin{Vmatrix} S' & 0 \\ 0 & S'' \end{Vmatrix} \cdot \underbrace{\|\theta\| \cdot \begin{vmatrix} E^x_{inc} \\ E^y_{inc} \end{vmatrix}}_{\begin{vmatrix} E'_{inc} \\ E''_{inc} \end{vmatrix}}}_{\begin{vmatrix} E'_{sc} \\ E''_{sc} \end{vmatrix}} \qquad (16)$$

$$\text{where } \|\theta\| = \begin{Vmatrix} \cos\theta & \text{sen}\theta \\ -\text{sen}\theta & \cos\theta \end{Vmatrix} .$$

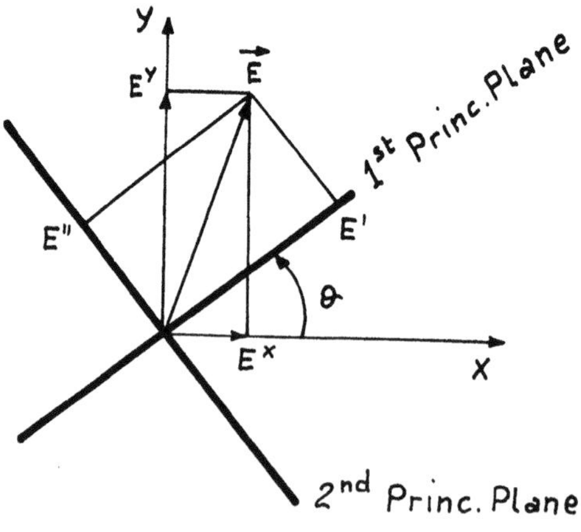

Fig. 3

The (16) can be rewritten:

$$\begin{vmatrix} E^x_{sc} \\ E^y_{sc} \end{vmatrix} = \frac{e^{-j\beta_o R}}{j\beta_o R} \begin{Vmatrix} S_{xx} & S_{xy} \\ S_{yx} & S_{yy} \end{Vmatrix} \begin{vmatrix} E^x_{inc} \\ E^y_{inc} \end{vmatrix} \tag{17}$$

where:

$$S_{xx} = \frac{S'+S''}{2} + \frac{S'-S''}{2} \cos 2\theta$$

$$S_{yy} = \frac{S'+S''}{2} - \frac{S'-S''}{2} \cos 2\theta \tag{18}$$

$$S_{xy} = S_{yx} = \frac{S'-S''}{2} \operatorname{sen} 2\theta$$

The (18) shows that cross-talk terms vanish for $\theta=0$ or $\pi/2$, being dependent on the 2θ angle: the relevant electrical parameter is the differential forward amplitude $S'-S''$.

Let us now consider the mean transfer properties of the dispersion.

Observe that the receiving point is practically influenced only by the particles in the first Fresnel zone; for this reason, at microwaves, only nearly-axial re-radiating contribution have to be taken into account.

This justifies the attention we devoted to the forward parameters even if traditionally the global parameter "extinction section" is involved in propagation problems.

The "extintion section" is a suitable parameter for attenuation studies only, due to its energetic nature. Conversely polarization problems need the knowledge of the phase properties as well, and so the more comprehensive parameter S has to be used.

The forward amplitude S and extintion section C_{ext} are bonded by the relationship

$$C_{ext} = \frac{\lambda^2}{\pi} \operatorname{Re} |S| \tag{19}$$

In the next step we will establish the equivalence between the dispersion and a lossy-dielectric medium with suitable characteristics.

Instead of a parameters' equivalence, the input-output electric field relationship will be directly computed. This equivalence preserve the mean properties of the medium and is based on the simple-scattering assumption.

Second order phoenomena like space coherency on the wavefront or multiple scattering effects are completely lost.

The logical steps in the search of medium equivalent parameters are the following:

a) Substitution of the real particles with a pair of orthogonal equivalent small dipoles, oscillating along the x and y directions.
Amperian moments are obtained by imposing the complete forward equivalence.

b) Substitution of the small concentrated dipoles contained in a slab of thickness δz, with a laminar electric current which preserve the electrical momentum per unit transverse area.

c) Evaluation of the δE_{RAIN} field strength variation caused by the laminar current. This variation has to be joined to the natural free space propagation one ($\delta E_{F\ S}$)

d) Evaluation of the overall field strength variation by integrating the elementar variations computed in the above point.

The equivalence described in the step a) is summarized in the figure 4, where the two amperian moments are denoted $(I\ell)^x$ and $(I\ell)^y$.

SCATTERED FIELD BY ELECTRICAL SMALL DIPOLES:

$$E^x_{sc} = \frac{-j\eta}{2\lambda R}(I\ell)^x\, e^{-j\beta_o R}$$

$$E^y_{sc} = \frac{-j\eta}{2\lambda R}(I\ell)^y\, e^{-j\beta_o R}$$

SCATTERED FIELD BY ASYMMETRIC PARTICLES

$$\begin{vmatrix} E^x_{sc} \\ E^y_{sc} \end{vmatrix} = \frac{e^{-j\beta_o R}}{j\beta_o R} \begin{Vmatrix} S_{xx} & S_{xy} \\ S_{yx} & S_{yy} \end{Vmatrix} \begin{vmatrix} E^x_{inc} \\ E^y_{inc} \end{vmatrix}$$

$$\begin{vmatrix} (I\ell)^x \\ (I\ell)^y \end{vmatrix} = \frac{4\pi}{\eta\beta_o^2} \begin{Vmatrix} S_{xx} & S_{xy} \\ S_{yx} & S_{yy} \end{Vmatrix} \begin{vmatrix} E^x_{inc} \\ E^y_{inc} \end{vmatrix} \qquad (20)$$

Fig. 4

Calling δJ^x and δJ^y the laminar current values, as indicated in the step b), we put the equivalence of electrical momentum, as summarized in the figure 5.

AMPERIAN MOMENTUM OF CONSTANT LAMINAR CURRENT	AMPERIAN MEAN MOMENTUM OF PARTICLES
$\delta J^x [A/m] \cdot 1[m^2]$	$\sum_{\text{UNIT VOLUME}} (I\ell)_i^x \cdot \delta z$
$\delta J^y [A/m] \cdot 1[m^2]$	$\sum_{\text{UNIT VOLUME}} (I\ell)_i^y \cdot \delta z.$

$$\begin{vmatrix} \delta J^x \\ \delta J^y \end{vmatrix} = \frac{4\pi\delta z}{\eta\beta_o^2} \begin{Vmatrix} \Sigma_1 S_{xx} & \Sigma_1 S_{xy} \\ \Sigma_1 S_{yx} & \Sigma_1 S_{yy} \end{Vmatrix} \begin{vmatrix} E^x_{inc} \\ E^y_{inc} \end{vmatrix} \qquad (21)$$

Fig. 5

The evaluation δE_{RAIN} is easily accomplished using the Maxwell equation to get the magnetic strength (and hence electric strength) variation due to the electrical currents evaluated above.

Step c) is then summarized in Fig. 6.

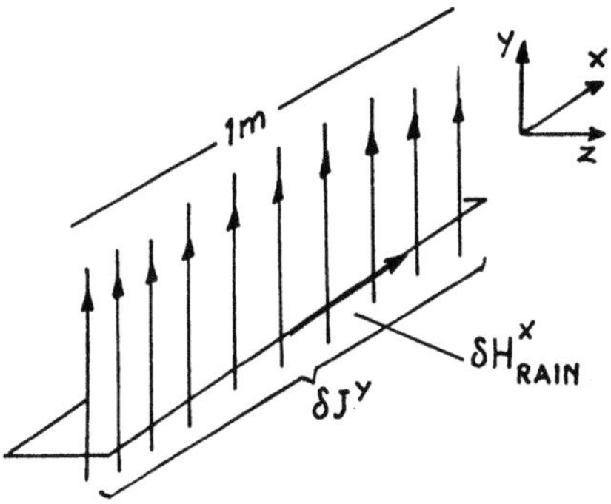

$$\delta H^x_{RAIN} = -\frac{1}{\eta}\,\delta E^y_{RAIN} = \frac{1}{2}\,\delta J^y$$

Fig. 6

$$\begin{vmatrix} \delta E^x_{RAIN} \\ \delta E^y_{RAIN} \end{vmatrix} = \frac{2\pi\delta z}{\beta_o^2} \begin{Vmatrix} -\Sigma_1 S_{xx} & -\Sigma S_{xy} \\ -\Sigma S_{yx} & -\Sigma S_{yy} \end{Vmatrix} \begin{vmatrix} E^x_{inc} \\ E^y_{inc} \end{vmatrix} \qquad (22)$$

The variation δE_{FS} due to the free space propagation is easily obtained from the law:

$$\vec{E}_{inc}(z+\delta z) = \vec{E}_{inc}(z)\ e^{-j\beta_o(z+\delta z)} \quad ;$$

for small values of $\delta z/\lambda$ we obtain:

$$\delta E^x_{FS} = -j\beta_o\ E^x_{inc}\ \delta z$$
$$\delta E^y_{FS} = -j\beta_o\ E^y_{inc}\ \delta z \qquad (23)$$

The overall variation is then: $\delta E = \delta E_{FS} + \delta E_{RAIN}$:its value is given by the system

$$\begin{vmatrix} \delta E^x \\ \delta E^y \end{vmatrix} = \delta z \begin{Vmatrix} -\gamma_x(z) & -M(z) \\ -M(z) & -\gamma_y(z) \end{Vmatrix} \begin{vmatrix} E^x \\ E^y \end{vmatrix} \qquad (24)$$

where

$$\gamma_x(z) = j\beta_o + \frac{2\pi}{\beta_o^2} \sum_1 S_{xx}$$
$$\gamma_y(z) = j\beta_o + \frac{2\pi}{\beta_o^2} \sum_1 S_{yy} \qquad (25)$$
$$M(z) = \frac{2\pi}{\beta_o^2} \cdot \sum_1 S_{xy}$$

Notice that $\gamma_x(z)$, $\gamma_y(z)$ and $M(z)$ are longitudinal variables, depending on the statistical mean properties of the slab δz.

We have now to carry out the resolution of the fundamental differential system (24),as in the point d)
The solution is particularly easy to find in the case of moderately depolarizing medium or in the case of homogeneous rain |Capsoni C - Paraboni A. - 1972|.
In the first case an error less than 2 dB is made up to 20 dB of overall depolarization.

The basic assumption of moderate depolarization capability allows the following semplifications:

a) if only a wave is transmitted (called main wave) this is virtually not influenced by the much smaller cross-wave.

b) Each contribution to the growing cross-wave propagates as in an isotropic medium from the coupling section to the end.

The solution is then represented in figs. 7 and 8 for linear and circular polarization. If both the horthogonal waves are transmitted the resulting scheme can be simply doubled and the outputs summed, provided that the signals are uncorrelated, as it happens in practice.

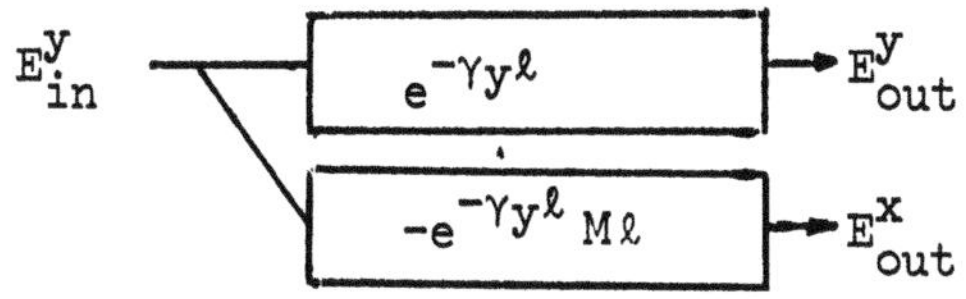

Fig. 7

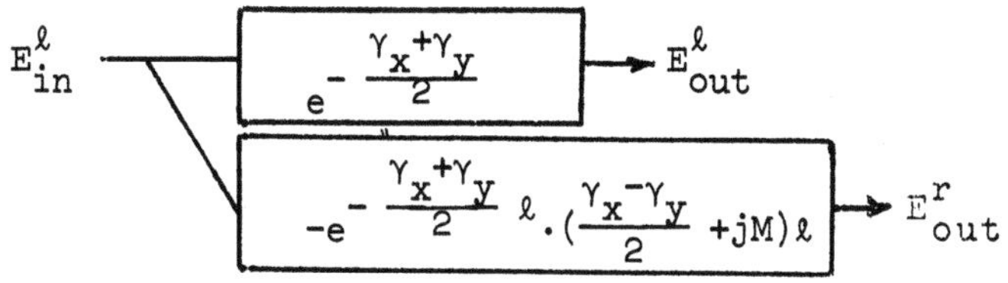

Fig. 8

The cross-talk ratios are:

$$\delta_{xy} = -M\,\ell \qquad \delta_{r\ell} = -\left(\frac{\gamma_x-\gamma_y}{2} + jM\right)\ell \tag{27}$$

By using the elementar scattering parameters we get

$$\delta_{xy} = -\frac{2\pi}{\beta_o^2} \sum_{\ell\cdot 1} \frac{S'_i - S''_i}{2} \,\mathrm{sen}\, 2\theta_i$$

$$\delta_{r\ell} = -\frac{2\pi}{\beta_o^2} \sum_{\ell\cdot 1} \frac{S'_i - S''_i}{2} \, e^{j2\theta_i} \tag{28}$$

where the sum is extended to the whole path length ℓ and per unit area.

Notice that the overall differential propagation constant is similarly given by:

$$\frac{\gamma_x-\gamma_y}{2}\,\ell = \frac{2\pi}{\beta_o^2} \sum_{\ell\cdot 1} \frac{S'_i - S''_i}{2} \cos 2\theta_i \tag{29}$$

Unless the δ's (or $\gamma_x-\gamma_y$) are directly measured, their knowledge would require to know the statistical properties of the physical parameters S'-S" and θ of the particles.
These parameters depend on the ellipticity degree and the canting angle of the droplets.

Today very few data about these factors are available. Nevertheless the relationships (27), (28) and (29) are useful because they express simple bonds existing among integral parameters which can be separately measured.

The (27) can be rewritten

$$\delta_{r\ell} = -\left(\frac{\gamma_x-\gamma_y}{2}\ell - j\,\delta_{xy}\right) \tag{30}$$

Similarly we have

$$\delta_{\ell r} = -\left(\frac{\gamma_x-\gamma_y}{2}\ell + j\,\delta_{xy}\right) \tag{31}$$

The above description requires then two complex parameters to be measured: each pair of parameters among the four seen in the (30) and (31) (say $\delta_{r\ell}$, $\delta_{\ell r}$, δ_{xy}, and $\gamma_x-\gamma_y$) is equally suitable; this description is complete in the sense that the whole information concerning depolarization in any type of polarization is obtainable |A. Attisani, C. Capsoni, A. Paraboni - 1973|.

Two descriptive parameters are then required instead of three as in section 4.

6. GEOMETRICAL REPRESENTATION OF THE DEPOLARIZING PROPERTIES OF THE MEDIUM.

The single particle can be represented in a complex plane by a couple of phasors as indicated in the Fig. 9.
By summing all the contributions separately, we get, unless the constant $-2\pi/\beta^2_0$, the relevant parameters $\delta_{r\ell}$, δ_{xy}, $\gamma_x-\gamma_y$ according to the (28) and (29). This is shown in Fig. 10.

Notice that in general the originary orthogonality is lost. There can be, therefore, two significant cases in which the orthogonality still holds:

- The various contributions (S'-S"/2) have the same argument, i. e. for each particle is Im(S'-S")/Re(S'-S") = constant.

- The droplets have all the same canting angle, or they can be supposed to be like that.

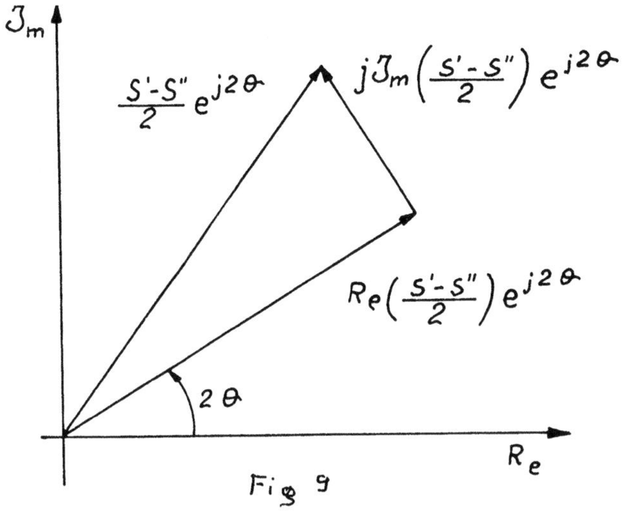
$\Im_m$
$\frac{S'-S''}{2} e^{j2\theta}$
$j\Im_m\left(\frac{S'-S''}{2}\right) e^{j2\theta}$
$R_e\left(\frac{S'-S''}{2}\right) e^{j2\theta}$
2θ
R_e

Fig 9

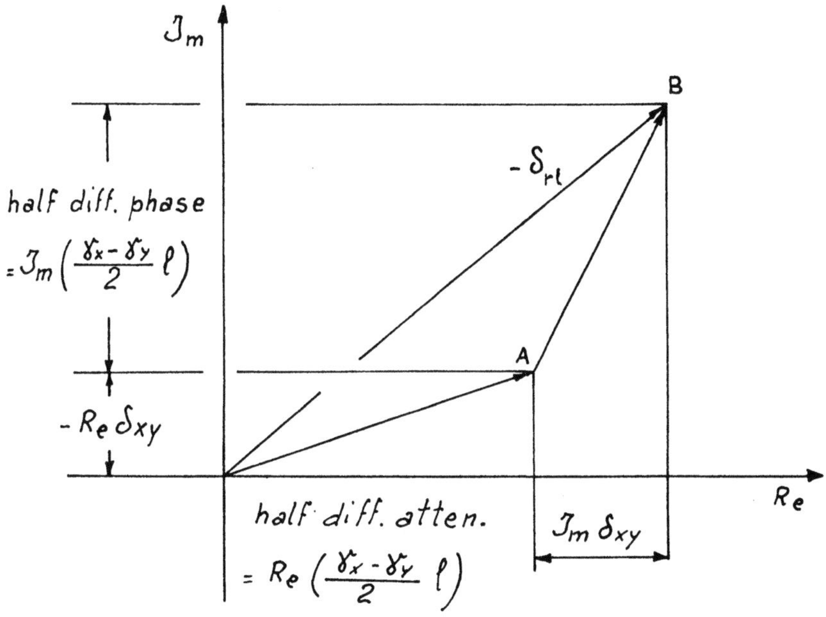
$\Im_m$
B
$-\delta_{rl}$
half diff. phase
$= \Im_m\left(\frac{\gamma_x - \gamma_y}{2} \ell\right)$
A
$-R_e\,\delta_{xy}$
R_e
half diff. atten.
$= R_e\left(\frac{\gamma_x - \gamma_y}{2} \ell\right)$
$\Im_m\,\delta_{xy}$

Fig. 10

In these two cases two orthogonal non-depolarizing planes still exist for the whole medium: they still will be named principal planes.

Along these two planes propagation of two linearly polarized waves takes place independently. In this case only three real parameters are necessary to describe fully the medium:differential propagation and phase along the principal planes and tilt angle of these with respect to the transmitted waves. This appears to be the maximum semplification in the medium description.

The rigorous solution is easily obtained with reference to the principal planes by solving the incoming waves along the principal planes: of course this is possible only if the principal planes exist.

Proceeding as in the section 5, with the use of rotational matrices, the overall transfer matrix in the case of linear polarization is given by:

$$\begin{vmatrix} E^x_{out} \\ E^y_{out} \end{vmatrix} = \begin{Vmatrix} \dfrac{e^{-\gamma' \ell}+e^{-\gamma'' \ell}}{2} + \dfrac{e^{-\gamma' \ell}-e^{-\gamma'' \ell}}{2} \cos 2\theta & \dfrac{e^{-\gamma' \ell}-e^{-\gamma'' \ell}}{2} \operatorname{sen} 2\theta \\ \dfrac{e^{-\gamma' \ell}-e^{-\gamma'' \ell}}{2} \operatorname{sen} 2\theta & \dfrac{e^{-\gamma' \ell}+e^{-\gamma'' \ell}}{2} - \dfrac{e^{-\gamma' \ell}-e^{-\gamma'' \ell}}{2} \cos 2\theta \end{Vmatrix} \begin{vmatrix} E^x_{in} \\ E^y_{in} \end{vmatrix} \tag{32}$$

Misagreement between the rigorous cross-talk values from the (32) and the approximate but more general and simple ones obtained above, is easily computed; it turns out to be negligible in most cases of practical interest.

7. MEDIUM MEASUREMENTS

It has been shown that the medium is completely identified by the phasor diagram of Fig. 10. (In the case of left circular polarization a similar diagram can be easily plotted). Up to date complete measurements of the medium have not yet performed, to our knowledge.

Particularly remarkable are:

-Differential attenuation measurements |Thomas D.J. - 1970|
- Linear depolarization meas. (in modulus only)| Watson P.A., Goodall F., Arbabi M. - 1973 ; Evans B.G.,Troughton J.-1973|.

-Droplets canting angles meas. |Saunders M.J. 1971|
- Maximum differential attenuation theoretical computations |Oguchi T. 1970| These latter appear to be applicable only to excessively semplified rain models.
-Maximum differential phase from differential attenuation data, using the Hilbert functional transformation |Evans B.G., Troughton J. 1973|.

As it can be seen no one of these data set is complete.Recently circular polarization measurements (left and right) have been proposed |Attisani A., Mauri M., 1973 |then it would be possible to get, from the (30) and (31) the parameters γ_x-γ_y and δ_{xy} and, in the case of orthogonal phasors in fig.10,the tilt angle of the principal planes and the differential propagation constant along these planes (attenuation and phase).

8. FUTURE TRENDS

There is today a fundamental lack of knowledge about the following areas:
-equioriented raindrops models validity
-correlation between the droplets canting angles and their ellipticity
-wind effects on depolarization
-time fluctuation rate
-coherence bandwidth of the phoenomena.

Given the importance of the subject it is easily guessed that in the near future polarization phoenomena will be widely investigated, both theoretically and experimentally. Suitably equipped radio links have been recently installed in Italy, Switzerland, and U.K.. Many others will be soon installed by post offices where the frequency re-use technique is foreseen.

Extensive theoretical results about the scattering by ellypsoids, wet hailstones, ecc. could be very useful if canting angles distributions would be reasonably known in typical wind condition.

At last orthogonality restoring techniques must be remembered, by which a sort of space equalization, based on real-time measurements, is performed |Chu T.S. 1971; 1973|.

If such techniques will be successful it is believed that time fluctuation rate and coherence bandwidth will become primary importance parameters.

REFERENCES

Attisani A., Mauri M.: "Identification of physical properties concerning non-spherical scatterers dispersion, with respect to the depolarization caused on centimetric and millimetric waves". To be published on "Alta Frequenza" Italy.

Barge B.L. "Hail detection with a polarization diversity radar" Stormy Weather Group Scientific Report MW 71 1972.

Beckman P. "The depolarization of electromagnetic waves" The Golem Press 1968.

Capsoni C., Paraboni A. "Depolarization of an electromagnetic wave traveling through a stratified aerosol of non-spherical scatterers" 18 AGARD Tech. Meeting on Telecommunication Aspects on Frequencies between 10 and 100 GHz Sept. 1972.

Chu T.S. "Restoring the orthogonality of two polarizations in radiocommunication systems" B.S.T.J. Nov. 1971 n. 9, March 1973 n. 3.

Evans B.G., Troughton J. "Linear and circular cross-polarization statistics" IEE Conference on Propagation of radio waves at frequencies above 10 GHz, April 1973, London.

Evans B.G., Troughton J. "Calculation of cross-polarization due to precipitation" IEE Conference on propagation of radio waves at frequencies above 10 GHz, April 1973, London

Oguchi T. "Attenuation of electromagnetic wave due to rain with distorted raindrops (Part II)" Journal of Radio Research Lab. (Tokio) Vol. II n° 53 Jan. 1964.

Saunders M.J. "Cross polarization at 18 GHz and 30 GHz due to rain" IEEE A.P. Trans. Vol. 19, n° 2, March 1972

Thomas D.J. "Cross polarization discrimination in microwave radio transmission due to rain" USNC/URSI Fall Meeting Columbus, Ohio 1970.

Watson P.A.., Goodall F., Arbabi M. "Linear cross-polarization and attenuation measurements at 11 and 36 GHz" IEE Conference on propagation of radio waves at frequencies above 10 GHz, April 1973 - London.

ANALYSIS OF THE INFLUENCE OF THE VARIOUS PARAMETERS ON THE ATTENUATION-RAIN RATE RELATION

F. Fedi and P. Mandarini

Fondazione Ugo Bordoni, Roma, Italy

1. INTRODUCTION

It is well known how important is the part played by the attenuation-rain rate relation in forecasting the reliability of terrestrial radio-links at frequencies above 10 GHz by using the meteorological data available or obtainable in the various localities.

On the basis of the strict solution of the problem of the scattering of electromagnetic waves by a dielectric sphere, obtained by Mie in 1908 (Ref. 1), Ryde (Ref. 2), around 1945, calculated the theoretical relation between the attenuation and the rain rate.

Since tha date, numerous experiments have been carried out to verify the validity of the theory that is commonly known as the Mie-Ryde theory. Unfortunately the agreement between the measured values and the values predicted by the theory has not always proved satisfactory and the conclusions reached by the various experimentors in some cases appear contradictory. In the light of present-day knowledge, it is still not entirely clear whether it is the Mie-Ryde theory that should be questioned or whether the disagreement between the experimental data and the theoretical data should be ascribed to the inadequacies of the experimental techniques used.

In the present work an analysis is carried out on the possible causes of error that might justify the discrepancy between the measured values and those predicted by the theory. In the first part the Mie-Ryde theory is briefly reviewed. In the second part the validity of certain hypotheses is examined, as well as the validity of the experimental values of the parameters assumed as the basis of the calculation of the theoretical relation, together with the influence of their variations. The third part is devoted to an examination of the influence of the methods by which the experimental cheks are carried out: some results of a quantitative analysis is of the errors that depend on the parameters characterising the experiment are shown.

2. MIE-RYDE THEORY

In passing through regions containing precipitation, microwave radiation will be attenuated through two mechanisms: part of the energy will be absorbed within the raindrops and converted into heat and part of the energy will be scattered.

In the case of a plane monochromatic wave incident upon a spherical raindrop embedded in a non-dissipative medium, the total energy W removed per unit time by the drop may be written:

$$W = SQ\left[D, \lambda, \eta(\Theta, \lambda)\right] \quad , \qquad (1)$$

where S is the magnitude of the incident Poynting vector and Q is the extinction cross-section of the drop, which may be interpreted as the sum of a scattering cross-section and an absorption cross-section. The extinction cross-section is a rather complicated function of the drop diameter D, of the wavelength λ and of the complex index of refraction of water η , which, in turn, is a function of the temperature Θ and of λ .

On the assumption of independent single scattering(*), the total energy removed by N identical particles is N times the energy removed by a single particle. Therefore in passing through a layer dx in thickness the decrease in S is:

$$-\frac{dS(x)}{dx} = S(x) \int_0^\infty n(D,x)\, Q\, dD \ , \qquad (2)$$

where n(D,x) dD gives the average number of drops per unit volume with diameter lying between D and D+dD, and n(D,x)dD gives the total average number of drops per unit volume.

(*) Independent scattering means scattering by well-defined and separate particles and implies that the distance between them is sufficiently large to allow each particle to form its own scattering pattern undisturbed by the presence of the others. Single scattering implies the absence of multiple scattering. Each particle is exposed only to the incident radiation and not to the radiation scattered by the other particles.
These are allowable assumptions when the distance between the drops is greater than about five times their diameter, as normally seems to be the case.

Integrating (2) along a section of length d, we have:

$$S(d) = S(0)\, e^{-\int_0^d \gamma(x)dx} \qquad (3)$$

where:

$$\gamma(x) = \int_0^\infty n(D,x)\, Q\, dD \quad , \qquad (4)$$

and is commonly defined as the extiction coefficient.

If it is assumed that n(D,x) does not depend on x:

$$n(D,x) = n(D) \quad , \qquad (5)$$

the attenuation in dB/unit length can be expressed by:

$$a = 4.34\,\gamma = 4.34 \int_0^\infty n(D)\, QdD. \qquad (6)$$

Generally, it is preferred to associate the attenuation with the rain rate, which is the meteorological parameter most commonly used to characterise the precipitations. It is defined as the volume of water reaching the ground per unit time and is given by:

$$p = \frac{\pi}{6} \int_0^\infty n(D)v(D)D^3 dD \quad , \qquad (7)$$

where v(D) is the terminal velocity in still air of the raindrop of diameter D.

From (6) and (7) it may be seen that a and p can be related to each other once we know the distribution n(D), which, however, is not a conveniently measurable parameter. In general, an experimental measurement is made of the distribution m(D), defined as the fraction of the total volume of water reaching the ground due to drops of diameter D. This distribution can be related to n(D) and p by means of the expression:

$$m(D)\, dD = \frac{\frac{\pi}{6} n(D)v(D)D^3}{p}\, dD. \qquad (8)$$

It is obvious that:

$$\int_0^\infty m(D)\, dD = 1. \qquad (9)$$

By means of (6), (7) and (8), the attenuation per unit length can therefore be espressed by:

$$a = \frac{6 \times 4.34}{\pi}\, p \int_0^\infty \frac{Q[D, \lambda, \eta(\Theta, \lambda)]}{v(D)\, D^3}\, m(D)dD. \qquad (10)$$

From (10) it may be noted that:
- once λ, Θ and p are fixed, the attenuation per unit length may be calculated from a knowledge of $\eta(\Theta, \lambda)$, v(D) and m(D);

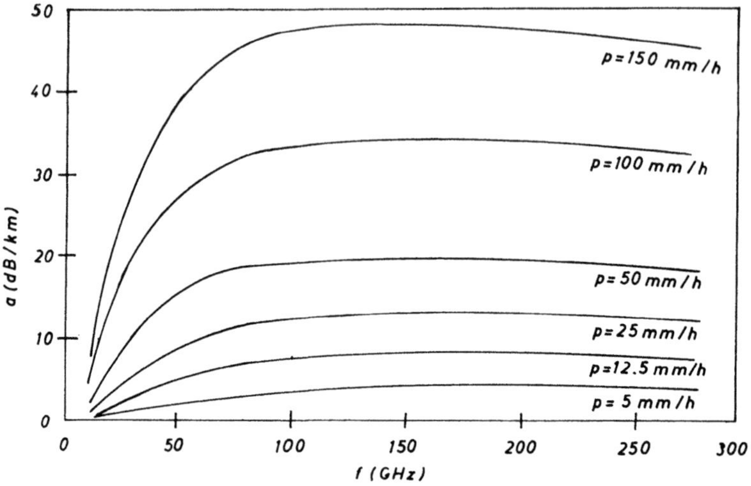

Fig. 1: Attenuation per unit length versus frequency

There are a considerable number of causes that could account for the lack of agreement between the experimental data and the theoretical data, and it is not always easy to assess the errors that could arise from each of them.

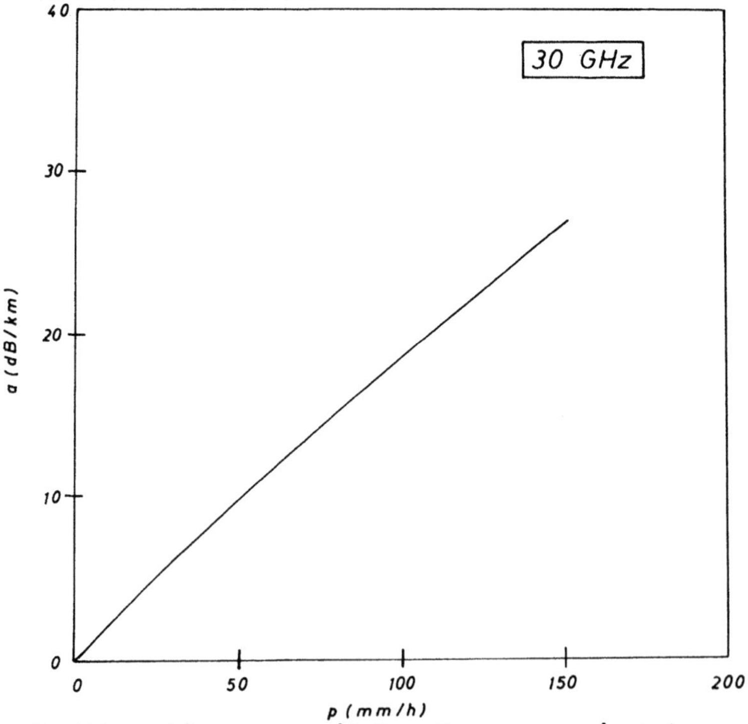

Fig. 2: Attenuation per unit length versus rain rate

- in general, a is not necessarily proportional to p, since also m(D) is generally associated with p. Unless a given rain rate corresponds always to a definite drop-size distribution and vice-versa, a is not even a single-valued function of p;
- for the wavelength considered, the more constant the ratio $Q(D)/v(D)D^3$ in the range of the diameters expected for the raindrops, the more linear will be the relation a-p and the more independent it will be of the drop-size distribution.

Fig. 1 shows the manner in which a(dB/Km) behaves in relation to the frequency f(GHz) for various values of p(mm/h). The curves have been calculated for Θ = 20°C, on the basis of the values of $\eta(\Theta, \lambda)$, m(D) and v(D) measured experimentally by Saxton-Lane (Ref. 3), Laws-Parsons (Ref. 4) and Gunn-Kinzer (Ref. 5) respectively. As may be seen, for the various rain rates, the attenuation increases rapidly with the frequency, reaches a maximum and then remains fairly constant.

The behaviour of the attenuation a in relation to p for f = 30 GHz is shown in Fig. 2. Again, the curve refers to Θ=20°C and has been calculated on the basis of the same values of η, m and v mentioned above.

In view of the importance of the relation a-p for forecasting the reliability of terrestrial communication systems on the basis of the meteorological data available or obtainable in the various localities, numerous experiments have been carried out at various frequencies in order to verify the validity of the Mei-Ryde theory. They normally consisted in attenuation measurements made in natural rain over surface paths and simultaneous rain rate measurements made along the path with a certain number of rain-gauges.

In general, the agreement between measured and predicted values was not completely satisfactory: measured attenuation data were found to be very scattered around the regression curve and generally higher than those predicted. In 1965 Medhurst (Ref. 6), in his excellent review of the experimental data obtained hitherto, concluded that "the applicability of the theory to the practical rainfall situation cannot be said to be demonstrated".

In the experiments in which an attempt was made to minimise the shortcomings in the experimental techniques, data scatter was much less. However, in some of them the measured values of attenuation, even if they show virtually no scatter, are found to be about twice as large as the calculated values (Ref. 7). On the other hand, other measurements, made recently, seem to confirm the data predicted by the theory (Ref. 8).

There are a considerable number of causes that could account for the lack agreement between the experimental data and the theoretical data, and it is not always easy to assess the errors that could arise from each of them.

An initial series of causes of error is concerned with: the validity of the experimental values assumed for m(D), $\eta(\Theta,\lambda)$ and v(D) in the calculation of the relation a-p; the variations in $\eta(\Theta,\lambda)$ and v(D) due to temperature and wind velocity respectively; and the validity of the hypotheses underlying the Mie-Ryde theory.

A second series of causes of error depends on the parameters that characterise the experiment carried out to ascertain the validity of the theory (path length; frequency used; number, collecting surface and integration time of the raingauges) and on the methods used in the data processing.

A third series of causes is concerned with the errors due to the radio-electrical and meteorological instruments used in the experiments.

In the following pages the first two series of causes of error will be examined and an attempt will be made to assess the effect of each cause taken separately. On the other hand, no account will be taken of errors due to the various instruments used. In general, the reference frequency will be taken as 30 GHz, since experiments are at present being carried out in Italy at that frequency (Ref. 9) with the object of ascertaining the validity of the Mie-Ryde theory; however, the conclusions can easily be extended to other frequencies.

3. VALIDITY OF THE HYPOTHESES AND EXPERIMENTAL DATA ASSUMED IN THE THEORY. INFLUENCE OF DROP-SIZE DISTRIBUTION, WIND AND TEMPERATURE.

In this section, in the light of our present knowledge, we shall examine the validity of the experimental values generally assumed for m(D), $\eta(\Theta,\lambda)$ and v(D) for the calculation of the relation a-p, as well as the influence on this relation of the variations in $\eta(\Theta,\lambda)$ and v(D) due to the temperature and wind velocity respectively, and the validity of certain hypotheses underlying the Mie-Ryde theory.

3.1. Drop-size distribution

One of the raindrop-size distributions most commonly used for the calculation of the relation a-p is undoubtedly that proposed by Laws-Parsons. This distribution was derived from samples obtained by exposing in the rain pans of dry sifted flour: a drop striking the flour formed a small ball of dough which was hardened by

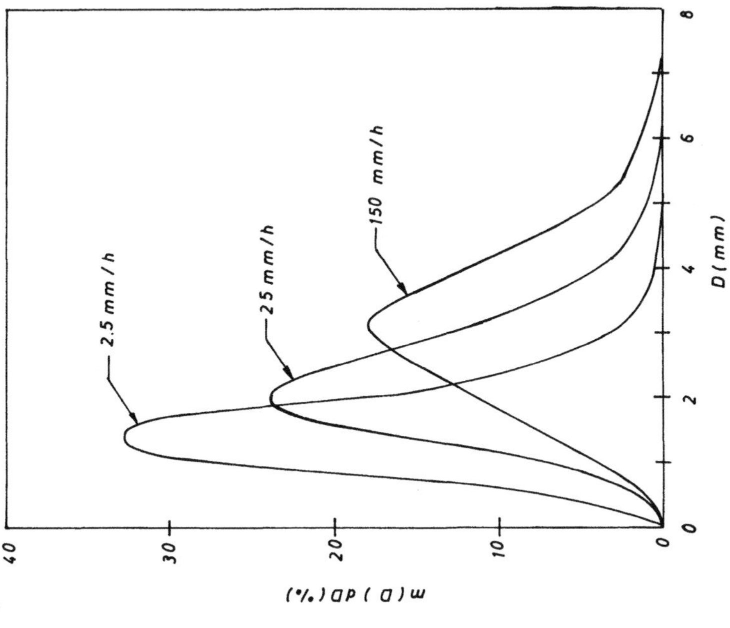

Fig. 3 – Laws – Parsons raindrop size distribution

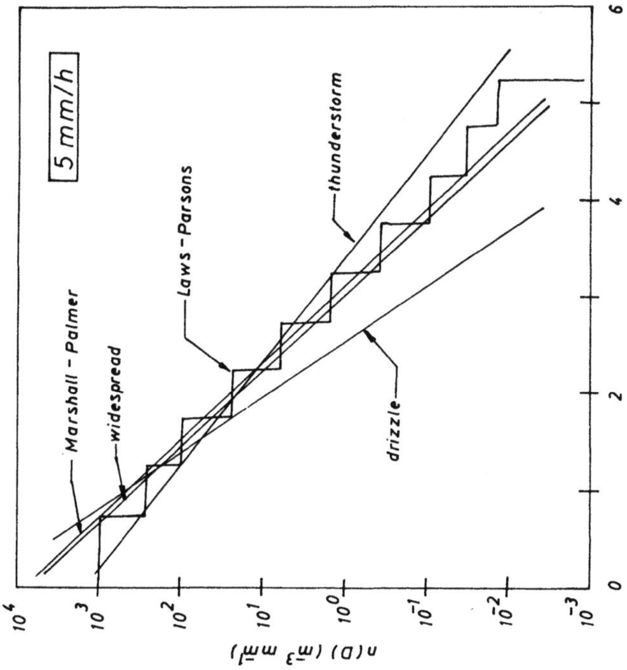

Fig. 4 – Various raindrop – size distributions

drying and examined at leisure. Fig. 3 shows three distributions of m(D)dD (%) corresponding to three different rain rates. It may be noted that the diameters of the raindrops can vary from a fraction of a millimetre to about 7 mm and that their distribution varies as p varies: in particular, as p increases, so also does the number of larger drops.

Subsequently Marshall-Palmer (Ref. 10) and later Joss et al. (Ref.11) proposed distributions of an exponential type that give n(D) directly. All these distributions have the same expression:

$$n(D) = n_o \quad e^{-bD} \quad , \tag{11}$$

where n_o is a constant and b is a parameter obtainable by means of the function v(D) through (7). In particular, b is found to be associated with p through the relation:

$$b \;=\; c\, p^{-0.21} \quad . \tag{12}$$

Whereas Marshall-Palmer proposed a single distribution, Joss et al. proposed three different distribution according to the characteristics of the rainfalls: Drizzle, Widespread, Thunderstorm.

Figs 4 and 5 show these distributions for two values of p , equal to 5 and 150 mm/h respectively. The figures also show the histogram obtained from the Laws-Parsons distributions by means of the velocity v(D) proposed by Gunn-Kinzer. As may be seen, there is fairly good agreement between the various distributions, except in the case of Drizzle and of Thunderstorm. Compared to the Laws-Parsons distribution, Drizzle assumes fewer large drops, while Thunderstorm assumes a greater number of large drops.

It is therefore as well to ask ourselves what variations in a, for the various values of p, can be expected, assuming one or other of the above distributions as a basis for the calculations. Fig. 6 shows the results obtained at 30 GHz. Compared with the Laws-Parsons distribution, the Marshall-Palmer distribution gives values of a that are almost coincident; Widespread gives slightly higher values; Thunderstorm gives lower values. This can easily be explained on the basis of (10): in fact, at 30 GHz the maximum values of the ratio Q(D)/v(D)D3 occur for diameters of about 2 mm, for which the Thunderstorm distribution assumes fewer drops compared with the Laws-Parsons distribution. Fig. 6 also shows the maximum and minimum limiting values theoretically predictable for a, assuming (Ref. 6) a monodisperse drop-size distribution corresponding to the diameter that makes the ratio Q(D)/v(D)D3 maximum or minimum respectively.

From these considerations it seems reasonable to conclude that the influence of the drop-size distribution on the relation a-p is not excessive. As already noted, this influence is found to be less,

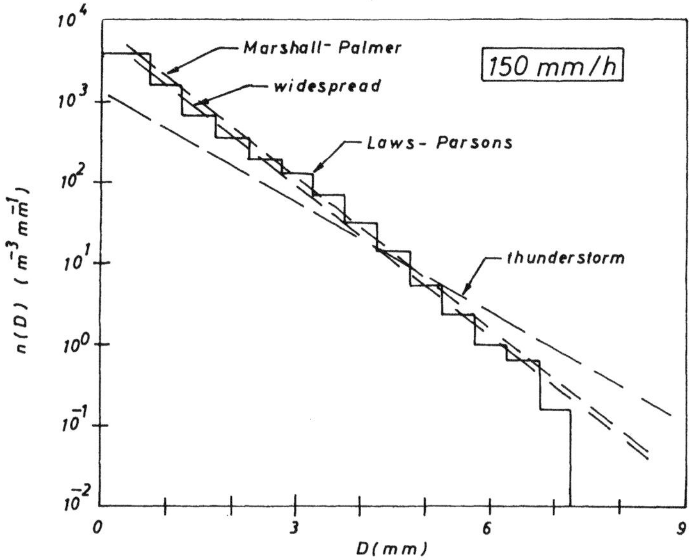

Fig. 5 – Various raindrop – size distributions

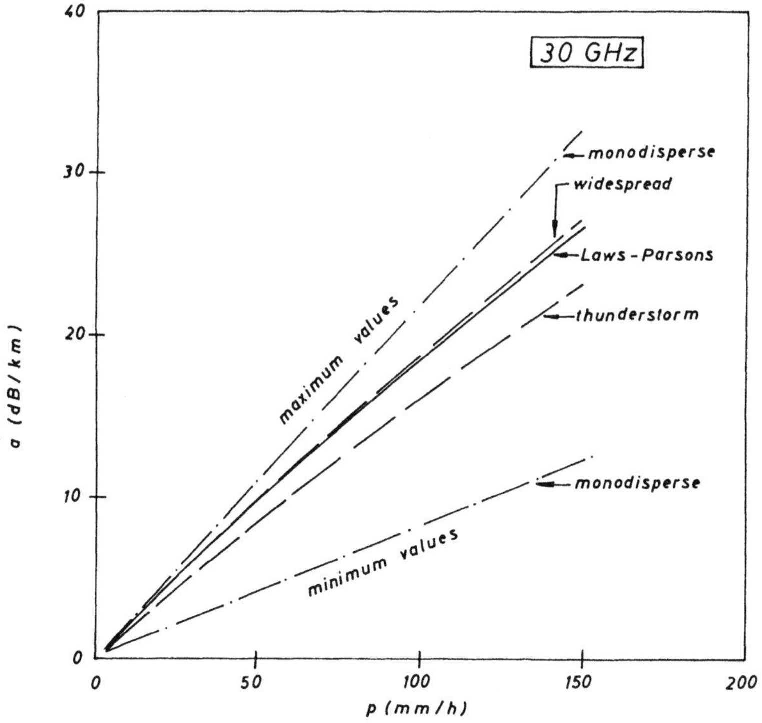

Fig. 6 – Influence of raindrop – size distribution

the less the ratio $Q(D/v(D)D^3$ varies in relation to the diameter, for the wavelength considered. From an examination of the variability of this ratio, it seems that the influence of the drop-size distribution on the relation a-p is a minimum in the frequency range 20-40 GHz.

However, in order to reduce the uncertainties arising from the variations in drop-size distribution in the experimental verification of the Mie-Ryde theory, it seems advisable, at the same time, to measure this distribution. An indirect method for checking the variability in the drop-size distribution would seem to be to measure, at the same time, on the same link, the supplementary attenuations at two different frequencies (Ref. 12).

3.2. Index of refraction of water. Influence of temperature.

Let us now see whether the values relative to index of refraction of water generally assumed for calculating the attenuation per unit length are modified or not, in the light of the most recent knowledge on the subject.

Recently (Ref. 13) an empirical model has been proposed for the calculation of the index of refraction of water, for wavelengths of between 10^{-4} and 10^9 cm and for temperatures of up to 50°C, based on an elaborate fitting of the available measured data. Fig. 7 shows the real and imaginary parts of the index of refraction of water, calculated by means of the above model, for values of λ between 0.1 and 10 cm and for Θ = 20°C. It may be noted that the calculated values coincide exactly with the experimental values obtained by Saxton-Lane in 1952 (Ref. 3) and that the results obtained with the model proposed by these authors in the same year are in excellent agreement with the curve shown in Fig. 7.

The same figure shows, for λ = 1 cm, the values calculated by means of a method proposed in 1946 (Ref. 14) and used by many authors for the calculation of the attenuation a (Refs. 6, 15). The agreement with the data presented is so good that the values of attenuation calculated by these authors can be considered sufficiently precise.

As the temperature varies, so does the index of refraction of water. Consequently also the values of the theoretical attenuation depend on the temperature. However, examination of the values of a calculated by Ryde (Ref. 2) for various frequencies and temperatures shows that the variations in a, at least in the range 10° - 30°C and for frequencies of around 30 GHz, are fairly restricted.

On the basis of the values calculated by Ryde, a rough estimate has been made of the possible variations in the theoretical va-

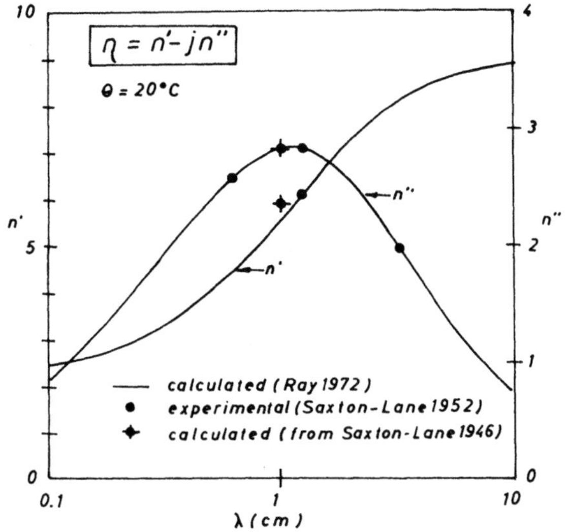

Fig. 7 – Index of refraction of water

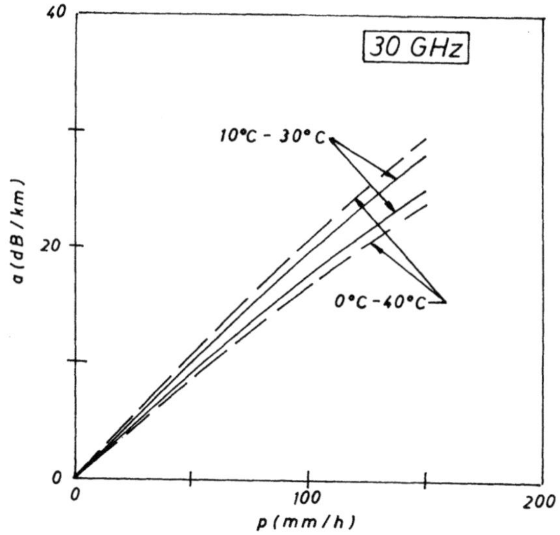

Fig. 8 – Influence of temperature

lues of the attenuation for the two temperature ranges 10° - 30°C and 0° - 40°C and for f = 30 GHz: these variations are shown in Fig. 8. Although the variations in a with the temperature are not excessive, it would however seem convenient, in order to increase the precision of the data obtained in the experimental checks of the Mie-Ryde theory, to take measurements of the temperature at the same time.

3.3. Terminal velocity of the raindrops in still air. Influence of the vertical component of the wind velocity.

Certain authors (Ref. 6) assume for the terminal velocity of raindrops in still air the values proposed by Gunn-Kinzer in 1949 and shown in Fig. 9. As may be seen for the range of diameters normally considered, this velocity varies from about 2 to about 9m/sec. Other authors (Ref. 15), on the other hand, assume the values proposed by Best in 1944 (Ref. 16), also shown in Fig. 9, although in 1950 (Ref. 17), Best himself reported certain data obtained by Davie that seem to be in excellent agreement with the values proposed by Gunn-Kinzer. However, the differences between the two curves are so slight that the attenuations calculated by these authors are found to be approximately coincident.

In the calculation of the theoretical relation a-p, as has been seen, it is assumed that the drops fall through still air. However, this assumption is rarely respected in actual rainfall situations, on account of the effect of the wind. In many situations the wind velocity may have a vertical component. In such cases, if it is assumed that the distribution n(D) does not vary, then a does not vary but p may undergo considerable variations. For example, a vertical component of the wind velocity in an upward direction has the effect of reducing the terminal velocity of the drops and hence causing an error by reducing the measured value of the rainfall intensity at the ground.

Fig. 10 shows the differences from the theoretical values, calculated on the assumption of still air, that can be expected, in the presence of a vertical wind component assumed constant, equal to 1 and 2 m/sec and directed downwards and upwards, respectively. As was to be expected, in the case of a component directed upwards, for example, measurement of the rain-rate at the ground gives too low a value, and hence the measured values of the attenuation will tend to be higher than the values calculated in the case of still air. The contrary occurs for a vertical component directed downwards.

The considerable departures from the theoretical values shown in Fig. 10, even if certain assumptions are made to simplify the calculation with respect to the complexity of the phenomenon, make it advisable to measure the wind velocity also in the vertical plane

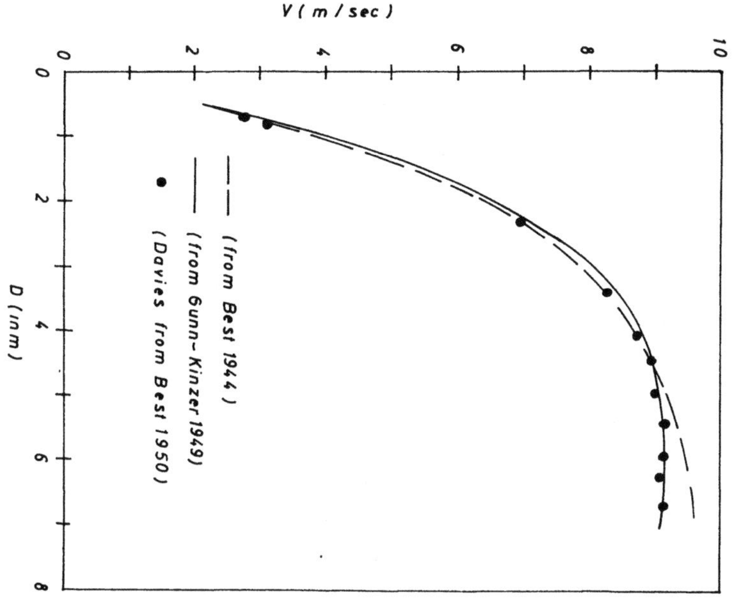

Fig. 9 – Terminal velocity of the drops

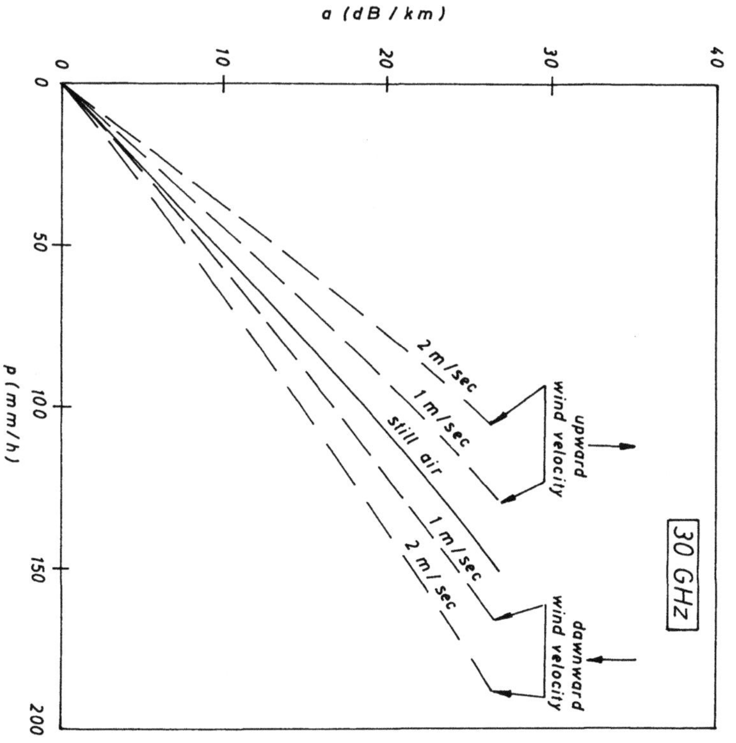

Fig. 10 – Influence of vertical wind speed

in experiments aimed at verifying the Mie-Ryde theory. Particular caution should be adopted, however, in processing data obtained in conditions of strong wind, not only for the reasons indicated above but also on account of the effect of turbulence - not discussed here - which could be produced by the wind on the collecting surface of the raingauges.

3.4. Non-spherical shape of the raindrops.

Among the assumptions underlying the Mie-Ryde theory, one of the most questionable seems to be that concerning the spherical shape of the raindrops. Experimental evidence (Ref. 18) has shown that the raindrops tend to be flattened and that this flattening increases as the size of the drop increases. Consequently, the attenuation tends to be sensitive to polarization.

On the basis of the experimental data regarding the variation in the deformation of the drops in relation to their dimensions, it has been possible to calculate the difference in the attenuation undergone by two waves polarised orthogonally at the same frequency (Ref. 19). From the results obtained it appears that for horizontally polarised waves the attenuation is always greater than for vertical polarised waves and that the former is very close to what would occur in the case of spherical drops. Fig. 11 shows certain data relative to 30 GHz calculated recently. (Ref. 20). In the figure it is assumed that the attenuation for horizontal polarization a_H coincides with that obtained on the assumption of spherical drops a.

From what has been said above it would therefore seem advisable to measure, at the same time and at the same frequency, both the attenuation for horizontal polarization and that for vertical polarization, in the experiments aimed at verifying the validity of the theory. This would make it possible to estimate to what extent the differences between the measured values and the values calculated on the assumption of spherical drops are to be ascribed to the non-spherical shape of the raindrops.

4. INFLUENCE OF THE PARAMETERS OF THE EXPERIMENT AND OF THE METHODS USED IN THE DATA PROCESSING.

An analysis has been carried out recently (Ref. 21) on the causes of error that depend on the methods used for experimentally checking the validity of the Mie-Ryde theory. An attempt has also been made to assess the extent of these errors which are mainly due to the integration time and to the spacing of the raingauges along the path. Expecially for high values of p, the former causes p to be underestimated; the latter causes p to be overestimated.

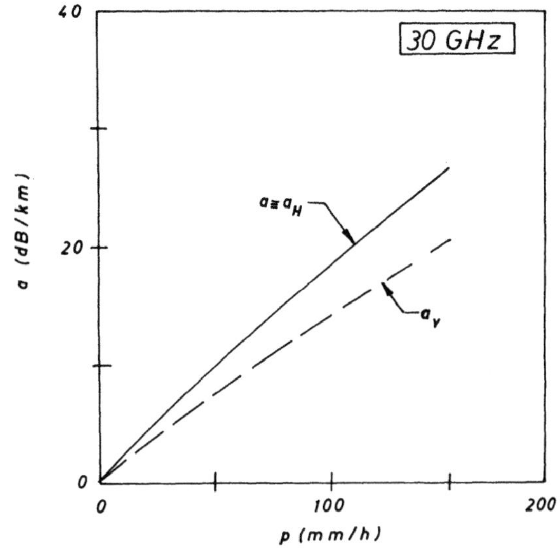

Fig. 11 – Attenuation for horizontally and vertically polarized waves

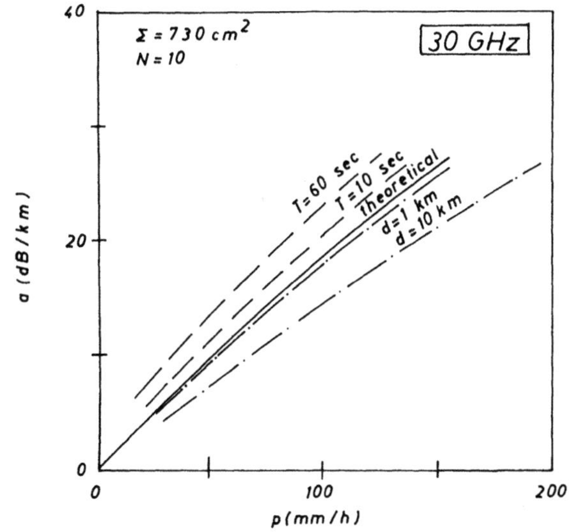

Fig. 12 – Influence of integration time and spacing of the raingauges

It ensues that measured values of attenuation tend to be higher than those predicted by the theory if the error due to the integration time is considered; lower if the error due to the spacing is considered.

An estimate has been made (Ref. 21) of the spread of the values that may be expected in the experimental cheks of the theoretical attenuation-rain rate relation at 30 GHz due to the integration time and the spacing of the raingauges. In particular, this spread has been estimated for two values of the integration time, equal to 10 and 60 sec. respectively, and for two values of the length of the link, 1 and 10 km, along which 10 raingauges are located, with a collecting area of 730 sq. cm. The results obtained are shown in Fig. 12.

From the figure it may be observed that a spacing of about 100 m between the raingauges and an integration time lower than 10 sec, should give a fairly restricted spread of values. On the other hand, if the spacing between the raingauges is kept below 100 m and the integration time is increased to values above 10 sec, the error due to the latter seems to exceed by a long way the error due to the spacing between the raingauges. These conclusions are found to be in agreement with the data obtained experimentally by Norbury (Ref. 22).

A value higher than about 100 m for the spacing of the raingaùges or a value higher than about 10 sec for the integration time considerably increases the spread of values that may be expected, at least with the assumptions made. However, while the error due to the spacing between the raingauges does not appear to be reducible (unless the number of raingauges is increased in a given length of link), the error introduced b- an excessive integration time of the raingauges appears to be considerably reducible by means of an integration of the attenuation, carried out in the same interval of time.

5. CONCLUSIONS

On the basis of the observations made in the present paper, it seems reasonable to draw the following conclusions.

In the light of present-day knowledge, it seems that the values normally assumed for the index of refraction of water, the drop-size distribution and the terminal velocity of the drops are sufficiently reliable for the calculation of the attenuation-rain rate relation on the basis of the Mie-Ryde theory.

The experimental verification of this theory, however, is anything but simple in view of the numerous sources of error that may be present.

A first series of causes of error consists in variations in temperature and raindrop-size distribution, the non-spherical shape of the raindrops and, above all, the presence of wind during the experiments. In order to limit the effects of these errors and possibly to sort out in a suitable manner the data obtained, it seems convenient in the experimental cheks to take measurements, simultaneously, of: temperature; wind direction and velocity, also in the vertical plane; and raindrop-size distribution. If this last cannot be measured directly, it seems in any case advisable to try to estimate its variations at least indirectly, by measuring, at the same time and on the same link, the attenuations encountered at two different frequencies. Lastly, in order to estimate to what extent the differences between the measured values and those predicted by the theory are to be ascribed to the non-spherical shape of the raindrops, it appears convenient to measure, at the same time, the difference between the attenuations undergône by two waves at the same frequency, polarised horizontally and vertically.

Other considerable causes of error consist in an excessive distance between the raingauges used in the experiment or in their excessive integration time. At a first analysis it appears that a spacing closer than about 100 m and an integration time of between 1 and 10 sec should ensure a fairly restricted spread of values. In the case where the integration time of the raingauges is higher, it seems that the resulting spread of values may be reduced by the procedure of effecting an integration of the attenuation in the same interval of time.

Finally, even if the subject has not been discussed in the present paper, it seems as well to mention the considerable influence that can be assumed by the errors due to the radio-electric and meteorological instruments used in the experiments.

REFERENCES

(1) Mie, G.C., "Beitrüge zur optik trüber medien speziell kolloidaler metallosungen" Ann. Physik, Vol. 25, n. 3, March 1908.

(2) Ryde, J.W., Ryde, D., "Attenuation of centimeter and millimeter waves by rain, hail, fogs and clouds" Rept. 8670, General Electric Co. Research Labs., Wembley, May 1945.

(3) Lane, J.A., Saxton, J.A., "Dielectric dispersion in pure polar liquids at very high radio frequencies" Parts I and II, Proc. of the Royal Society, A, Vol. 213, 1952.

(4) Laws, J.Q., Parsons, D.A., "The relation of raindrop-size to intensity" Trans. Am. Geophysical Union, Vol. 24, 1943.

(5) Gunn, R., Kinzer, G.D., "The terminal velocity of fall for water droplets in stagnant air" Journal of Meteor., Vol. 6, August 1949.

(6) Medhurst, R.G., "Rainfall attenuation of centimeter waves: comparison of theory and measurement" I.E.E.E. Trans. on AP, Vol. 18, July 1965.

(7) Anderson, L.J. et al., "Attenuation of 1.25 centimeter radiation through rain" P.I.R.E., Vol. 35, April 1954.

(8) Norbury, J.R., White, W.J.K., "Microwave attenuation at 35.8 GHz due to rainfall" Electronics letters, Vol. 8, n. 4, Feb. 1972.

(9) Fedi, F., "Criteria and expected accuracy of the measurements envisaged in the research programme under way in Italy" AGARD Symposium on Telecommunication aspects on frequencies between 10 and 100 GHz" Gausdal (Norway), September 1972.

(10) Marshall, J.S., Palmer, W.M.K., "The distribution of raindrops with size" J. Met., Vol. 5, 1948.

(11) Joss, J. et al., "The variation of raindrop size distribution at Locarno" Proc. Int. Conf. Cloud Physics, Toronto, 1968.

(12) Semplak, R.A., "Dual frequency measurements of rain induced microwave attenuation on a 2.6 km propagation path" B.S.T.J., Vol. 50, n. 8, October 1971.

(13) Ray, P.S., "Broadband Complex refractive indices of ice and water" Appl. Opt., 11 August 1972.

(14) Saxton, J.A., Lane, J.A., "Meteorological factors in radiowave propagation" The Physical Society, London, 1946.

(15) Setzer, D.E., "Computed transmission through rain at microwave and visible frequencies" B.S.T.J., October 1970.

(16) Best, A.C., "Water in the Atmosphere" from Kerr D.E. Propagation of short radio waves, New York: Dover Publ. Inc.

(17) Best, A.C., "Empirical formulae for the terminal velocity of water drops falling through the atmosphere" Quart. J.R.Met. Soc., Vol. 76, July 1950.

(18) Magono, C., "Shape of water drops falling in stagnant air" Journal of Met., February 1954.

(19) Oguchi, T., "Attenuation of e.m. wave due to rain with distorted raindrops" J. of Radio Research Lab., September 1960, January 1964.

(20) Morrison, J.A., Cross, M.J., Chu, T.S., "Rain-induced differential attenuation and differential shift at microwaves frequencies" B.S.T.J., April 1973.

(21) Fedi, F., Mandarini, P., "Errors inherent to the experimental testing of the Mie-Ryde theory". (To be published).

(22) Norbury, J.R., White, W.J.K., "Correlation between measurements of rainfall and microwave attenuation at 36 GHz" IEE Conf. on Propagation of radio waves at frequencies above 10 GHz - London April 1973.

QUANTITATIVE ANALYSIS OF PRECIPITATIONS THROUGH RADAR TECHNIQUES

G. Drufuca and A. Paraboni

Centro Telecomunicazioni Spaziali CNR
Istituto di Elettronica ed Elettrotecnica
Politecnico di Milano, Italy

1. FOREWORD

The quantitative evaluation of precipitations through radar observations is a technical problem of formidable complexity. A complete review of the problems involved would be a very hard task. Because of the limited space that is allowed, the title of the present work has little modesty as we wish to satisfy both the contradicting exigencies of synthesis and depth. A choice is necessary: therefore we decided for a certain degree of completeness, sending the reader to the specialistic works for the particular problems. This choice is justified by the individual experiences of the authors which have been limited to the use of radar as a suorce of data to be directly applied, rather than consisting in a specialistic research on the instrument itself or on the theory of the measurement. This work does not present deep analysis nor original results, but it attempts to give a panoramic view of the measurements, of the precision, of the limits, of the main difficulties to be encountered and possibly of the methods to overcome them.

2. PRECIPITATION MEASUREMENTS AND APPLICATIONS

Today the applications of radar meteorology cover a large number of fields: cloud physics, storm physics, mesoscale analysis, hydrology forecasting, storm avoidance, attenuation studies, interference studies and so on.

Each of these applications has its own peculiarities and in many cases the quantity directly measured reflectivity, has to be converted in a different quantity: tipically rainfall rate, water content, specific attenuation at a given frequency and so on.

These conversions are possible only if the drop size distribution is known or more generally if the size distribution of the meteors is known. In practice this function is not known perfectly because it can have very fast spatial variability. This is a characteristic common to all remote sensing devices. Nevertheless the radar being a remote sensing device offers the possibility of having a complete, spatial and temporal, view of the process under observation that other conventional instruments, although more precise, cannot offer. In fact it is the only instrument able to perform measurements of precipitations aloft. For these reasons the problem of the interpretation of radar data is not: "in which measure are the radar data reliable?" but rather: "in which way is it possible to collect and elaborate radar data in order to take advantage from the enormous quantity of information available?".

The answers to this question are tipically "problem oriented". Some example can be clarifying. For instance is hydrology it does not matter so much how much rain is fallen in a given point in a short time but how much water is fallen on a large area for a relative long time; averaging over large areas and times is luckily one of the more effective ways to reduce the confidence limits of point measurements. In an other case, forecasting for instance, it could be more useful to know the shape and the trajectory of cells rather than the absolute value of the precipitation rate.

A typical application, considered to be difficult respect to precision, consists in the prediction of microwave attenuation. In this case the final result consists in statistics of attenuation, possibly extended down to extreme values of probability, for various configurations of the communication path. In this scheme the errors encountered in the single realization have importance only in the measure with which they can influence the probability distribution: in this example also, the exploitation of an enormous sample eliminates all but the systematic errors. Reliable results can than be obtained with the auxiliary use of an instrument that can be more precise even if much poorer in quantity of information (satellite, radiolink, radiometer, raingauge), using the results of this instrument to adjust some parameter of the distribution curves. This procedure has been followed by Strickland (1972) for satellite links and by Drufuca (1973) for ground links.

What has been said does not eliminate the necessity of knowing as deeply as possible the causes for error and of trying, in relation to the marginal advantages of the improvement and its cost, to improve the single measurements.

The causes of errors are examined in the following in the general context of the theory of radar meteorology.

3. RADAR EQUATIONS

Let us consider a radar set measuring the power returned from spherical particles composed entirely of water or ice and randomly distributed in space.

For a single scatter we can write:

$$P_r = \frac{P_t\, G^2\, \lambda^2}{(4\pi)^3\, r^4}\; \sigma_i \qquad (1)$$

where P_r received power

P_t transmitted power

G antenna gain

λ wavelength

σ_i backscattering cross section of a single scatter.

For an array of scatterers illuminated at the same time an average has to be made because the phase of each scatterer has to be considered. If the received power is averaged over a large number of independent arrays we write

$$\bar{P}_r = \frac{P_t\, G^2 \lambda^2}{(4\pi)^3 r^4} \int_0^n{}_i\; \sigma_i \qquad (2)$$

where the summation is carried on over the illuminated volume.

The illuminated volume can be expressed in terms of antenna beamwidths and pulse length

$$V = \pi \left(r\, \frac{\theta}{2}\right) \left(r\, \frac{\phi}{2}\right) \frac{h}{2}$$

where θ and ϕ are the horizontal and vertical beamwidths and h is the pulse length.

We have:

$$\bar{P}_r = \frac{P_t\, G^2 \lambda^2 \theta \phi\, h}{512\, \pi^2 r^2} \int_{vol} \sigma_i \qquad (3)$$

where $\int_{vol} \sigma_i$ represents a summation over a unit volume and is called "radar reflectivity" η .

Eq. (3) has a more realistic form due to Probert-Jones (1962) which takes into account the radiation pattern of paraboloidal antennas with circular reflectors:

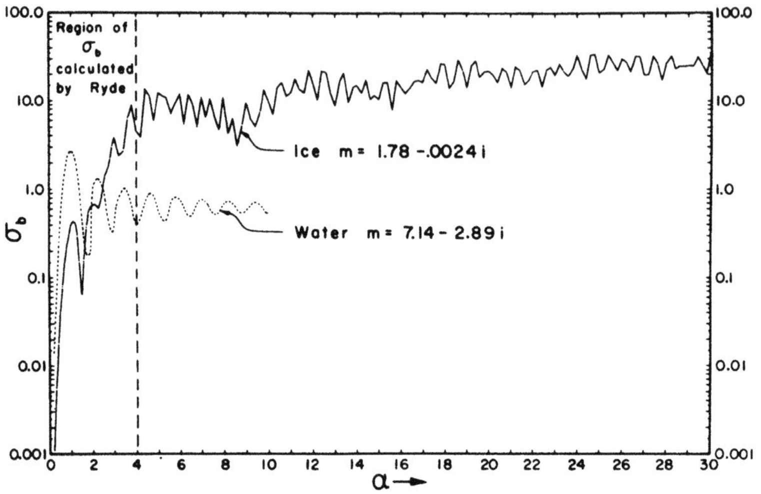

Fig. 1 Calculated values of the normalized backscattering cross section ($\sigma_b = \sigma/\pi a^2$) for water and ice spheres. The water curve applies at a temperature of 0°C and a wavelength of 3.21 cm. The ice curve is valid for wavelengths from 1 to 10 cm. From Herman and Battan (1961*a*).

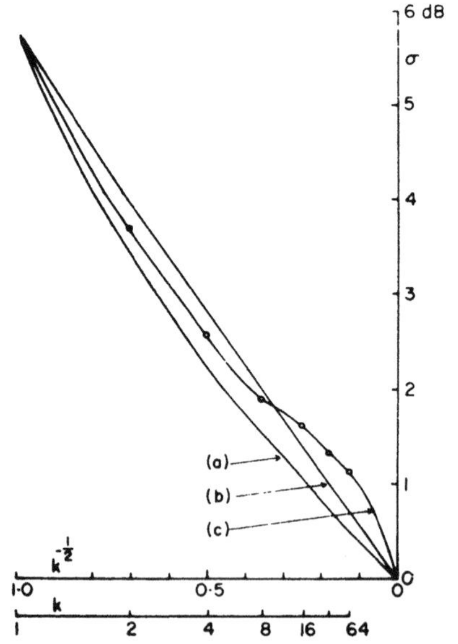

FIG. 2 Standard deviation of probability distribution of target level for groups of *k* data combined by averaging intensities (a), averaging intensity levels (b), and peak reading (c), based on data taken from Smith (1964).

$$\bar{P}_r = \frac{P_t\ G^2 \lambda^2 \theta \phi\ h}{512(2\ln 2)\ \pi^2 r^2} \int_{vol} \sigma_i \qquad (4)$$

ore more expressively:

$$\bar{P}_r = \left|\frac{c}{1024\ \pi^2 \ln 2}\right| \left|P_r\ \tau \lambda^2\ G^2 \theta \phi\right| \left|\frac{\eta}{r^2}\right| \qquad (5)$$

where c is the speed of light.

The terms in the first brackets are constants, in the second brackets we have radar parameters and, in the third, target parameters.

For the scattering of a plane wave by spherical particles a general theory was developed by Mie (1908). The back scattering cross section of a sphere is:

$$\sigma = \frac{\pi a^2}{\alpha^2} \left| \int_1^{\infty}{}_n \quad (-1)^n \quad (2n+1) \quad (a_n - b_n) \right|^2$$

where a is the particle radius and $\alpha = \frac{2\pi a}{\lambda}$, a_n and b_n represent the scatter from magnetic and electric dipoles, quadrupoles, etc., induced in the particle by the incident wave. a_n and b_n are expressed in terms of spherical Bessel functions and Henkel functions of the second kind with arguments α and m (complex index of refraction).

$$m = n - ik.$$

Fig. 1, from Herman and Battan (1961), shows calculated values of the cross section for water and ice spheres at 3.21 cm (0°c).

If the radius of the particle is small ($\alpha \ll 1$):

$$\sigma_i = \frac{\pi^5}{\lambda^4} \left|K\right|^2 \quad D_i^{\,6} \qquad \text{(Rayleigh approximation)}$$

where D_i is the diameter of the particle and

$$K = \frac{m^2-1}{m^2+1} \ .$$

In Table 1 Battan (1973) shows the relevant parameters for water and in Table 2 Battan (1973) for ice. The data are taken from Gunn and East (1954).

TABLE 1 The Components of the Complex Index of Refraction, $|K|^2$, and the Imaginary Part of $(-K)$ of Water as Functions of Temperature and Wavelength

Quantity	Temperature (°C)	Wavelength (Cm)			
		10	3.21	1.24	0.62
n	20	8.88	8.14	6.15	4.44
	10	9.02	7.80	5.45	3.94
	0	8.99	7.14	4.75	3.45
	−8	....	6.48	4.15	3.10
κ	20	0.63	2.00	2.86	2.59
	10	0.90	2.44	2.90	2.37
	0	1.47	2.89	2.77	2.04
	−8	....	....	2.55	1.77
$\|K\|^2$	20	0.928	0.9275	0.9193	0.8926
	10	0.9313	0.9282	0.9152	0.8726
	0	0.9340	0.9300	0.9055	0.8312
	−8	....	....	0.8902	0.7921
$\mathrm{Im}(-K)$	20	0.00474	0.01883	0.0471	0.0915
	10	0.00688	0.0247	0.0615	0.1142
	0	0.01102	0.0335	0.0807	0.1441
	−8	....	....	0.1036	0.1713

SOURCE: Gunn and East 1954.

Table 2

The Components of the Complex Index of Refraction, $|K|^2$, and the Imaginary Part of $-K$ of Ice as Functions of Temperature

Quantity	Temperature (°C)	Value
n	All temperatures when $\rho = 0.92$ gm/cm^3	1.78
κ	0	2.4×10^{-3}
	−10	7.9×10^{-4}
	−20	5.5×10^{-4}
$\|K\|^2$	All temperatures when $\rho = 1$	0.197
$\mathrm{Im}(-K)$	0	9.6×10^{-4}
	−10	3.2×10^{-4}
	−20	2.2×10^{-4}

SOURCE: Gunn and East 1954.

From Battan 1973

In the case of small particles it is possible to write:

$$\bar{P}_r = \left|\frac{\pi^3 c}{1024 \ln 2}\right| \left|\frac{P_t \tau G^2 \theta \phi}{\lambda^2}\right| \left| |K|^2 \frac{Z}{r^2} \right| \qquad (6)$$

where $Z = \int_{vol} D_j^6$ is the "radar reflectivity factor", then $\eta = \frac{\pi^5}{\lambda^4} |K|^2 Z$.

Z is independent of the wavelength, thus allowing comparison of measuremets made with different radar systems. Its MKS units are $m^6/m^3 = m^3$, because the diameters of the particles are usually measured in millimeters values of Z are usually given in mm^6/m^3.

It is common to use a logarithmic scale for Z. The logical base for such a scale is 1 mm^6/m^3. Then:

$$Z_{dB} = 10 \log \left| Z/ (1mm^6/m^3) \right| .$$

As an example due to Smith (1969) suppose the precipitation target to be a shower with all drops 1 mm in diameter, with a concentration of 10^3 drops per cubic meter. In this case $Z = 10^{-15}$ m^6/m^3 or 10^3 mm^6/m^3 or 30 dB_Z, η (with a wavelength of 10 cm) would be 2.85×10^{-9} m^2/m^3.

If the drops were all of 5 mm diameter, with the same concentration of 10^3 drops per cubic meter, Z would be more than 10^7 mm^6/m^3 ($Z_{dB} = 70$).

Values of Z in excess of 10^7 are encountered in thunderstorms, but the concentration of 10^3, 5 mm drops per cubic meter seems unrealistically high; such a value of Z probably indicates presence of hail. At the other end of the scale, values of Z in drizzle, with drop diameters of the order of 0.2 mm can be as small as 0.01 mm^6/m^3 ($Z_{dB} = -20$).

Equation (6) is valid provided that a given set of assumptions are satisfied (P. Smith (1969)), they are:

a) The scatterers are homogeneous dielectric spheres, with diameters small compared to the wavelength.
b) The illuminated volume is filled with scatterers.
c) Z is uniform throut the illuminated volume.
d) The scatterers all have the same dielectric constant; that is they are all water drops or all ice spheres.
e) Multiple scatter is negligible.
f) The main lobe of the antenna pattern can be described by a gaussian function.
g) Attenuation along the propagation path between the radar and the target is negligible.
h) The incident and back-scattered waves are linearly polarized.

Whenever any of these assumptions are invalid Equation (6) has to be modified.

Equation (6) can be inverted:

$$Z = \frac{1024 \ln 2}{\pi^3 c} \left| \frac{\lambda^2}{P_t \, \tau \, G^2} \right| \left| \frac{\bar{P}_r \, r^2}{|K|^2} \right| \qquad (7)$$

As long as the assumptions made are satisfied, the value of Z measured by radar using equation (7) is a characteristic parameter of the precipitation target as liquid water content or rainfall rate. In cases when the assumptions break down eq. (7) can be used to define an "equivalent radar reflectivity factor" Z_e which is what the radar sees. It is not simply related to the drop diameters and it is in general dependent of the wavelength so that many advantages of its use are lost, although it provides a way of coping with inhomogeneous, non-spherical and large scatterers.

4. RANDOM PHASE FLUCTUATION

Eq. (7) requires the averaging of the received backscattered power P_r.

This averaging is necessary to overcome the effects due to fluctuation of the backscattered power from a random array of abjects such as precipitation particles. Because the fluctiation arises from the changes in the phase of the reflections from the individual particles they are usually called "random phase" fluctuations.

The problem of estimation of the average power and then of the average reflectivity has been studied by various authors: Marshall and Hitschfeld (1953), W allace (1953), Kodaira (1960), Smith (1964) Rogers (1971). In the case of spatially uniform reflectivity, Fig. 2 taken from Marshall (1971) shows the standard deviation of the estimate of the average level (reflectivity or received power expressed in dB) in function of the number of independent samples and for different methods of averaging.

If the reflectivity field is not uniform the standard fluctuation theory will not strictly apply as shown by Rogers (1971); Fig. 3, taken from his work, shows the effect for two models of spatial distribution of reflectivity. Model 1 assumes uniform distribution function of the reflectivity level in the volume over which the average is made and model 2 assumes uniform distribution function of the reflectivity itself; in this figure k is a measure of the reflectivity gradient. For instance if the volume is 1 km long, k = 20 (in dB) means that there is a reflectivity gradient of

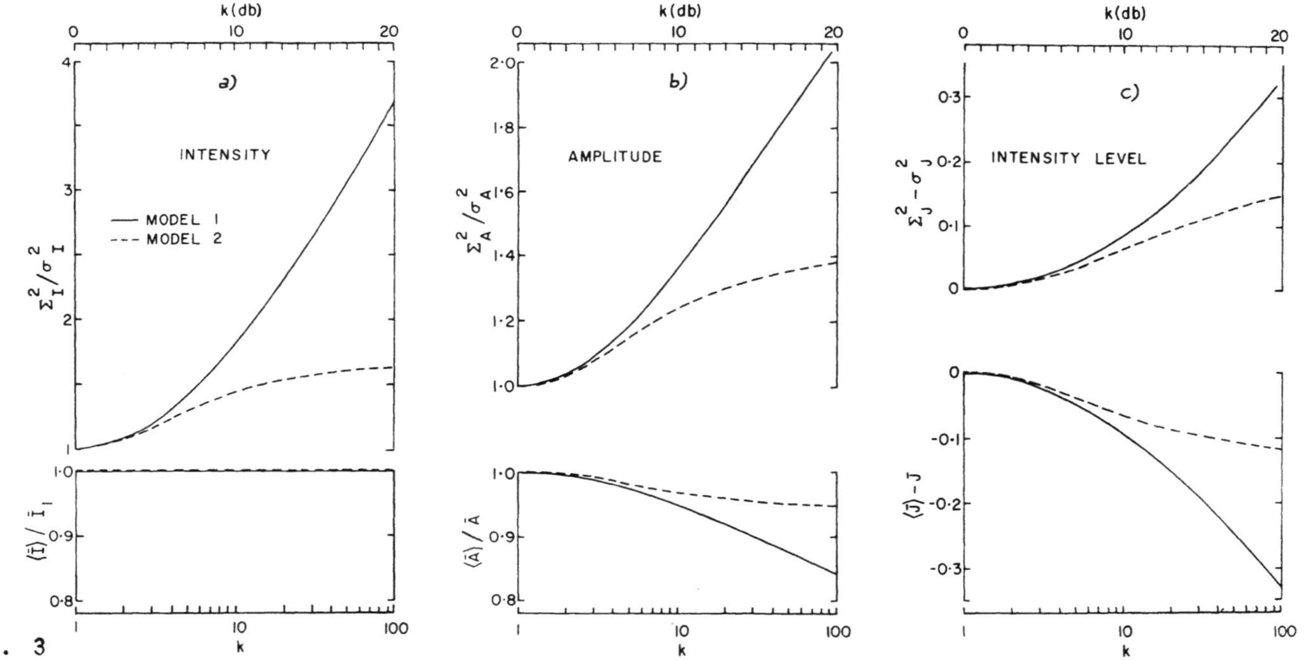

Fig. 3

a) The effect of non uniform reflectivity on the mean value and variance of signal intensity. Plotted as functions of the non-uniformity factor k are the ratio of variances (above) and the ratio of means (below) for signals of the same mean reflectivity with and without nonuniformities.

b) Same as a), except for signal amplitude

c) Same as a), except for signal intensity level.

from Rogers 1969

20 dB_Z/km.

5. DISCUSSION OF THE RADAR EQUATION

The equation (7) is valid provided that a set of assumptions are satisfied: if not, further considerations and corrections have to be made.

Condition a) is satisfied if the diameter of the particle is such that $D<0.07\lambda$ for water at 18°c (Gunn and East, 1954). Ryde (1946) showed that for ice the Rayleigh approximation can be used for hail for diameters such that $D<0.16\ \lambda$.

Conditions b) and c) are self explanatory, although caution has to be used in interpreting the results of averages made over reflectivity or reflectivity levels.

Condition d) is not always satisfied; often in the same volume it is possible to have raindrops and melting snow, or raindrops and wet hail. A good review of this complicate problem is found in the recent book by L.J. Battan (1973).

Condition f) is easily controllable; if not, the actual antenna pattern has to be considered.

Condition g) is usually satisfied for 10 cm radars. It is not for 3 cm radars. In this case for any rain of more than moderate intensity a quantitative analysis is hopeless.

Several attempts: Atlas and Banks (1951), Hitschfeld and Bordan (1954), Rao and Shaw (1968), have been made to compensate for the rain attenuation but without success. The main reason being that a compensating technique requires, among other things, a calibration of the radar set in the order of 1%, which is impracticable.

The work of McCormick and Hendry (1970) and Barge (1970) indicates that condition h) is generally satisfied at frequencies less than 16 GHz.

6. MORE TROUBLES

Other problems affect the performances of a meteorological radar. Among them, of great practical importance, are the calibration and the influence of ground clutter.

Equation (7) can be written as:

$$Z = C\ \overline{P}_r\ r^2$$

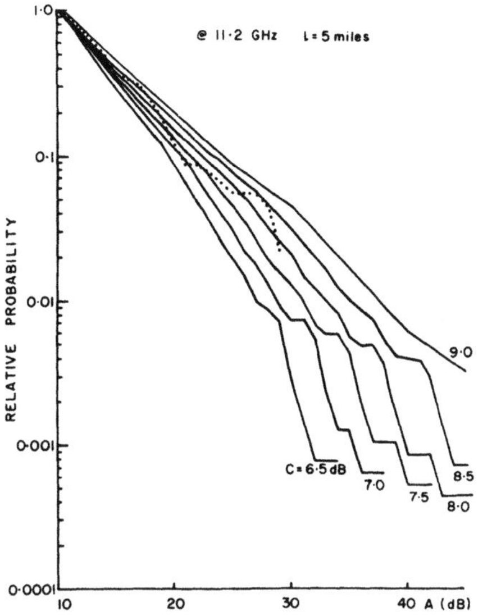

Fig. 4 . Relative probabilities that a given attenuation is exceeded (continuous lines) as obtained for different values of the calibration constant (in a relative scale). The dotted curve is obtained from the experimental link. All curves are normalized to 1 for 10 dB of attenuation.

from Drufuca 1973)

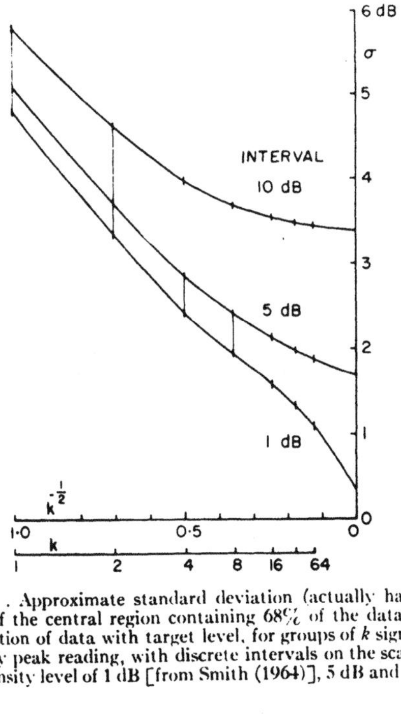

FIG 5 . Approximate standard deviation (actually half of the width of the central region containing 68% of the data) for the distribution of data with target level, for groups of k signals combined by peak reading, with discrete intervals on the scale of target intensity level of 1 dB [from Smith (1964)], 5 dB and 10 dB.

C is a calibration constant to be determined for each radar set. Smith (1968) describes and proposes several techniques to accomplish a good calibration.

Except for very particular kind of radar operation, as in Joss and others (1968), it is difficult to accomplish a radar calibration with an accuracy better than ± 1 dB, ± 2 dB being a more realistic figure. Among the various reasons one is particularly out of control, that is the raindrops distribution aloft.

Calibration is particularly important when, for communication purposes, attenuation is computed from reflectivity measurements.

For instance Fig. 4 from Drufuca (1973) shows the dependence of the statistics of ground links at 11.2 GHz (5 miles length) on the radar calibration constant. It appears that only a calibration " a posteriori" that is a comparison between observed and computed long-term statistics can cope with the problem.

Calibration with the aid of raingauges is possible but it must be made with great caution due to the different sampling volumes of the two instruments, differences both in size and location. Careful statistical considerations have to be made as illustrated by Joss and Waldwogel (1968).

For what regards ground clutter it appears possible to reduce it from the observations using the particular spectral properties of the phenomena as proposed by Paraboni (1971) and Schaffner (1971) At the moment the only available remedy is to point the antenna with a sufficient elevation angle.

Finally the radar data are stored, usually in quantized levels. How fine and therefore how expensive this quantization has to be? The answer to this question depends on the method of processing the data.

Fig. 5 taken from Marshall (1971) shows for an integration performed by peak reading the performances of different discrete intervals versus number of independent samples.

BIBLIOGRAPHY

Atlas, D. and Banks, H.C. - The interpretation of microwave reflections from rainfall - J. Met. 8:271-82, 1951.

Barge, B.L. - Polarization observations in Alberta - Preprints of the 14th Radar Met. Conference. Am.Met.Soc. Boston Mass., 1970.

Battan, L.J. - Radar Observation of the Atmosphere - The University of Chicago Press, 39,40. 1973.

Drufuca, G. - Rain attenuation statistics for frequencies above 10 GHz from radar observations. - Proceedings of the Conference on Propagation of Radio waves at frequencies above 10 GHz. IEE n. 98, 1973, London.
---------- - Rain attenuation studies. - Scientific Report MW-77. Stormy wheather group. McGill University, p. 3, 1973.
Gunn, K.L.S. and East, T.W.R. - The microwave properties of precipitation particles. - Quart. J. Roy. Meteor. Soc., 80:522-45, 1954.
Hermann, B.M. and Battan, L.J. - Calculations of the backscattering of microwaves from ice spheres. - Quart. J. Roy. Meteor. Soc., 87:223-30, 1961.
Hitschfeld, W. and Bordan, J. - Errors inherent in the radar measurement of rainfall at attenuating wavelengths. - J. Meteor., 11:58-67, 1954.
Joss, J., Thams, J.C., and Maldvogel, A. - The accuracy of daily rainfall measurements by radar. - Proceedings of the 13th Radar Met. Conference Am. Met. Soc., Boston Mass., 1968.
---------- - and Waldwogel, A. - Raindrop Size Distributions and Sampling Size Errors. - Journal of the Atmospheric Sciences, 26:566-69, 1969.
Kodaira, N. - The characteristic of the averaged echo intensity received by the logarithmic I.F. amplifier. - Proc. Eight Weath. Radar Conf., pp. 255-61, Boston: Amer.Meteor.Soc., 1960.
McCormick, G.C. and Hendry, A. - The study of precipitation backscatter at 1.8 cm with a polarization diversity radar. - Preprints of the 14th Radar Met. Conference Am.Met.Soc. Boston, Mass., 1970.
Marshall, J.S. - Peak reading and thresholding in processing radar weather data. - J. of Appl. Meteor., 10,6 pp. 1213-23, 1971.
----------- and Hitschfeld, W. - Interpretation of the fluctuating echo from randomly distribute scatterers. - Pt. I. Canadian J. Physics 31:962-94, 1953.
Mie, G. - Beiträge zur Optik trüber Medien, speziell kolloidaler metallösungen (Contribution to the optics of suspended media, specifically colloidal metal suspensions). - Ann.Phys. 25:377-445, 1908.
Paraboni, A. - Principio di funzionamento di un sistema anticlutter per radar meteorologico (Principles of an anticlutter system for meteorological radar). - Internal Report 71-1, Istituto di Elettronica, Politecnico di Milano, Milano, 1971.
Probert-Jones, J.R. - The radar equation in meteorology. - Quart. J. Roy. Meteor. Soc., 88:485-95, 1962.
Rao, K.M. and Shaw, R.W. - Attenuation parameters from a radar map. - Proceedings of the 13th Radar Met. Conference, Am.Met.Soc., Boston, Mass., 1968.
Rogers, R.R. - The effect of variable target reflectivity on weather radar measurements. - Quart. J. Roy. Meteor. Soc., 97:154-67; 1971.

Ryde, J.W. - The attenuation and radar echoes produced at centimeter wavelengths by various meteorological phoenomena. - Meteorological Factors in radio wave propagation, pp. 169-88, London, Physical Soc., 1946.

Schaffner, M. - Ground echo rejection. - Private Communication, 1973.

Smith, P.L.Jr. - Interpretation of the fluctuating echo from randomly distributed scatterers - Pt. III, Scientific Report MW 39. Stormy wheather group, McGill University, 1964.

--------- - Calibration of weather radars; proceedings of the 13th Radar Met. Conference, Am.Met.Soc., Boston, Mass., 1968.

--------- - Project M. A course of seminars held at McGill University, 1969.

Strickland, J.I. - Comparison of direct and indirect measurements of precipitation attenuation at 15.3 GHz. - AGARD Conference Proceedings n. 107. Telecommunication aspects of frequencies between 10 and 100 GHz. 1972.

Wallace, P.R. - Interpretation of the fluctuating echo from randomly distributed scatterers. Pt. II, Canadian J. Physics, 31:995-1009, 1953.

REPORTS OF WORKING GROUPS

REPORT OF WORKING GROUP A

SOME GENERAL COMMENTS ON THE EXCHANGE AND STABILITY PROCESSES IN THE SURFACE LAYER

G. Emmanuel, D. Hay, T. Kester, G. Kruspe, P. Matthews
G. Schayes, S. Wickerts, A. Zancla

It is quite apparent that there does not exist an adequate theory which is capable in describing the physical processes involved in the air-sea interaction problem of the marine boundary layer. Most workers in the field use, and with some success, the Monin-Obukhov similarity formulation which is, at best, only an approximation.

The concept behind the Monin-Obukhov formulation is based on the assumption that the turbulent fluxes of momentum, sensible heat and moisture are constant (or quasi-constant) with height. The stability of the flow is specified by the dimensionless parameter z/L, where z is the height and L is defined by

$$L \equiv - \frac{[\overline{u'w'}]^{3/2} T}{kg \overline{w'T'}}$$

where T is the temperature, k is the von Karman constant, g is the acceleration due to gravity, and u'w' refer, respectively, to the fluxes of momentum and heat. To take into consideration the effect of moisture in the marine boundary layer, the temperature T is replaced by the virtual temperature t_V.

During the last few years, the main effort of meteorological experimentation in the surface layer above the sea has been aimed at the determination of the turbulent fluxes of momentum, heat and moisture. This has been done by using one of several techniques available: the eddy correlation technique, the aerodynamic profile method, or the energy dissipation technique.

Under the conditions of the similarity formulation, the structure of the boundary layer and the intensity of turbulence within this layer can be related to the stability parameters z/L or the Richardson number (Lumley and Panofsky, 1964), properly modified to take into account the buoyancy effect of moisture. Except for conditions of free convection, for which a "convective similarity theory" exists (Wyngaard et al, 1971; Arya, 1972), or under conditions of strongly stable situations, we can assume for the calculation of the "gradient fluxes" the well known approximations for the stability function as given by the Monin-Obukhov formulation.

We must, however, keep in mind that the surface layer is not isolated: it is being affected by the sea-air interface itself and also of higher layers. Such higher layers may represent strata of laminar or turbulent flow. The latter is found to occur within hydrodynamically unstable layers in which the Richardson number is ≤ 0.25 'Emmanuel, 1972). In fact, theoretical considerations, laboratory observations, as well as observations made of the clear atmosphere, suggest that clear air turbulence may be attributed to a form of dynamic instability commonly referred to as Kelvin-Helmholtz (K-H) instability. It should be emphasized, however, that this problem is far from being completely, or even satisfactorily, resolved.

In the clear atmosphere, and in particular within the planetary boundary layer, K-H instability occurs within a stratum characterized by strong thermal stability and large vertical wind shear. Initially, this type of dynamic instability appears as amplifying waves which, in time, "roll" and "break" into turbulent flow on a rang of smaller scales. The onset of this type of instability within a layer of thickness dz is determined by the value of the correspondin layer Richardson number

$$Ri = g\beta \left(\frac{dU}{dZ}\right)^{-2}$$

where $\beta \equiv \frac{1}{\theta}\frac{d\theta}{dZ}$ represents the thermal stability. To date, it has been shown that $Ri \leq 0.25$ is a necessary but not sufficient condition for the onset of K-H instability. Furthermore, it has been shown that the stability criterion may be expressed as

$$\alpha D = \left[\frac{1}{2} \pm \left(\frac{1}{4} - Ri\right)^{\frac{1}{2}}\right]^{\frac{1}{2}}$$

where $\alpha = \frac{2\pi}{\lambda}$ wavenumber, and D is the layer thickness. Another way of expressing this criterion is

$$\left[\frac{1}{2} - \left(\frac{1}{4} - Ri\right)^{\frac{1}{2}}\right]^{\frac{1}{2}} < \alpha D < \left[\frac{1}{2} + \left(\frac{1}{4} - Ri\right)^{\frac{1}{2}}\right]^{\frac{1}{2}}$$

which states that for $Ri > 0.25$ all disturbances of dimensionless wavenumber αD will be stable, while for $Ri < 0.25$ only those disturbances will be unstable whose dimensionless wavenumbers fall inside the inequality.

It should be evident then that what is needed is an integrated concept of the entire boundary layer including the effects from the surface itself. Parameterization techniques should prove quite fruitful, however, more measurements are needed at different sites and under different states of the oceanic wave-wind field.

A physical process that is of interest to both the radio-meteorologist and the scientist studying air-sea interaction is the vertical flux of pollutant gases above water. Examples of such gases are carbon dioxide, carbon monoxide, nitrogen oxides and sulphur oxides. In general, the transfer process is complicated within the water, where the gas may decompose or react with other components to produce new compounds that are not readily traced. Hence, the flux measurement is made more readily in the air above the water.

The flux measurement in the air assumes for simplicity that the pollutant gas is a conservative quantity (a passive additive). Then the vertical transfer of this gas by atmospheric turbulence is the vertical flux (number of gas particles passing through unit horizontal area per unit time):

$$F_G = \varrho \; \overline{w'c'}$$

according to the eddy correlation technique, where ϱ is air density (assumed approximately constant), w' is the fluctuating component of vertical wind speed, and c' is the fluctuating component of pollutant gas concentration (per unit mass of air). The averaging is carried out over a period containing a large number of fluctuations.

Techniques for the measurement of w' and c' are of immediate concern here. Conventional sonic anemometers or bivane anemometers with rapid response are generally adequate for observing w'. Two methods of measuring c' are suggested for further consideration. One of these is the ventilated millimeter-wavelength cavity, that is operated at the resonance frequency of the pollutant gas. The other involves Raman scatter by the pollutant gas at wavelengths in the ultra violet and optical regions. Each of these requires further examination, concerning sensitivity and speed of response. It is important to note that matched response-times are required in measuring w' and c' by this method.

For the special case of water vapour flux, measurements using the microwave refractometer and the capacitor refractometer are being studued currently. These and other techniques are described by McBean G.A. and Miyake M., 1972, Martin H.C., 1972 and Pond S. et al, 1971.

REFERENCES

Arya, S.P.S., 1972: Free Convection Similarity and Measurements in Flows with and without Shear. J. Atmos.Sci., nO 5, vol. 29,

Emmanuel, C.B., 1972: Richardson number profiles through shear instability wave regions observed in the lower planetary boundary layer. Paper presented at IUCRM Colloq. on Waves and Turbulence, La Jolla, Calif., June 6-15. (To be published in Boundary Layer Meteorology, 1973).

Lumley, J.L., and Panofsky, H.A., 1964: The Structure of Atmospheric Turbulence, Interscience, New York.

Martin, H.C., 1972: Humidity and temperature microstructure near the ground. Quart. J. Roy. Met. So., 98, p 440, Apr.

McBean, G.A., and Miyake, M., 1972: Turbulent transfer mechanism in the atmospheric surface layer. Quart. J. Roy. Met. So., 98, p 383, Apr.

Pond, S., Phelps, G.T., Paquin, J.E., McBean, G. and Stewart, R.W. 1971: Measurements of the Turbulent Fluxes of Momentum, Moisture and Sensible Heat over the Ocean. J. Atmos. Sci. 28, p 901, Sept.

Wyngaard, J.C. et al, 1971: Local free convection, similarity, and the budgets of shear stress and heat flux. J. Atmos. Sc., 28 1171-1182.

REPORT OF WORKING GROUP B

CONCERNING EVAPORATION AND INTERMITTENCY

B.R. Bean, J. Batesti, A. Bernardini, J. Boisen,
R. Bolgiano, I. Ghidini, L. Ladell, K.J. McCormick,
A. Nania, G. Racca, S. Rotheram, H.D. Seehars, S.A. Stage
A. Tangerud

A. EVAPORATION

The group considered that approximately one-third of the earth's surface is water (± 23° North and South of the equator). In this region solar energy is stored by absorption by the sea and in the latent heat of evaporation. On the average about 0.5 cm per day evaporates from this surface, thus providing about 300 calories per cm to fuel the earth's weather machinery. It is this key to the weather that is the subject of intense research by the world's meteorological research laboratories as to how this fuel gets distributed.

It appears from the Barbados Oceanographic and Meteorological Experiment (BOMEX) that a pattern is beginning to emerge of turbulent flux over the open tropical sea. The time series of evaporation shows a "ramp"-like structure when observed with the wind and a "box"-like structure when observed crosswind (see Bean and Emmanuel, this volume). The spectral properties are quite different for the two aspects of examining the fluxes. Table I shows that

Table I

Length Corresponding to Spectral Peak

Height	Downwind	Crosswind
100 ft	2 km	0.25 km
500 ft	3.3 km	1 km

a very much elongated pattern emerges such as roll vortices as the driving mechanism with plumes or bubbles embedded in the flow.

Note: the downwind pattern is the one observed by towers, captive balloons, etc. This, however, is for atmospheric stability near neutral ($0 < -Z/L < 0.35$). For $-Z/L \leq 0.075$ the flow goes into free convection (i.e. random rather than organized convection) and the difference between downwind and crosswind disappears.

B. INTERMITTENCY

The discussion of intermittency centered around its definition. It was clear that the process could be defined in either a statistical or physical sense. Among statistical descriptions could be included:

- 4th moment, rms and 3rd moment
- spectral properties
- deviations from the mean
- coherency analysis
- moving averages
- "bursts" of energy within "bursts".

Physical processes giving rise to intermittent "bursts" in the time series could include:

- convective plumes advecting past the observing point
- breaking of internal waves
- laminar-turbulent transitions
- fossil turbulence
- perturbation of stable layers by convective plumes.

From consideration of the above the group concluded that delineation of one physical process from another was difficult at best when utilizing standard meteorological observing techniques, even including the most sophisticated of turbulence devices. Experience has shown that the use of high powered radars, FM-CW radars and acoustic sounders are of inestimable aid in guiding the interpretation of such standard observations.

GPSR Compliance

The European Union's (EU) General Product Safety Regulation (GPSR) is a set of rules that requires consumer products to be safe and our obligations to ensure this.

If you have any concerns about our products, you can contact us on ProductSafety@springernature.com

In case Publisher is established outside the EU, the EU authorized representative is:

Springer Nature Customer Service Center GmbH
Europaplatz 3
69115 Heidelberg, Germany

Batch number: 10723233

Printed by Printforce, the Netherlands